Association of Reptilian and Amphibian Veterinarians Conference (ARAV 2018)

Held at ExoticsCon 2018

Atlanta, Georgia, USA
23 - 26 September 2018

ISBN: 978-1-5108-7528-9

Printed from e-media with permission by:

Curran Associates, Inc.
57 Morehouse Lane
Red Hook, NY 12571

Some format issues inherent in the e-media version may also appear in this print version.

Printed by Curran Associates, Inc. (2019)

For permission requests, please contact Association of Reptilian & Amphibian Veterinarians at the address below.

Association of Reptilian & Amphibian Veterinarians
5003 SW 41st Blvd
Gainesville FL 32608-4930
USA

Phone: 480-703-4941

info@arav.org

Additional copies of this publication are available from:

Curran Associates, Inc.
57 Morehouse Lane
Red Hook, NY 12571 USA
Phone: 845-758-0400
Fax: 845-758-2633
Email: curran@proceedings.com
Web: www.proceedings.com

EXOTICSCON 2018 PROCEEDINGS

September 22-27 (Exhibits September 23-26) • Hilton Atlanta Hotel, Atlanta, GA

EXOTICSCON.ORG

BUILDING EXOTICS EXCELLENCE.
ONE CITY. ONE CONFERENCE.

BROUGHT TO YOU BY

Association of Avian Veterinarians

AEMV
ASSOCIATION OF EXOTIC MAMMAL VETERINARIANS

ARAV
ASSOCIATION OF REPTILIAN AND AMPHIBIAN VETERINARIANS

WELCOME TO EXOTICSCON 2018

On behalf of the Board of Directors for the Association of Avian Veterinarians (AAV), Association of Exotic Mammal Veterinarians (AEMV), and the Association of Reptilian and Amphibian Veterinarians (ARAV), we welcome you to Atlanta, Georgia.

The Trilateral Conference Planning Committee is excited to bring you quality education and amazing networking opportunities. We hope you'll make the most of the program the committee worked tirelessly to assemble. From educational opportunities like hands-on workshops, roundtables, and lecture sessions, to special and social events, we know there is something for everyone.

We have the privilege to work with various professionals to make this conference possible and would like acknowledge the following groups and individuals who helped make this conference possible:

- **Auction chairs**, for undertaking this monumental task and insuring the future of all the organizations' research grant programs:
    - AAV: Katherine Baine, DVM, DABVP (Avian Practice) and Brynn McCleery, DVM, DABVP (Avian Practice)
    - AEMV: Teresa Bousquet, DVM
    - ARAV: Nelson Bricker, DVM & Kourtney Grimm, DVM
- **The *volunteer* abstract and proceedings reviewers** for ensuring high-quality, accurate, and evidence-based, scientific programing and proceedings.
- **Session chairs** for their time and expertise to ensure the conference proceeds smoothly and attendees have the best educational experience possible.
- **Hands-on workshop instructors** who develop exciting hands-on experiences to convey new clinical techniques for attendees to implement in their practices. Their commitment and dedication does not go unnoticed.
- **Presenters**, for their time, energy, expertise, and willingness to share their knowledge and further the profession with their passion for exotic animals.
- **Executive directors** who provide leadership and guidance through the conference planning process, and for their devotion to our organizations and adopting our missions and passions as their own:
    - AAV: Robert Groskin, DVM
    - AEMV: Mandy Taylor, MBA, APR
    - ARAV: Dana Varble, DVM and (incoming) Wendy Dox
- **Conference planning team with NAVC** for their help with budgets, registration, A/V, and other logistics and details impacting the conference:
    - Pete Scott, MBA, CAE, APR, HABC
    - Joe Sorrentino, MBA, CMP
    - Sheri Matheny
    - Jarisa Esquivel

Additionally, thank you to the **lead conference planner Marisa Hackemann**, for taking on and tackling the overwhelming task of coordinating, aligning, creating compromise, and promoting the individual and collective needs of three devoted and passionate associations as well as our facilities and contractors to create and shape a record-setting conference.

Last but certainly not least, we thank our members and attendees for your faith and loyalty as well as your contributions to our associations. AAV, AEMV, and ARAV exist because of our members. Over the years, you have helped ExoticsCon grow in both size and scope. Your feedback continues to shape future planning and we look forward to your feedback. We hope you enjoy ExoticsCon 2018!

Sincerely,

Michelle Hawkins, VMD, DABVP (Avian Practice), *AAV Conference Chair*
Lauren Powers, DVM, DABVP (Avian Practice), DABVP (Exotic Companion Mammal Practice), *AAV*
Yvonne R.A. van Zeeland, DVM, MVR, PhD, DipECZM (Avian) DipECZM (Small Mammal), CPBC, *AAV*
Byron de la Navarre, DVM, *AEMV Conference Chair*
Melissa Kling, DVM, *AEMV*
Elizabeth Mackey, DVM, *AEMV*
Tim Tristan, DVM, DABVP (Reptile and Amphibian Practice), *ARAV Conference Chair*
Nicole Johnson, DVM, *ARAV*
Mike Corcoran, DVM, CertAqV, DABVP (Reptile and Amphibian Practice), *ARAV*

Introduction

We invite you to enjoy the benefits provided by these Proceedings of ExoticsCon 2018, a combined conference of the Association of Avian Veterinarians (AAV), Association of Exotic Mammal Veterinarians (AEMV), and the Association of Reptilian and Amphibian Veterinarians (ARAV). The 2018 combined conference provides access to leading specialty veterinary researchers and practitioners world-wide sharing their experience and knowledge in person and through these proceedings, which are furnished to members of the three associations.

AAV, AEMV and ARAV recognize and thank the following committees and staff for their time and hard work in assuring a quality program and publication:

ExoticsCon Trilateral Conference Committee

Michelle Hawkins, VMD, DABVP (Avian Practice), *AAV Conference Chair*
Lauren Powers, DVM, DABVP (Avian Practice), DABVP (Exotic Companion Mammal Practice), *AAV*
Yvonne R.A. van Zeeland, DVM, MVR, PhD, DipECZM (Avian) DipECZM (Small Mammal), CPBC, *AAV*
Byron de la Navarre, DVM, *AEMV Conference Chair*
Melissa Kling, DVM, *AEMV*
Elizabeth Mackey, DVM, *AEMV*
Tim Tristan, DVM, DABVP (Reptile and Amphibian Practice), *ARAV Conference Chair*
Nicole Johnson, DVM, *ARAV*
Mike Corcoran, DVM, DABVP (Reptile and Amphibian Practice), CertAqV, *ARAV*
Marisa Hackemann, *NAVC Meeting Planner*
Pete Scott, MBA, CAE, APR, HABC, *NAVC Chief Product Officer*

AAV Executive and Conference Committees

Lauren V. Powers, DVM, DABVP (Avian Practice), DABVP (Exotic Companion Mammal Practice), *President, Conference Committee Member*
Yvonne R.A. van Zeeland, DVM, MVR, PhD, DipECZM (Avian) DipECZM (Small Mammal), CPBC, *President-Elect, Conference Committee Member*
Michelle Hawkins, VMD, DABVP (Avian Practice), *Immediate Past President, Conference Committee Chair*
Anna Osofsky, DVM, DABVP (Avian Practice), *Education Committee Co-Chair*
Geoff Olsen, DVM, DABVP (Avian Practice), *Education Committee Co-Chair*
Steve Metz, DVM, *AAV Hands-On Workshop Coordinator*
Laurel Degernes, DVM, DABVP (Avian Practice), *Hands-On Workshop Coordinator*
Michael Lutz, DVM, *Treasurer*
Jesse Fallon, MS, DVM, *Advisory Council Representative to the Executive Board*
Robert Groskin, DVM, *Executive Director*

AEMV Executive Committee

Charly Pignon, DVM, DipECZM (Small Mammal), *President*
Nicola Di Girolamo, DMV, MSc (EBHC), GPCert(ExAP), PhD, DipECZM (Herpetology), *President-Elect*
David Hannon, DVM, DABVP (Avian Practice), *Vice President*
Laurel Harris, DVM, *Secretary*
Melissa Kling, DVM, *Treasurer*
Byron de la Navarre, DVM, *Immediate Past President*
Elizabeth Mackey, DVM, *Member at Large*
Jennifer Graham, DVM, DABVP (Avian Practice), DABVP (Exotic Companion Mammal Practice), Dipl ACZM, *Member at Large*
Teresa Bousquet, DVM, *Member at Large*
Mandy Taylor, MBA, APR, *Executive Director*

ARAV Executive Committee

Nicole Johnson, DVM, *President*
Tim Tristan, DVM, DABVP (Reptile and Amphibian Practice), *President-Elect*
Mike Corcoran, DVM, DABVP (Reptile and Amphibian Practice), CertAqV, *Vice President*
Kimberlee Wojick, DVM, *Secretary*
Stacey Leonatti Wilkinson, DVM, DABVP (Reptile and Amphibian Practice), *Treasurer*
Susan Horton, DVM, *Immediate Past President*
Louisa M. Asseo, DVM, DABVP (Canine and Feline Practice), *Member at Large*
Brad Waffa, MSPH, DVM, *Member at Large*
Matt Allender, DVM, MS, PhD, Dipl ACZM, *Member at Large*
Anke Stoehr, DVM, *Member at Large*
Dana Varble, DVM, *Executive Director, outgoing*
Wendy Dox, *Incoming Executive Director, incoming*

Association of Avian Veterinarians

AEMV
ASSOCIATION OF EXOTIC MAMMAL VETERINARIANS

ARAV
ASSOCIATION OF REPTILIAN AND AMPHIBIAN VETERINARIANS

Veterinary Review and Editing Committee

Technical Editor
Charlotte Kirk Baer, MS

Conference Staff — North American Veterinary Community
Pete Scott, MBA, CAE, APR, HABC, Chief Product Officer
Joe Sorrentino, MBA, CMP, Vice President of Event Operations
Dana Varble, DVM, Vice President of Veterinary Education
Marisa Hackemann, Senior Director, Industry Services
Mandy Taylor, MBA, APR, Executive Director, Industry Services
Wendy Dox, Executive Director, Industry Services
Sheri Matheny, Manager, Customer Service
Jarisa Esquivel, Event Manager, Operations

Proceedings Layout and Design — BrightLizard, Inc.
Elizabeth Smith

Continuing Education Accreditation

CE accreditation for ExoticsCon 2018 is sponsored by the American College of Zoological Medicine.

Veterinary technician CE accreditation was provided by RACE. The Association of Veterinary Technicians in Clinical Practice (AVTCP) has also reviewed and verified CE hours that are approved for Veterinary Technician Specialist certification.

Conference Sponsors

Platinum Sponsors

Silver Sponsor

Disclaimer

The material appearing in this publication comes exclusively from the authors and contributors identified in each manuscript. The techniques and procedures discussed reflect the personal knowledge and experience of the authors and contributors, and demonstrate their views of the methods that may be used for these medical procedures. The procedures demonstrated do not incorporate all known techniques, are not exclusive, and other techniques and technology may also be available. Any questions or requests for additional information concerning any of the manuscripts should be addressed directly to the authors.

AAV, AEMV and ARAV do not research, review, or otherwise verify any of the information contained in this publication. Opinions expressed in this publication are those of the authors and contributors and not necessarily those of AAV, AEMV, or ARAV. AAV, AEMV and ARAV are not responsible for errors or for opinions expressed in this publication.

AAV, AEMV and ARAV expressly disclaim any warranties or guarantees, expressed or implied, and shall not be liable for damages of any kind in connection with the material, information, techniques, or procedures set forth in this publication.

Manuscripts contained in these conference proceedings vary in length and format based on author intentions for possible future publication. Authors are given the option to publish their complete manuscript herein as the final, official publication of the manuscript. Authors who wish to retain the right to publish a complete manuscript in another professional publication at a later date have limited their submission herein to an abstract only.

For permission to reproduce or order copies of this publication, contact the respective association office:

- AAV Office, www.aav.org, PO Box 9, Teaneck, New Jersey 07666, USA, e-mail pubs@aav.org.
- AEMV Office, www.aemv.org, North American Veterinary Community, 622 E. Washington Street, Suite 300, Orlando, Florida 32801 e-mail info@aemv.org.
- ARAV Office, www.arav.org, North American Veterinary Community, 5003 SW 41st Boulevard, Gainesville, Florida 32608-4930, USA, telephone 480-703-4941 , email info@arav.org.

For information on ExoticsCon educational conferences, visit http://www.exoticscon.org. For information on membership and other association programs visit the association websites listed above.

Table of Contents

Hybrid Sessions

Section 1: Hybrid Wellness Session

Section 2: Hybrid Technician Session

Section 3: Plenary Session

Section 4: Hybrid Aquatics Session

Section 5: Hybrid Mixed Bag Session

Section 6: Hybrid Master Classes

Section 7: AAV Hot Topics Session

Evaluation of the Thermal Antinociceptive Effects and Pharmacokinetics of Hydromorphone Hydrochloride in Orange-Winged Amazon Parrots (*Amazona amazonica*) (Session #2215)
Pharmacokinetics of Oral Tapentadol in Hispaniolan Amazon Parrots (*Amazona ventralis*) (Session #2409)
Age-Dependent Carrier Status and Clinical Disposition of Avian Bornavirus in Experimentally Infected Cockatiels (Session #2095)
Effects of Doxycycline, Polidocanol, and Absolute Ethanol on the Subcutaneous Tissues of Chickens (Session #2212)
Intravenous and Intraosseous Regional Limb Perfusion in an Avian Model (Session #2145)
The Use of Pentoxifylline in Birds (Session #2288)
Advancing Communication with Birds: Can They Learn to Read? (Session #2265)
Epidemiology, Epizootiology, and Management of Chlamydiosis in an Interactive Lorikeet (*Trichoglossus* sp.) Zoologic Park Aviary (Session #2325)

Section 8: AAV Clinical Potpourri Session

Broudifacoum Toxicity in a Flock of Morocco Racing Pigeons (*Columba livia domestica*) (Session #2277)
Unusual Pathologic Brain Changes Associated with Severe Zinc Toxicosis in an Amazon Parrot (Session #2170)
Peripheral Neuropathy Secondary to Lead Intoxication in Psittacine Species (Session #2245)
Acute Meloxicam Overdose and Treatment in a Sun Conure (*Aratinga solstitialis*) (Session #2129)
Congenital Megalocornea in Blue-Throated Conures (*Pyrrhura cruentata*) with MicroCT Analysis and Histopathology (Session #2273)
Use of Computed Tomography in the Diagnosis of Complicated Skull Fractures in Two Parrots (Session #2345)
Psittacine Beak and Feather Disease Virus Infection in a Golden Conure (*Guaruba guarouba*) (Session #2230)
Three Cases of Suspected Exocrine Pancreatic Insufficiency in Psittacine Birds (Session #2174)

Section 9: AAV Master Classes Track 1

What Parrots Want?!? Avian Foraging and Environmental Enrichment (Session #2297)
Understanding and Appreciating Avian Cognition to Address Behavioral Issues (Session #2051)

Section 10: AAV Master Classes Track 2

Section 11: AAV Master Classes Track 3

Section 12: AAV Master Classes Track 4

Section 13: AAV Oncology Session

Section 14: AAV Hormones, Behavior, and Nutrition Session

Section 15: AAV Research Session

Section 16: AAV Surgery and Anesthesia Session

Section 17: AAV Wildlife Session

Section 18: AAV Posters

Xerosis is a Common Cause of Feather Destruction Behavior in Eclectus and other Parrots (Session #2001)
Application of DuoDERM® Patch as an Alternative to Close Surgical Wounds in Birds (Session #2151)
A Retrospective Look at Risk Factors Associated with Wildlife Rescue Presentations of North American Strigiformes (Session #2238)
Ascorbic Acid Content in Avian Pellet Diets and Their Effectiveness as an Ascorbic Acid Source (Session #2388)

Association of Exotic Mammal Veterinarians

Section 19: AEMV Diagnostics Session

Acquired Dental Disease in Guinea Pigs (*Cavia porcellus*) (Session #2204)
Diagnostic Accuracy of Abscess Capsule Culture Compared to Pus Culture for Isolation of Bacteria in Rabbit Dental Abscesses (Session #2395)
Polycystic Kidneys with Ureter Mineralization in a Guinea Pig (Session #2377)

Section 20: AEMV Oro-Facial Conditions Session

Use of Thermoreversible Gel for the Prevention of Perioperative Corneal Ulcers in Rabbits (Session #2117)
Corneal Diseases in Rabbits and Rodents: Taking Care of Those Great Big Beautiful Eyes (Session #2157)
Ototoxic Effect of Baytril Otic® Applied to the Middle Ear of Rabbits (Session #2050)
Surgical Management of Ear-Origin Abscesses in Rabbits and Rats (Session #2134)
A Retrospective Study of Intranasal Tumors in Rabbits: Six Cases (2007-2017) (Session #2076)

Section 21: AEMV Oncology Session

Are Mammary Gland Tumors in Rats Correlated with the Presence of Prolactinoma? (Session #2290)
Metronomic Chemotherapy in a Pet Rat (*Rattus norvegicus*) with Subcutaneous Fibrosarcoma (Session #2256)
Survival Times of Lymphoma Treatment Protocols in Domestic Ferrets (*Mustela putorius furo*): A Retrospective Study (Session #2123)
Successeful Treatment of Lymphoma by Radiation Therapy in Gerbil (*Meriones unguiculatus*) (Session #2179)
Diagnosis and Treatment of Chronic Lymphocytic Leukemia in a Pet Rabbit (*Oryctolagus cuniculus*) (Session #2314)
Cutaneous Xanthomas with Ocular and Cardiovascular Involvement in a Lop Rabbit (Session #2045)

Section 22: AEMV Master Classes Track 1

Section 23: AEMV Master Classes Track 2

Section 24: AEMV Internal Medicine Session

Section 25: AEMV Rabbit Session

Section 26: AEMV Infectious Diseases Session

Section 27: AEMV Ferret Medicine Session

Section 28: AEMV Surgery Session

Section 29: AEMV Anesthesia and Analgesia Session

Section 30: AEMV Urology Session

Section 31: AEMV Posters

Association of Reptilian and Amphibian Veterinarians

Section 32: ARAV Reproductive Session

Section 33: ARAV Master Classes

Section 34: ARAV Medicine and Surgery Session

Section 35: ARAV Diagnostics Session

Section 36: ARAV Anesthesia Session

Section 37: ARAV Case Reports Session

Section 38: ARAV Nutrition and Gastroenterology Session

Section 39: ARAV Poster

Author Index

Stacy Burdick, DMV, Dipl ACVIM, Fellow IR – page 755
Mark E. Burgess, DVM – page 569, 767, 881
Kristine E. Burgess, DVM, Dipl ACVIM (Oncology) – page 581
Veronica Burresi, DRN – page 509
Steven J. Bursian, PhD – page 485, 487
Nicholas A. Buscaglia, DVM – page 463
Karina Butler-Perez, BS – page 975, 981, 983
Dave Cacela, BS – page 487
Sarah R. Camlic, BA – page 817
Alvin C. Camus, DVM, PhD – page 97
James W. Carpenter, MS, DVM, Dipl ACZM – page 187, 433, 453
Francisco Javier Duque Carrasco, DVM, PhD – page 781
Daniel Calvo Carrasco, LV, CertAVP (ZooMed), MRCVS – page 515, 757, 773, 781
Sharon A. Center, DVM, Dipl ACVIM – page 729
Sabine Chahory, DVM, PhD, Dip ECVO – page 557
Ya-Pei Chang, BVM, MVM, DipECVN – page 589
Hui-Wen Chang, DVM, PhD – page 391
Lucile Chassang, DVM, IPSAV (Médecine zoologique) – page 579, 793
Simon Chavez, MS, MVM, CVT – page 67
Javier Ojeda Chávez, MVZ, MSc – page 739
Yi-An Chen, DVM – page 391
Sue Chen, DVM, DABVP (Avian Practice) – page 131
Ivan-Chen Cheng, DVM, PhD – page 391
Valérie Chetboul, PhD, Pr vet med, Dipl ECVIM-CA – page 587, 961
Chau-Hwa Chi, DVM, PhD – page 473, 809
Linus Li-Hsung Chiang, BVM, MVSc – page 803
Marty Chu, DVM – page 713, 775
Paula Ciembor, DVM, PhD – page 205
P. Ciembor, DVM, PhD – page 221
William Claude, Dr vet med – page 961
Ashley Clayton – page 445
Christal R. Clements, DVM – page 221
Susan Clubb, DVM – page 197
Camille Coisnon, DVM – page 793
Ophelie Cojean, DMV, IPSAV – page 933
Sara Collins, DVM – page 493
Jessica Comolli, DVM – page 97, 99, 921
Mike Corcoran, DVM, DABVP (Reptile and Amphibian Practice), CertAqV – page 957
Pamela Coy, Da Chen, MS, PhD – page 479
Carolyn Cray, PhD – page 431, 717
Jeannette Cremer, DVM, Dr med vet, DACVAA, CVA – page 101
Ivan Crotaz, MRCVS – page 779
Jennifer Cunha, JD – page 197
Fred L. Cunningham, DVM, PhD – page 485, 487
Daniel C. Cutler, DVM – page 567
Robert D. Dahlhausen, DVM, MS – page 273
Cécile Damoiseaux – page 587
Gigi S. Davidson, BSPH, DICVP – page 89
Jeann Leal de Araujo, GTS – page 397
Ricardo de Matos, LMV, MSc, DABVP (Avian Practice), DipECZM (Avian), DipECZM (Small Mammal) – page 463
Loes de Nies, BSc – page 787
Karen M. Dean, PhD – page 485, 487
Adeline Decambron, DVM, PhD, Resident ECVS – page 765
Gianluca Deli, DVM, GPCert(ExAP) – page 509, 811, 987
Isabelle Desprez, DVM, Resident ECZM (Small Mammal) – page 557, 587, 765, 799, 961
Nicola Di Girolamo, DMV, MSc(EBHC), GPCert(ExAP), PhD, DipECZM (Herpetology) – page 549, 659, 707, 795, 825, 935
Orlando Diaz-Figueroa, DVM, MS, DABVP (Avian Practice) – page 993

Britton J. Grasperge, DVM, PhD, DACVP – page 427, 993
Cheryl B. Greenacre, DVM, DABVP (Avian Practice), DABVP (Exotic Companion Mammal Practice) – page 35, 219, 333, 647
Clayton Greenway, DVM – page 731
J. Gregory Massey, DVM – page 89
Christopher R. Gregory, DVM, PhD – page 109, 205, 221, 507
Erik Groeneveld, BSc – page 91
Mackenzie Hallman, DVM, Dipl ACVR – page 513
Samer S. Hameed, BVMS, MSc, PhD – page 83
Chiara Hampton, DVM, MS – page 827
Katie C. Hanson-Dorr, MS – page 485 ,487
Kendal E. Harr, DVM, MS, DACVP – page 117, 249, 485, 487, 591
Don J. Harris, DVM – page 287
Tara M. Harrison, DVM, MPVM, Dipl ACZM, Dipl ACVPM, DipECZM (Zoo Health Management) – page 389
Scott H. Harrison, PhD – page 389
Stefan Harsch, DVM – page 445
Carolynn Harvey, DVM – page 721
Sayed M. Hassan, PhDÊ – page 209
Michelle G. Hawkins, VMD, DABVP (Avian Practice) – page 711
Kate A. Healy, MS – page 485, 487
Darryl J. Heard, BSc, BVMS, PhD, Dipl ACZM – page 959
J. Jill Heatley, DVM, MS, DABVP (Avian Practice), DABVP (Reptilian and Amphibian Practice), Dipl ACZM – page 397, 423, 429, 845
Joanna Hedley, BVM&S, DZooMed (Reptilian), DipECZM (Herpetology), MRCVS – page 913
Tom Hellebuyck, Dr med vet, DipECZM (Herpetology) – page 893
Coenraad Hendriksen, DVM, PhD – page 787
Melody Hennigh, RVT, KPA-CTP – page 47
Christiane Herden, Dipl ECVP – page 189
Sonia M. Hernandez, DVM, PhD, Dipl ACZM – page 991
Sybille Herzog, Dr med vet – page 189
Laurie Hess, DVM, DABVP (Avian Practice) – page 393
Aaron RGS Hess, MD, PhD – page 445
Julia M. Hill, DVM – page 491
E. Hofmeister, DVM, MA, DACVAA, DECVAA – page 19
Jan Hooimeijer, DVM, CPBC – page 239
Lisa A. Hoopes, PhD – page 991
Sharman M. Hoppes, DVM, DABVP (Avian Practice) – page 217, 449
Katherine E. Horak, PhD – page 487
Kendall Houlihan, DVM – page 11
Elizabeth W. Howerth, DVM, PhD, DACVP – page 205, 567
Li Hsung Chiang, BVSc, MSc – page 999
Li-Hsin Huang, DVM – page 589
Wei-Hsiang Huang, DVM, PhD – page 589
Clémence Hurtrel, DVM – page 799
Minh Huynh, DVM, MRCVS, DipECZM (Avian), Dipl ACZM – page 161, 579, 793
Timothy H. Hyndman, BVMS, PhD – page 929, 955
Shashikant Jadhav, MVSc – page 445
Mark Jankowski, MSPH, PhD – page 479
Chian-Ren Jeng, DVM, PhD – page 965
Carmen Jerry, DVM, DACVP – page 221
Brian Jochems, DVM – page 429
Cathy Johnson-Delaney, DVM – page 21
Dan H. Johnson, DVM, DABVP (Exotic Companion Mammal Practice) – page 663
Cia L. Johnson, DVM, MS – page 11
Michael Jones, DVM, DABVP (Avian Practice) – page 219
Vickie Joseph, DVM, DABVP (Avian Practice) – page 143, 209, 481
SeungWoo Jung, DVM, MS, PhD, Dipl ACVIM – page 493
Claudia Kabakchiev, DVM – page 455, 771, 789

Ketsia Kahambwe – page 695
Ian Kanda, RVT – page 49, 451, 759, 911
Chu-Lin Kang, DVM – page 965
Lorraine G. Karpinski, VMD, DACVO – page 491
Philip Kass, DVM, MS, MPVM, PhD – page 711
Spencer Kehoe, DVM – page 99, 921
Katharine Kierski, BS – page 967
Marja Kik, DVM, PhD, DipECZM (Herpetology) – page 91, 577
Shannon G.M. Kirejczyk, DVM, MPH – page 567
Matti Kiupel, DVM, MS, PhD, DACVP – page 393
Marcel Klaassen, PhD – page 989
Kevin Kleinow, DVM, PhD – page 101
Heather Knych, DVM, PhD, DACVCP – page 185, 711
Petra Koelle, DVM, Dr med vet, PrivDoz, Cert Spec for Fish including Ornamental Fish, Cert Spec for Reptiles, Cert Spec for Pet Nutrition consultation – page 985
Rüdiger Korbel, Prof, Dr med vet, DipECZM (Avian) – page 345, 437
Mark Kramer, DVM – page 911
Yuan-Peng Kuan, DVM, MS – page 391, 589, 965
Butch KuKanich, DVM, PhD, DACVCP – page 187, 433, 919
Elise E. B. LaDouceur, DVM, DACVP – page 971
Eve Laloy – page 587
Louise Lam – page 935
Isabelle Langlois, DMV, DABVP (Avian Practice) – page 753, 755
James D. Lauderdale, PhD – page 945
Stefania Lauzi, DVM, PhD – page 17
Kévin Le Boédec, DVM, Dipl ACVIM, Dipl ECVIM – page 793
Matthew Leach, PhD – page 787
Amber Lee, BVSc, DABVP (Avian Practice) – page 215
Yen Tao Lee, BVSc, MSc – page 803, 809, 999
Ber-In Lee, DVM, MS – page 965
E. Lepri, DVM, PhD, DipECVP – page 795
Wen-Ta Li, DVM, MS – page 803
Marla Lichtenberger, DVM, DACVECC – page 305, 653, 749
Michael Lierz, DipECZM, DZooMed, DipECVPS – page 189
Jane E. Link, MS – page 485, 487
Chen-Hsuan Liu, DVM, PhD – page 965
Joshua Llinas BVSc, MVS – page 955
Jason Locklin, PhD – page 109
M. Lodi, DVM, PhD – page 811
Maria Belen Moreno Lopez, MRCVS – page 515, 757, 781
Walter Lorenz, PhD – page 507
Silvia Lovati, DRN – page 509
Emanuele Lubian, DVM, GPCert(ExAP) – page 509
E. Lubian, DVM, GPCert(ExAP) – page 811
Emanuele Lubian, DVM, GPCert(ExAP) – page 813, 821, 987
Édouard Maccolini, DMV, IPSAV – page 933
John Mackie BVSc, PhD, DACVP – page 955
Geraldine C. Magnin, PhD, Dipl ABT – page 433
Shachar Malka, DVM, DABVP (Avian Practice) – page 585
Maria Manou – page 705
Christoph Mans, Dr med vet, Dipl ACZM, DipECZM (Zoo Health Management) – page 93, 213, 337, 447, 511, 797, 877, 923, 939, 941, 967
P. Maratkanova, DVM – page 583
Jeffrey Marr, MS, PE – page 441
Rachel E. Marschang, PD, Dr med vet, DipECZM (Herpetology), FTÄ Mikrobiologie, ZB Reptilien – page 929
Rob Marshall, BVSc, MACVSc (Avian Health) – page 405, 413, 519
Anna Martel-Arquette, DVM – page 213, 511, 797

Anne Martin, PhD – page 721
Wendy Massey – page 89
Karina Mathes, DVM, Dr med vet, Cert Spec in Reptiles and Amphibians, DipECZM (Herpetology) – page 953
Crystal Matt, BS – page 513
Giorgia Matteucci, PhD – page 549
Jöerg Mayer, DVM, MS, DABVP (Exotic Companion Mammal Practice), DipECZM (Small Mammal), Dipl ACZM – page 19, 99, 567, 921
Vincent Mayousse – page 705
Stephanie McCain, DVM, Dipl ACZM – page 205
Brynn McCleery, DVM, DABVP (Avian Practice) – page 783
Emily McConnelly, DVM – page 559
Julianne McCready, DVM – page 783
Andrew K. McFadden, BS – page 487
JoAnn McGee, PhD – page 441
Alicia McLaughlin, DVM – page 249
Rita McManamon, DVM – page 221
Dave McRuer, DVM, Dipl ACVPM – page 529
Erica Mede, CVT – page 69, 71
Caroline Meek – page 89
Marie Melin, DVM – page 557
Douglas B. Menke, PhD – page 945
Blake Metcalf – page 695
Ericka Meyers, DVM – page 909
Sarah Miller, DVM – page 971
Whitney L. Miller, DVM, MBA, Dipl ACVPM – page 111, 113
Mark A. Mitchell, DVM, MS, PhD, DipECZM (Herpetology) – page 817, 819, 829, 975, 981, 983, 995
Yasutsugu Miwa, DVM, PhD – page 573
Laure Molitor, MS, DVM – page 995
Rob Moore, PhD – page 989
Jennilee Morrison, DVM – page 457, 517
Stephanie Munyon, DVM – page 215
Jeffrey MB Musser, DVM, PhD, DABVP (Dairy Practice) – page 397
Makoto Nakata, DVM, PhD – page 573
G. Nardini, DVM, PhD, DipECZM – page 795
Guinevere Nease – page 389
Dmitriy Neginskiy, DVM – page 733
Peggy Nelson, PhD – page 441
Robert D. Ness, DVM – page 149, 551
Aleisha Nesset, DVM – page 461
Javier G. Nevarez, DVM, PhD, Dipl ACZM, DipECZM (Herpetology) – page 403, 827, 915, 993
Hoa Nguyen-Phuc, PhD – page 479
Alexandra Nicolier, DVM, Dipl ECVP – page 579, 793
Roger Nilsen, MS – page 507
Brendan Noonan, DVM, DABVP (Avian Practice) – page 223
Terry M. Norton, DVM, Dipl ACZM – page 87, 991
Tailai O'Brien, BBus (HRM & BMGT), Cert IV Assessment and Workplace Training, Cert III Captive Animal Management – page 405, 413, 519
Tim O'Connell, MA, PhD – page 513
Ryan O'Shea, CVT – page 65
Susan E. Orosz, PhD, DVM, DABVP (Avian Practice), DipECZM (Avian) – page 273, 305
Robert J. Ossiboff, DVM, PhD, DACVP – page 889
Seth C. Oster, DVM – page 329, 493, 737
Sarah Ozawa, DVM – page 711
Clément Paillusseau, Dr vet med – page 961
Lily Parkinson, DVM – page 923, 967
Riccardo Pascuzzo, PhD – page 787
Peggy Passavin, Dr vet med – page 961

Karen Russell, DVM, PhD, DACVP (Clinical Pathology) – page 429
Riccardo Russo, DVM, GPCert(ExAP) – page 813
David Sanchez-Migallon Guzman, LV, MS, DipECZM (Avian), DipECZM (Small Mammal), Dipl ACZM – page 185, 461
Susan Sanchez, BSc, MSc, PhD, CBiol, FSB – page 963
Vittorio Saponaro, PhD, Dr vet med – page 961
Emi Sasaki, DVM – page 211
Andrea Savarese – page 211
Alyssa M. Scagnelli, DVM – page 101, 191
Matthew Schexnayder, DVM – page 993
Lionel Schilliger, Dr vet med, DipECZM (Herpetology), DABVP (Reptile and Amphibian Practice) – page 961
Marion Schmicke, JProf Dr med Vet – page 703
Nico J. Schoemaker, DVM, PhD, DipECZM (Small Mammal), DipECZM (Avian), DABVP (Avian Practice) – page 91, 195, 297, 577, 607, 687, 779, 787
Renée Schott, DVM, CWR – page 27, 167, 475, 489
Celine Schulze, Dr med vet – page 437
M. Scott Echols, DVM, DABVP (Avian Practice) – page 373
Erin M. Scott, VMD, DACVO – page 217, 449
Lynne M. Seibert, DVM, DVM, MS, PhD, DACVB – page 261, 597
P. Selleri, DVM, PhD, DipECZM – page 795
Yun Shan Chiu, BVSc, MSc – page 999
Gillian C. Shaw, DVM, MS, PhD, DACVP – page 217
Zoe Shelton – page 695
Naomi Shimizu, DVM, MRCVS – page 773
Susan Shriner, PhD – page 487
Regina Sigg, Dr med vet – page 437
Richard Sim, DVM – page 205
Jules Simard, med vet – page 893
Elisabeth Simone-Freilicher, DVM, DABVP (Avian Practice) – page 223
Shane Simpson, BVSc(Hons), GCM(VP) – page 949
Ameet Singh, DVM, DVSc, Dipl ACVS – page 771, 789
Devinn Sinnott, DVM – page 461
Adrianna Skarbek, MRCVS – page 757, 773
Kurt K. Sladky, MS, DVM, Dipl ACZM, DipECZM (Herpetology), DipECZM (Zoo Health Management) – page 797
Alison Smith, DVM, DACVAA – page 783
Ryan M. Smith, DVM, DACVECC – page 827
Domingo Casamian Sorrosal, DVM CertSAM, DVC, DipECVIM-CA, MRCVS – page 781
Jeny Soto, BS – page 431
Brian Speer, DVM, DABVP (Avian Practice), DipECZM (Avian) – page 373, 453
Nicole I. Stacy, DVM, Dr Med Vet, DACVP (Clinical) – page 969
Scott J. Stahl, DVM, DABVP (Avian Practice) – page 851
Fouzia Stambouli – page 587
Anne Staudenmaier, VMD – page 223
Nicole Stec-Phillips, DVM – page 911
Brittany N. Stevens, DVM – page 491
Georgina V. Stewart, BVSc – page 567
Mathew Stewart, DVM – page 211, 827
Melanie Stodel, MSc – page 787
Eva Stoffels, DVM, PhD – page 539
Anke C. Stöhr, DVM – page 805, 915
Rhett W. Stout, DVM, PhD, DACLAM – page 403
Anneliese Strunk, DVM, DABVP (Avian Practice) – page 117
Leonardo Susta, DVM, PhD, DACVP – page 423
Kellyn M. Sweeley, DVM – page 943
Simon Swift, MA, VetMB, CertSAC, Dipl ECVIM-CA (Cardiology), MRCVS – page 959
Lara Sypniewski, DVM, DABVP (Canine and Feline Practice) – page 911
Zoltan Szabo, DrMedVet, DABVP (Exotic Companion Mammal Practice), DABVP (Avian Practice), GpCert(ExAP), MSc, MRCVS – page 627

Section 1

Hybrid Wellness Session

Cathy Johnson-Delaney, DVM
Session Chair

Caring for the Caregivers: Mental Health and Well-Being in the Veterinary Profession

Jennifer E Graham, DVM, DABVP (Avian Practice), DABVP (Exotic Companion Mammal Practice), Dipl ACZM,
Genevieve Rajewski

Session #2064

Affiliation: From the Department of Clinical Sciences (Graham), Office of Marketing and Communications (Rajewski), Cummings School of Veterinary Medicine at Tufts University, 200 Westboro Rd, North Grafton, MA 01536, USA.

Abstract: Physical, emotional, and intellectual challenges, combined with the traits of perfectionism and introversion shared by many in the profession—may contribute to depression, anxiety, and even suicide among some veterinary professionals. This master class will provide an overview of these issues with a focus on initiatives to improve mental health and a sense of well-being among veterinary students and professionals. Specific examples of wellness initiatives at the Cummings School are highlighted along with discussion of wellbeing and peer assistance programs available through AVMA and other organizations.

Why Worry?

According to the first-ever mental health survey of veterinarians in the United States, published October 2015 in the Journal of the American Veterinary Medical Association (JAVMA), one in 11 US veterinarians has experienced serious psychological distress, and one in six has contemplated suicide since graduating.[1] What's behind these startling numbers? Experts say that for many veterinarians, practice carries environmental risk factors for suicide that are exacerbated by a culture and personality types that are common among veterinary professionals. The good news is that with implementation of measures to help veterinarians cope with practice-related stressors and reducing barriers in seeking mental health treatment, suicide risk among veterinarians may be reduced.

Startling statistics

The vast majority of veterinarians surveyed in the JAVMA study said they were pleased with their occupation. "One of the key findings was that 80 percent agreed or strongly agreed they were happy being a veterinarian," said Randall Nett, co-author of the study and a physician working for the Public Health Service at the US Centers for Disease Control and Prevention. "But we must take steps to make sure those who are suffering get the support and mental health services they need."

Nett stressed that only a small percentage of veterinarians are at risk of suicide, but pointed to studies that show vets are anywhere from 1.6 to 4 times more likely to kill themselves than are members of the general population.[2,3] Determining precise rates by profession is difficult—in part because suicide is believed to be vastly underreported as a cause of death—but data collected in 2015 by the National Institute for Occupational Safety and Health indicated that the suicide rate among veterinarians is the fourth-highest among all professions, trailing only medical doctors, dentists, and police officers.

Suicide in veterinarians likely occurs because of the complex interaction of job-related stresses and personality characteristics common to people who go into veterinary medicine, Nett said. In the JAVMA survey, veterinarians overwhelmingly identified the demands of practice as their key stressor: the long hours and large caseloads, the financial and management responsibilities, the grief related to animal deaths, and the ethical challenges posed by the need to euthanize healthy animals or to help clients who lack the means to pay for treatment. Nett said he was inspired to conduct the JAVMA survey in large part because he's married to a veterinarian. "She does critical care in emergency veterinary medicine," he said, "and I've seen her come home from days at work when she's had to do eight euthanasias and just been completely wrecked, emotionally and physically and mentally by that. I've seen a workload where she's supposed to work a 12-hour day but she ends up working 18, 19, 20 hours because there's so much work to be done."

But it's not just the demands of the job, said Barry Feldman, the director of psychiatry programs in public safety and an assistant professor of psychiatry at the University of Massachusetts Medical School. Veterinarians tend to have a rigid cognitive style, Feldman said, one that's intolerant of making mistakes—they're perfectionists. "You often see this cognitive style in the medical professions, where there are lots of bright people who were used to doing well in school," he said. "When they start to struggle, that can feel almost catastrophic."

Indeed, veterinary students are also at risk. In a nationwide survey conducted through the Student Chapter of the American Veterinary Medical Association (SCAVMA), 60 percent of veterinary students reported experiencing depression, with two-thirds of them saying that the bouts lasted for longer than two weeks—the clinical definition of depression. Five percent of the students said they'd gone so far as to formulate a suicide plan. "So of the 3,888 students who answered our survey, 194 said yes, they had put a plan in place to kill themselves," said Michael McEntire, a vet student in Texas A&M's Class of 2017, who helped convene a yearlong SCAVMA wellness taskforce after learning that one of his classmates had tried to kill herself.

Many vets and vet students work under tremendous pressure, and some of them are predisposed to take failure—or the perception of it—especially hard. The final piece of the puzzle is that veterinarians also have easy access to a means to actually do themselves harm: the drugs used for animal euthanasia. "You find that when veterinarians die from suicide, they more often die from self-poisoning compared with the general population, which makes sense," Nett said. "Even compared with physicians, they really have unparalleled access to lethal drugs and the knowledge of how to use them. Our study also found that, compared with US adults, veterinarians actually reported fewer suicide attempts, and unfortunately that's probably because they're more successful at completing suicide."

Not talking about suicide has dangerous consequences said Feldman, who specializes in suicide intervention and prevention. "There's a pervasive myth that asking someone if they're depressed or having thoughts of suicide plants the seed for suicide. But research absolutely refutes that," said Feldman, who spoke at the invitation of Tufts adjunct associate professor Michael McGuill, V91, at the American Veterinary Medical Association conference in Boston. "Telling someone that you're worried about them and asking directly if they're thinking about ending their life can actually serve to help get them into the mental health arena. I find it really encouraging that there's finally a lot of very open dialogue about opiate addiction and opiate overdoses. We need the same kind of discussion around suicide."

That's especially important—and challenging—when it comes to vets. Nett said that the JAVMA study found that veterinarians, particularly those suffering from significant psychological distress, were less likely than the rest of the population to believe that people are generally caring toward those with mental illness. "So the people who need the help the most, as far as mental health treatment goes, appeared to have the highest degrees of perception of stigma against those with mental illness," he said.

The Cummings community's response

To begin changing the perception of stigma against those with mental illness, Cummings School students and faculty have made it their mission to promote mental health and to openly discuss their personal struggles. Two behavioral health counselors and one career counselor have been available to meet with students on campus in the recently established Cummings Support Center, and licensed counselors are also available 24/7 on a confidential phone line. Cummings School also supports its faculty, staff, residents and interns with an on-campus behavioral health counselor, as well as additional in-person counseling and a 24/7 hotline through Tufts' Employee Assistance Program.

Over the past academic years, students have also organized several events to address mental health. For instance, Pamela Bay, V18, invited a nationally known vet to give two SCAVMA talks on burnout and compassion fatigue. And as part of the first wellness week ever held on campus, Stephanie Chubb and Nina Bartling, both V18, organized a panel discussion in which faculty and other practicing veterinarians talked about how they've successfully coped with personal issues. During Diversity Week, meanwhile, a comedian performed a show about depression that was followed by a discussion. Heartened by the reaction on campus, the Tufts chapter of SCAVMA has established a permanent wellness committee to plan monthly events about mental health and positive coping strategies. Starting in the fall of 2016, Cummings School began offering The Healer's Art, a 15-hour selective for first- and second-year students led by Jennifer Graham, an associate professor and exotics veterinarian, and associate professor Flo Tseng, the director of Tufts Wildlife Clinic. The program (http://www.rachelremen.com/learn/medical-education-work/the-healers-art/), taught at 90-plus medical schools and a small, but growing number of veterinary schools, encourages students struggling with the stresses of the profession to focus on what drew them to it in the first place, and to make peace with the impossibility of curing every patient. The Tufts chapter of SCAVMA worked over the past with the student body to draft a student "bill of rights" that's designed to ensure a work-life balance for students during the clinical year—a period, the SCAVMA study found, during which self-reported rates of depression, suicide ideation and self-harm increased among veterinary students. Faculty is now using the document to craft new guidelines for clinical rotations.

Cummings Thrive

A recent student-led crowdfunding project successfully raised funds to provide monthly wellness workshops for students and launched the creation of Cummings Thrive. Cummings Thrive is a monthly/bi-weekly workshop intended to introduce participants (most focus on 1st year students) to important concepts to help them 'survive and thrive in their lives'. Some of the workshops have included topics covering conflict management, team building and communication, and self-awareness. As part of this initiative, the participants have created an open community where they can rely on support for the rest of their careers. A Facebook group and email listserv provide an ongoing dialogue on wellness topics covered in the workshops.

Wellness button campaign

A wellness button campaign is another initiative created to provide student, staff, house officer, and faculty support at the Cummings School. Individuals undergo QPR (Question, Persuade, Refer - https://qprinstitute.com) training in suicide prevention. Those who undergo training are then issued a button to wear to indicate they are a 'safe space' for individuals who feel they need a support system. Members of the campaign include licensed health care professionals and campus police to provide immediate assistance in crisis situations.

Leadership and doctoring course

Lastly, a leadership and doctoring course working committee has been created to meet on a regular schedule to discuss business, leadership, professionalism, communications, and wellness components of the Cummings curriculum and further development of these important areas. A day-long leadership seminar for 3rd year students entering their clinical rotation year is one example of the group's efforts. An orientation program for students entering the Cummings School, based on the Veterinary Leadership Experience leadership training program, is mandatory for incoming students of the veterinary school, ensuring exposure to these topics at the onset of their veterinary careers.

The goal of all these efforts is to change the culture of the veterinary profession. "As a student and then through my internship and residency, I didn't feel like I could show any signs of weakness or talk to anybody about it," Graham said. However, not long after Graham finished her residency, her mother died by suicide, a devastating loss that pushed her to seek help for her own feelings of depression "and a pretty intense struggle with my own sexual orientation." "I share all this openly now because it worries me that someone I could influence might be feeling like they can't reach out for help," said Graham.

In that spirit of open discussion about the challenges and stresses of being a veterinarian, Michael McEntire, aTexas A&M vet student, has worked with the SCAVMA to create a series of online video interviews in which vets and vet students who've struggled with their mental health discuss the importance of talking with others about feelings of depression. "We never think our very accomplished professors or classmates are going through anything, McEntire said. "It always seems like they've all got their acts together." According to Dr. Michele Gaspar, a feline veterinarian and licensed professional counselor from Chicago, veterinary schools should consider ways to develop "antidotes to perfectionism," teaching students the self-compassion that builds confidence, competence, and resilience.[4] Many veterinary schools across the country have wellness programs with this focus in mind.

What about me?

Social support is a protective factor for psychological health related to occupational stressors.[5] In the 2015 JAVMA study, poorer psychological health was associated with: 1) not being married or in a committed relationship, 2) being separated or divorced, 3) not having children, and 4) not being a member of a veterinary medical association. Less experienced veterinarians were also reported to have poorer psychological health than experienced colleagues. According to Tracy Witte, an associate professor of Psychology at Auburn University, "Fortunately, most veterinary professionals will not experience suicidal behavior, but what stood out was work-related stress, with work overload being the major stressor of the profession…when asked about where future efforts should be focused in their state, most respondents listed education about risk factors for mental health and substance abuse as well as stress management. I think it is important that we are focusing on people who are suffering and directing them to resources. But this is just one part of wellness, and we need to teach stress management on top of that".[6]

"Our wellness has to be in developing our students and our profession to courageously be able to hold others' pain," according to Michele Gaspar. A step outlined in the 4th Annual Veterinary health and Wellness Summit is the formation of more Finding Meaning in Veterinary Medicine groups.

Finding Meaning in Veterinary Medicine (FMVM)

An offshoot of the Healer's Art course, FMVM is based on Finding Meaning in Medicine (FMM). FMM are self-led physician groups meeting monthly, independent of institutional support or involvement, using a storytelling approach to uncover and deepen a sense of connection and community, and find greater satisfaction and meaning in their daily work and their profession. FMVM provide an ongoing conversation about important experiences, issues, and ideas that emerge from work as a veterinarian that individuals may not have the opportunity to reflect on and discuss with others. "It's not a therapy or support group. Each person shares, and the rest just listen. It's generous listening with an open mind and heart to whatever they're bringing forth," according to Dr. Laurie Fonken, PhD, psychological counselor at Colorado State University's veterinary college. "It's not fixing, giving advice, or saying, 'That happened to me.' It's holding a space for each to share." Groups of practices or students or state VMAs can easily form their own FMVM groups. Contact Laurie.Fonken@colostate.edu for more information.

Stress management

The Mayo Clinic suggests four strategies for coping with stress: avoid, alter, accept, adapt. While it is ideal to avoid or alter stressful situations, acceptance and adaptation are helpful when stressful situations are unavoidable. A good guide is the stress management checklist:[7]

- Exercise regularly
- Get sufficient sleep
- Plan for and follow a balanced diet
- Drink alcohol in moderation
- Maintain a healthy weight
- Balance work and play
- Delegate tasks appropriately. Don't try to do everything yourself
- Spend quality time with friends and family
- Resolve conflict in a timely and respectful manner
- Take psychological health seriously. Seek professional assistance if bothered by feelings of depression or other signs of mental illness
- Never underestimate the health benefits of playing with your pet

Thinking of life in terms of 'sinkers' and 'balloons' can be a helpful concept.[8] Sinkers are negative things that pull us down in daily life while balloons life us up. Balloon filling is unique to each person and they ultimately fill our physical, mental, emotional, and spiritual connections. For example, a physical connection balloon may be going out for a run. Running through a forest or on a beach may fill mental, emotional, and spiritual needs as well. Whether it be yoga, expressive writing, affirmations, listening to music, or 15 minutes of time spent alone, keeping our balloons full is a key to resiliency in times of stress.

Meditation and breathing exercises are exceptionally effective ways to manage stress. Our breath is something we have with us always and simply taking a moment for slow, deep diaphragmatic breaths is beneficial. Deep breathing has been shown to decrease oxygen consumption, decrease heart rate, and decrease blood pressure, as well as increase theta wave amplitude in EEG recordings, increasing parasympathetic activity resulting in alertness and reinvigoration.[9] Meditation is accessible to everyone and there are a myriad of applications and devices making meditation mainstream.

Resources

General

- AVMA Wellbeing and Peer Assistance:
  https://www.avma.org/ProfessionalDevelopment/PeerAndWellness/Pages/default.aspx

  This site contains helpful information on work and compassion fatigue, stress management, physical health, financial well-being, and work-life balance, among others. AVMA members can set up a workplace wellbeing program and are able to sign up for a pilot QPR training program to learn to identify and refer at-risk colleagues. The Professional Quality of Life Assessment tool can help individuals measure how they are being affected in key areas related to mental wellbeing, which can then help identify areas to focus self-care planning. Numerous wellbeing resources for veterinarians are available through this site.

- 1 Life Connected: http://1lifecc.com/

  This site contains information on life coaching, workshops, retreats, and ways to recognize your passion, embrace imperfection, and stay connected as a veterinarian.

- "Not One More Vet" is a private Facebook support group strictly for veterinarians. See www.nomv.org for more information.

Wellbeing

- Headspace: www.headspace.com

  Headspace is a meditation app with hundreds of themed sessions on everything from stress to sleep. Busy veterinarians can benefit from bite-sized meditations for hectic schedules and SOS exercises in case of sudden meltdowns.

- Some other meditation applications (there are many) include: Calm, The Mindfulness App, MINDBODY, buddhify, Insight Timer, Smiling Mind, Meditation Timer Pro, Sattva, Stop Breathe and Think, Simply Being.

- Your Healthiest Self: WellnessToolkits

  https://www.nih.gov/health-information/your-healthiest-self-wellness-toolkits

  NIH Wellness Toolkits highlight evidence-based tips for living well and improving your health. This is published by the Office of Communications and Public Liaison in the NIH Office of the Director.

Crisis

- If you are in crisis, call the National Suicide Prevention Lifeline at 1.800.273.8255, or dial 911 or go to a hospital emergency room.
- The American Foundation for Suicide Prevention explains risk factors and warning signs of suicide: bit.ly/AFSPrisks
- S.A.V.E. (Suicide Awareness in Veterinary Education) provides mental health information to vet students: www.vet.utk.edu/save/education
- Active Minds is a nonprofit organization that empowers students to speak openly about mental health to educate others and encourage help-seeking: www.activeminds.org
- The national Student American Veterinary Medical Association (SAVMA) has created a series of video interviews with veterinary students, faculty and clinicians about their struggles in veterinary school and in practice: https://www.youtube.com/playlist?list=PL4Qoyn0na9zvrXVwhFrXUrE3Jk-GybsHT
- The Survivors of Suicide website seeks to help those who have lost a loved one to suicide: www.survivorsofsuicide.com
- University of Tennessee Veterinary Social Work program – any veterinarian, not just Tennessee graduates, can contact a veterinary social worker at the University of Tennessee, Knoxville School of Veterinary Social Work. Call 1-865-755-8839 during business hours, or email vetsocialwork@utk.edu.

If someone you know is in crisis:[10]

1. Approach the person.
2. Ask how they are feeling.
3. Listen with care and concern.
4. Ask if they have suicidal thoughts.
5. If they do, find a crisis hotline and stay by their side as they make the call.
6. If they do but refuse to call, call a crisis hotline yourself for guidance. It is imperative to access professional support, direction, and services as needed.
7. Assure them that things can and will change.
8. Stay with them, and recruit the company of trusted others (as warranted), until you know the person is safe with the necessary supports in place.
9. When it is safe to leave, make specific plans to see them the next day so they have a reason to hang on for one more day.

Acknowledgments: Some content has been modified from Rajewski's article: http://sites.tufts.edu/vetmag/fall-2016/under-pressure/

References

1. Nett R, Witte T, Holzbauer, et al. Risk factors for suicide, attitudes toward mental illness, and practice-related stressors among US Veterinarians. *J Am Vet Med Assoc.* 2015;247(8):945-955.

2. Bartram DJ, Baldwin DS. Veterinary surgeons and suicide: a structured review of possible influences on increased risk. *Vet Rec.* 2010;166:388-397.

3. Jones-Fairnie H, Ferroni P, Silburn S, et al. Suicide in Australian veterinarians. *Aust Vet J.* 2008;86:114-116.

4. Larkin M. Wellness as an everyday affair. *JAVMA News.* Jan 15, 2017.

5. Michie S, Williams S. Reducing work related psychological ill health and sickness absence: A systematic literation review. *Occup Environ Med.* 2003;60:3-9.

6. Larkin M. Studies confirm poor well-being in veterinary professionals, students. *JAVMA News.* May 1, 2016.

7. American Veterinary Medical Association. AVMA Stress Management Checklist. Available at: https://www.avma.org/ProfessionalDevelopment/PeerAndWellness/Pages/wellness-checklist.aspx Accessed July 1, 2108.

8. Pope-Robinson K. The Bridge to Connection. In: *The Unspoken Life.* San Clemente, CA: 1 Life Connected Consulting;2017:13-32.

9. Jerath R, Edry J, Barnes V, Jerath V. Physiology of long pranayamic breathing: neural respiratory elements may provide a mechanism that explains how slow deep breathing shifts the autonomic nervous system. *Med Hypotheses.* 2006;67(3):566-71.

10. Stoewen D. Suicide in veterinary medicine: Let's talk about it. *Can Vet J.* 2015;56(1):89-92.

Interactive Assessment of Guinea Pig and Parakeet Welfare: You be the Judge

Cia L. Johnson, DVM, MS, and
Kendall Houlihan, DVM

Session #2154

Affiliation: From the Animal Welfare Division, American Veterinary Medical Association, 1931 N. Meacham Road, Schaumburg, IL 60173, USA.

Abstract: The Animal Welfare Judging and Assessment Contest (AWJAC) is an American Veterinary Medical Association (AVMA) program that aims to provide a unique educational experience through the opportunity to assess the welfare of animals in a variety of settings using science-based methods and reasoning. This session will be based on two scenarios, the first involving parakeets and the second about guinea pigs. Ensuring animal welfare is a human responsibility that includes consideration for all aspects of animal well-being, including proper housing, management, nutrition, disease prevention and treatment, responsible care, humane handling and, when necessary, humane euthanasia.

Seems straightforward, right? In reality, animal welfare assessment may be much more complex and challenging to assess; real-life situations may present advantages in one area and disadvantages in another, and an assessment often depends on weighing which elements of welfare are most important to the animals. Even well-respected animal welfare experts may disagree.

This session will provide attendees with realistic scenarios to help them appreciate the many factors that go into a thorough welfare assessment, identify areas for improvement, and provide sound, science-based opinions relating to the welfare of the particular species addressed in the scenarios.

Introduction

The public assumes veterinarians are the animal experts, and should be involved in all major animal-based industries and activities. So when major cases of abuse or neglect are exposed people are quick to ask: Where was the veterinarian? Why did the veterinarian let this happen? In an age where many previously respected professions have become mired in moral and ethical scandals the veterinary profession must demonstrate the expertise and the will to advocate for animals and promote their welfare. This duty relates not only to companion animals, or animals as our patients, but all of the diverse animal populations that are affected by human activities. As such, veterinarians must develop competencies relating to the science-based evaluation of animal welfare, and communicating about welfare to diverse audiences. As animal welfare issues become increasing contentious, the role of the veterinarian in negotiating how animals should be treated has never been more difficult or more important.

Animal Welfare

Animal welfare is difficult to define, although many have tried. The World Organization for Animal Health defines animal welfare:

> *"Animal welfare means how an animal is coping with the conditions in which it lives. An animal is in a good state of welfare if it is healthy, comfortable, well-nourished, safe, able to express innate behavior, and if it is not suffering from unpleasant states such as pain, fear and distress.... Animal welfare refers to the state of the animal, the treatment that an animal receives is covered by other terms such as animal care, animal husbandry, and humane treatment."*[1]

This definition is fairly cumbersome and difficult to understand. Another possible definition is given to us by Dr. David Fraser:

> *"Animal welfare is an everyday concept which people use to express their concerns about the quality of life of animals, especially as it is affected by human actions."*[2]

This definition seems to capture the complexity of animal welfare; however, we are left with the question: what is good quality of life? We can attempt to answer this by using a combination of animal centered outcomes and environmental measures.[3] One of the frameworks that assists us in evaluating these outcomes and measures is brought to us by Fraser et al.[2] and is referred to as the three concepts (aspects) approach to animal welfare: basic health and function (health, growth and productivity); affective state of the animal pain, suffering, contentment); and natural living (comparisons to wild or free-living counterparts). The diagram (Fig 1) represents three components considered critical to a holistic assessment of welfare. Under this model, welfare is seen as having different but overlapping pieces. Welfare is not only about the animal's health, how it feels, or whether it has a natural lifestyle. Instead these are all seen as aspects of what might be good or bad about how an animal is doing at a particular point in time. These components often overlap. For example, being able to chew is not only natural for a dog, but it also appears to be enjoyable, and is good for dental health. Welfare needs may also be separate or competing. For example, more opportunity to run and play outside might increase the chance that an animal is injured.

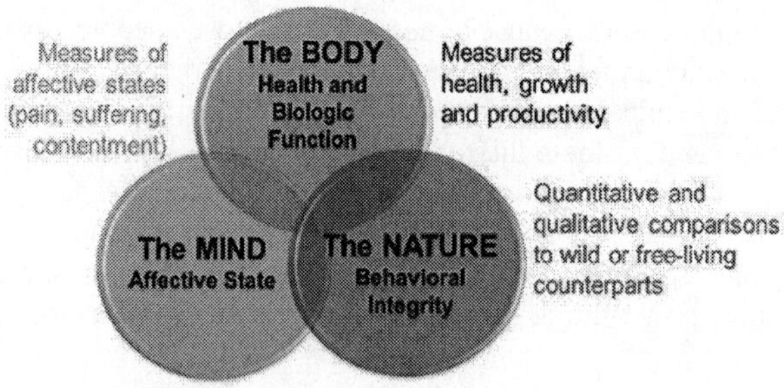

Figure 1. Diagram of the Three Concepts animal welfare assessment framework (Adapted from Fraser D, Weary DM, Pajor EA, et al. A scientific conception of animal welfare that reflects ethical concerns. *Anim Welf.* 1997;6:187-205).

As veterinarians, we tend to have an excellent handle on the health and biologic functioning side of things; however, since animal welfare encompasses the whole of the animal's experience, the whole of its quality of life, we must also scientifically assess the behavioral integrity (or natural life of the animal) and affective or emotional state of the animal.

When people make animal welfare decisions, they typically emphasize one of three attributes: how well the animal functions physically, how the animal feels, or how closely the environment or practice being evaluated corresponds to what a wild- or free-living member of the same species might experience. This is a reflection of what measures of animal welfare the person considers to be most important. Differences in emphasis can lead to different decisions about the environment or practice even when the same information is being evaluated. Regardless of the findings, how those findings are valued will determine which are given priority and which are implemented.

The Five Freedoms might also be used to evaluate animal welfare: freedom from hunger or thirst; freedom from discomfort; freedom from pain, injury or disease, freedom to express (most) normal behavior; and freedom from fear or distress.[4] A more recent framework developed by Mellor is known as the Five Domains.[5] Figure 2 is an abbreviated version of the Five Domains Model. It summarizes survival-related and situation-related factors and their associated physical/functional domains, and provides examples of negative or positive affects assigned to the mental domain. The overall affective experience in the mental domain equates to the welfare status of the animals. A "life worth living" minimizes an animal's negative experiences and at the same time provides opportunities to have positive experiences. Knowing that negative experiences, specifically the survival related ones, can never be completely eliminated. Foraging behavior is dependent on being hungry.

It is important to note that while the measures used in these frameworks may be objective, evaluating them can be quite subjective and is influenced by divergent ethical views.

Physical/Functional Domains							
Survival-Related Factors						*Situation-Related Factors*	
1: Nutrition		2: Environment		3: Health		4: Behaviour	
Negative	*Positive*	*Negative*	*Positive*	*Negative*	*Positive*	*Negative*	*Positive*
Restricted water & food; poor food quality	Enough water & food; balanced and varied diet	Uncomfortable or unpleasant physical features of environment	Physical environment comfortable or pleasant	Disease, injury and/or functional impairment	Healthy, fit and/or uninjured	Behavioural expression restricted	Able to express rewarding behaviours

Affective Experience Domains					
5: Mental State					
Negative Experiences			*Positive Experiences*		
Thirst Hunger Malnutrition malaise Chilling/overheating Hearing discomfort	Breathlessness Pain Debility, weakness Nausea, sickness Dizziness	Anger, frustration Boredom, helplessness Loneliness, depression Anxiety, fearfulness Panic, exhaustion	Drinking pleasures Taste pleasures Chewing pleasures Satiety Physical comforts	Vigour of good health & fitness Reward Goal-directed engagement	Calmness, in control Affectionate sociability Maternally rewarded Excited playfulness Sexually gratified
Welfare Status					

Figure 2. An abbreviated version of the Five Domains Model (Mellor DJ. Updating animal welfare thinking: Moving beyond the five freedoms towards a life worth living. *Animals.* 2016;6:21).

The Intersection of Animal Welfare and Ethics

Ethical judgment will impact animal welfare science due to its inherent subjectivity. An individual's ethical viewpoint will skew their evaluation of welfare; the direction of the bias will be dependent on what they value most. There are currently five major ethical theories that reflect the majority of the people's values and have the ability to affect assessment of animal welfare: 1) contractarian, 2) utilitarian, 3) animal rights, 4) contextual, and 5) respect for nature.[4]

Contractarians are self-serving and believe that being moral is in your best interest.[6] In the contractarian theory only human interests are important since only humans have the ability to enter into a 'contract' or agreement.[7] Utilitarianism defends that we should act in ways that bring about as much happiness as possible in the world. The animal rights theory states that all living creatures have inherent value[6-8] and that animals are not a vessel or resource for use by humans.[9] With contextual (relational) theory, the relationship a human has with the animal impacts the moral obligations the human has to the animal (ie, moral obligation differs based on context). Relational theorists believe that humans have a duty to assist those animals for which they are responsible for existence.[6,7] In the respect for nature viewpoint, species are to be preserved and represent a lifeline of biologic history.[9] This view attributes little value to what humans think about any particular species.

Understanding ethical viewpoints is a critical aspect of communication regarding animal welfare assessments.

Animal Welfare Judging and Assessment Contest

Beginning in 2001 as an idea to combine animal welfare science with traditional livestock judging,[10] the AWJAC has become a valuable event for undergraduate, graduate, veterinary students, and more recently AVMA member veterinarians. The AWJAC focuses on the welfare issues of all species, including agriculture, companion, laboratory, and exotic animals and brings together experts in welfare science, veterinarians, producers, and students. Most veterinary students believe animal welfare is an important part of their education, yet many believe their curriculum does not include enough about animal welfare.[11] The AWJAC offers participants the chance to gain and utilize knowledge in animal welfare science.

To achieve its mission of strengthening communication and reasoning skills, increasing knowledge of animal welfare assessment, and providing opportunities for networking the welfare science field, the AWJAC hosts an organized contest where participants evaluate several hypothetical/virtual scenarios relating to animal welfare and one live/on-site scenario to determine levels of animal welfare in animals observed across a variety of settings, including agriculture, biomedical research, companionship, and human assistance. Participants then present their findings and are evaluated by judges who are animal welfare experts. Preparation for the contest motivates participants to dive into the nuances of animal welfare assessment as well as species specific issues. The contest itself gives structure and, via the welfare expert judges, feedback to participants regarding their reasoning and communication. In addition, participants are provided with the opportunity to network with internationally renowned welfare experts.

On exit surveys performed from 2014-2016, 99% of respondents agreed or strongly agreed with the statement that the event was an overall valuable experience.[12] Respondents (99%) also agreed or strongly agreed that the event increased their animal welfare knowledge. The Contest has been very successful in providing opportunities to develop animal welfare knowledge, critical thinking, ethical decision-making, and communication skills. Due to this success, the AVMA is bringing the concept and structure of the event to members and allied organizations.

During this session attendees will review and assess two scenarios, the first involving parakeets and the second about guinea pigs. Animal welfare assessment frameworks will be applied as well as ethical theories. Attendees will be asked to fully participate in discussion and respectful debate.

Preparation

In order to prepare for a robust discussion about the welfare of parakeets and guinea pigs in the upcoming scenarios, please consider reviewing the resources, sample scenarios, and quizzes located on the AWJAC website at https://awjac.org/prepare.

References

1. OIE World Organization for Animal Health. A new definition for the Terrestrial Animal Health Code: animal welfare. Available at: http://www.oie.int/doc/ged/d5517.pdf. Accessed April 22, 2018.

2. Fraser D, Weary DM, Pajor EA, et al. A scientific conception of animal welfare that reflects ethical concerns. *Anim Welf.* 1997;6:187-205.

3. Hewson CJ. Can we assess animal welfare? *Can Vet J.* 2003;44(9):749-753.

4. Farm Animal Welfare Council. Farm animal welfare in Great Britain: Past, present and future. Available at: https://assets.publishing.service.gov.uk/government/uploads/system/uploads/attachment_data/file/319292/Farm_Animal_Welfare_in_Great_Britain_-_Past__Present_and_Future.pdf. Accessed April 22, 2018.

5. Mellor DJ. Updating animal welfare thinking: Moving beyond the five freedoms towards a life worth living. *Animals.* 2016;6:21.

6. Palmer C, Sandøe P. Animal ethics. In: Appleby MC, Mench JA, Olsson IAS, Hughes BO, eds. *Animal Welfare.* Cambridge, UK: CABI Publishing;2011:1-12.

7. Garner R. *A Theory of Justice for Animals: Animal Rights in a Nonideal World.* New York, NY: Oxford University Press;2013.

8. Regan T. The case for animal rights. In: Lafollette Hugh, ed. *Ethics in Practice.* Maldon, MA: Blackwell Publishing;2007:205-211.

9. Sandøe P, Christiansen SB. *Ethics of Animal Use.* Ames, IA: Blackwell Publishing;2008.

10. Broom DM. Animal welfare: An aspect of care, sustainability, and food quality required by the public. *J Vet Med Educ.* 2010;37(1):83-88.

11. Fraser D. Understanding animal welfare. *Acta Vet Scand.* 2008;50:S1.

12. Baumgardner R. Animal welfare judging and assessment contest: A student's perspective. *Am Vet Med Assoc.* 2018.

Use Of Thrombocyte-Leukocyte Rich Plasma to Treat Chronic Oral Cavity Disorder in Reptiles: Two Case Reports

Alessandro Vetere, DVM, GPCert(ExAP),
Edoardo Bardi, DVM, GPCert(ExAP),
Aquaro Vincenzo, DVM,
Giuliano Ravasio, DVM, PhD,
Davide Zani, DVM, PhD,
Marco Tecilla, DVM, PhD,
Paola Roccabianca, DVM, PhD, DECVP,
Stefania Lauzi, DVM, PhD,
Stefano Romussi, DVM, PhD

Session #2222

Affiliation: From Ambulatorio Veterinario Gaggiano, Via Galvani 42, Gaggiano, 20083, Italy (Vetere), Department of Veterinary Medicine, University of Milano, Via Ceolia 10, Milan, 20100, Italy (Bardi, Zani, Tecilla, Roccabianca, Lauzi, Romussi), Oasi WWF Policoro Herakleia, Policoro, Italy (Aquaro).

Oral cavity disorders are some of the most common conditions in pet reptiles. Due to their devious nature and development, animals often are presented in an advanced state of illness, with limited therapeutic options and poor outcomes. On the other hand, in recent years regenerative medicine has gained a strong interest in reptile practice as well, with transposition of techniques widely used in human and mammal medicine.[1-7] Specifically, platelet rich plasma (PRP) and derived products have been shown to be safe and promising tools to improve soft and bone tissue healing in this class. We present two cases of oral cavity disorders treated with heterologous thrombocyte-leukocyte rich plasma (TLRP): one ball python (*Python regius*) with chronic facial cellulitis and one veiled chameleon (*Chamaeleo calyptratus*) with severe maxillary osteolysis secondary to chronic periodontal disease. Complete blood work, microbiologic culture and antibiogram, and CT scan were performed. In the first case (ball python), surgical debridement was performed and the wound was irrigated with TLRP before closure. Wound healing was not significantly accelerated by TLRP administration, but post-surgical antibiotic therapy was avoided thanks to the antimicrobial properties of platelet concentrates, with absence of gross infection during all follow up. In the second case (veiled chamelen), non-healing osteomyelitis was treated with intralesional injection of TLRP, with a complete restoration of function achieved in about one month.

References

1. Hedley J. Anatomy and disorders of the oral cavity of reptiles and amphibians. *Vet Clin North Am Exot Anim Pract.* 2016;19(3):689-706.

2. Latney LV, McDermott C, Scott G, Soltero-Rivera MM, Beguesse K, Sánches MD, Lewis JR. Surgical management of maxillary and premaxillary osteomyelitis in a reticulated python (*Python reticulatus*). *J Am Vet Med Assoc.* 2016;248:1027-1033.

3. Mans C. Clinical update on diagnosis and management of disorders of the digestive system of reptiles. *J Exot Pet Med.* 2013;22:141-162.

4. Di Ianni F, Merli E, Burtini F, Conti V, Pelizzone I, Di Lecce R, Parmigiani E, Squassino GP, Del Bue M., Lucarelli E., Ramoni R., Grolli S. Preparation and application of an innovative thrombocyte/leukocyte-enriched plasma to promote tissue repair in chelonians. *PLoS ONE.* 2015;10(4):e0122595.

5. Pelizzone I, Di Ianni F. Use of Platelet Concentrate in a osteomyelitis in a *Boa constrictor imperator*. ICARE. 2017.

6. Intini G. The use of platelet-rich plasma in bone reconstruction therapy. *Biomaterials.* 2009;30(28): 4956- 4966.

7. Jia WT, Zhang CQ, Wang JQ, Feng Y, Ai ZS. The prophylactic effects of platelet-leucocyte gel in osteomyelitis: an experimental study in a rabbit model. *J Bone Joint Surg Br.* 2010;92(2):304-310.

Client Expectations and Preferences Regarding Veterinary Care of Exotic Pets

J. Mayer, DVM, MSc, DipECZM (Small Mammal),
DABVP (Exotic Companion Mammal Practice), Dipl ACZM,
E. Hofmeister, DVM, MA, DACVAA, DECVAA,
A. Bankert, PhD

Session #2049

Affiliation: From the Department of Small Animal Medicine and Surgery, College of Veterinary Medicine, University of Georgia, 2200 College Station Rd., Athens, GA 30605, USA (Mayer), Department of Surgery, College of Veterinary Medicine, Midwestern University, 5715 W Utopia Rd, Glendale, AZ 85308, USA (Hofmeister), Department of Political Science, School of Public and International Affairs, University of Georgia, 204 Candler Hall Athens, GA 30602, USA (Bankert).

The purpose of the present study is to determine client priorities and decision-making considerations in owners of exotic pets. An online survey was constructed and distributed via Facebook and pet owners' forums. Within a few months, responses from 1250 pet owners were recorded. The distribution regarding type of pet owned was as follows: small mammal (eg, rabbit, ferret) 39.10% (n = 547); reptile 25.23% (n = 353) and bird 35.67% (n = 499). The survey data reveal that, on average, the visit of an exotic pet owner is based mainly on the ability to be seen by a board certified specialist who offers a collaborative relationship to work up the case. During the initial exam and assessment, the clients would prefer to be seen directly by the most senior veterinarian even if this means longer waiting times. Moreover, the owners appear to prefer a more invasive approach, which has a potential for some risks but is more likely to identify the problem. Keeping the animal hospitalized for 2 days is preferred over home care. Most owners also preferred a surgical approach–even when it was significantly more expensive–to a more conservative approach as it might carry a better chance of success. Overall, our study demonstrates the need for a more empirical approach towards the evaluation of client preferences. Surveys are an important tool in this endeavor and can help veterinarians to improve the relationship with their clients.

Reality of Health of Exotics in the Pet Trade

Cathy Johnson-Delaney, DVM

Session #2090

Affiliation: From NW Zoological Supply, 3614 Smith Ave, Everett, WA 98201, USA.

Abstract: Puppy mills get all the press, but the health and status of small mammals and other exotics in the pet trade is of great concern. Mortality due to poor nutrition, husbandry practices and disease is higher than should be expected. Conditions found include cestodiasis, dermatophytosis, *Clostridium perfringens*, *Chlamydia caviae*, bordetellosis, and other bacterial and viral pathogens. Exotic pet veterinarians should be taking an active role in addressing these problems.

Introduction

Animals produced for the pet trade in large part are bred in large facilities where the emphasis is on numbers. In the United States, small mammals (with the exception of mice and rats) bred or distributed fall under the regulations of the Animal Welfare Act and breeders and dealers must be inspected by the U.S. Department of Agriculture (USDA). Most often animals include rabbits, ferrets, rodents, and exotic small mammals such as sugar gliders and hedgehogs. Reptilian, amphibian, and bird facilities are not inspected. Producing animals is done as economically as possible as the cost of production versus the sale value margin is low. Pet stores make their money largely from the sale of merchandise such as food, caging, and furnishings, not from the sale of the animals. Pet stores may not budget for veterinary care and lack staffing capable of providing medical care if such is required. One large pet store chain requires funds used for medical care to come from the budget needed for feeding animals. As this budget is tight to begin with, managers do the bare minimum. Several pet store chains have their own veterinary clinics on site and have those veterinarians attend to the exotic animals, although they have little training with those species. Veterinarians have been playing a marginalized role with most pet trade breeders and dealers. They often are called in only when there is a deficiency that needs correction, health certificates are needed for interstate transportation or sale, or if there has been a catastrophic disease outbreak and significant mortality. Many USDA inspectors are unfamiliar with exotic species and their specific nutrition and husbandry needs or disease conditions. They are only required to inspect such a facility on an annual basis, as is the designated attending veterinarian. Colony preventive health maintenance is largely one of herd health management, although such programs at many breeders and dealers are virtually nonexistent.

Animals, particularly rodents, often are over-produced as pet store orders vary over time. This results in a large number being euthanized, often simply because they have become too large or too old to be sold in the stores. While mice, rats, guinea pigs, hamsters, gerbils and rabbits may be euthanized and frozen to be sold as prey items, other species have no use as dead animals. Ill animals usually are culled. The morbidity and mortality rates are unknown, although from observation appear to be quite high. This is particularly true with imported reptiles such as bearded dragons. Many animals are shipped too young (barely weaned) and too small and succumb. Animals often are crowded in shipping crates and this stress, combined with underlying nutritional deficiencies and disease conditions, results in high morbidity and mortality. Inbreeding of some species has been noted as likely contributing to poor health and condition of the weanling animals.

The public expects animals purchased to be healthy. As many are to be children's pets, an animal purchased that becomes ill and dies shortly after purchase may result in complaints to the pet store. If the owner takes that animal to a veterinarian, costs usually exceed the actual purchase price of the animal. This prevents many owners from seeking veterinary care. Zoonotic disease risks also are a consideration. Common disease problems include dermatophytosis and *Salmonella* sp. Although not as commonly transmitted (or reported), other diseases found include *Rodentolepis nana* and *Giardia* sp.

The author is employed by a commercial pet trade breeder, dealer, and distributor and has been tasked with investigating disease problems found in animals from many sources here in the United States as well as animals being imported from sources overseas. The goal is to be able to develop pathogen-free populations for the pet trade, similar to the "specific pathogen-free" animals developed for laboratory animal medicine. Discussion with some breeders and their veterinarians has identified informational, nutritional and husbandry deficiencies and lack of effective disease diagnosis and preventive measures at those facilities. Frequently found disease conditions in selected species have been identified since June 2017 (Table 1).

Diagnostics

Many diseases can be diagnosed antemortem through fecal parasite examination; fungal culture; skin scrapings or tape tests (ectoparasites); bacterial cultures (aerobic and anaerobic); oral, nasal, rectal, lesion, or fecal swabs submitted for PCR; and blood testing primarily for viral screening. Necropsy samples of material within the gastrointestinal tract can be definitive for parasite identification and culture of pathogens particularly anaerobes such as *Clostridium* sp. Histopathology along with specific testing such as immunohistochemistry may be necessary to identify agents such as *Atadenovirus* sp. Screening of breeding colonies should be done on a routine basis using sentinel animals. In distribution facilities, animals from different vendors and shipments should not be mixed; animals that are ill upon arrival should be quarantined and appropriate diagnostics done.

Guinea Pigs

A large number of guinea pigs are being produced and raised in converted dairy barns. Many of these are not climate-controlled and ventilation is questionable. After discussing caging and diet with several producers as well as one of the veterinarians involved, it is clear that vitamin C levels are likely not adequate and there is little to no temperature or humidity controls within the barns. The guinea pigs often are given large-animal tetracycline or sulfa drug in their water, which explains the antibiotic resistance to these drugs found in the *Bordetella* and *Streptococcus* cultured. The frequent incidence of dermatophytosis may be exacerbated by poor husbandry, diet, bacterial and chlamydial infections as well as the age and size (often barely weaned and under 200 g body weight). Stress levels need to be reduced as part of the overall strategy to address the disease problems.

Many guinea pigs arrive with severe pneumonia and dermatophytosis and are euthanized. While the chlamydial infection responds to doxycycline (administered in the water (calculated to be 10 mg/kg PO q24h at 30 ml/100 g body weight, Doxyvet Liquid, Wagga Wagga NSW, Australia). They are wiped down using a paper towel soaked in F10 SC (diluted with water to a 1:250 strength, F10SC, Florida Hills, South Africa). Guinea pigs are then treated with selamectin (30 mg/kg topically, Revolution for cats, Zoetis, Madison, NJ, USA). They receive supplemental parsley as well as vitamin C in the drinking water. The vitamin C in the water is made fresh daily (mixed to a concentration 0.4 mg/ml drinking water, ascorbic acid powder, Bronson Laboratories, Lindan, UT, USA) Individual animals purchased from pet stores with respiratory disease may respond to fluoroquinolones, bronchodilators and nebulization in addition to further anti-chlamydial treatment using doxycycline.

Table 1. Disease agents and conditions found in selected species at one distributor in 2017.

Species	Health problem	Etiology
Guinea pig	Respiratory disease	*Bordetella bronchiseptica* *Chlamydia caviae* *Pasteurella pneumotropica* *Streptococcus pneumoniae, zooepidemicus* *Pseudomonas aeruginosa*
	Skin lesions, dermatitis, alopecia	Dermatophytosis (*Trichophyton* sp.) Pediculosis (*Gliricola* sp.) Hypovitaminosis C
	General weakness including poor dentition	Hypovitaminosis C Unknown nutritional status
Chinchilla	Skin lesions, dermatitis, alopecia	Dermatophytosis (*Trichophyton* sp.) Fight wounds
	Diarrhea	Bacterial dysbiosis (*Clostridium* sp., gram-negative enterics)
	Zoonotic concern	Cestodiasis (*Rodentolepis nana*). Found with/without diarrhea
Dwarf hamster	Skin lesions, dermatitis, alopecia	Dermatophytosis (*Trichophyton* sp.)
Syrian hamster	Diarrhea	Bacterial: *Clostridium perfringens* *Escherichia coli* *Klebsiella* sp. *Giardia* sp. *Lawsonia intracellularis* Stress (including weaning, shipping, crowding)
	Zoonotic concern	Cestodiasis (*Rodentolepis nana*). Found with/without diarrhea. Note: *Hymenolepis diminuta* also found
Ferret	Prolapsed rectum/anus with/without diarrhea	Post-demusking surgery, secondary bacterial infection *Coccidia* sp.
	Respiratory disease	Influenza
	Epistaxis, hemorrhage	Ferret hemorrhagic syndrome
	Deafness	Associated with blaze, panda coat colors; genetic Note: many have ear mites with no associate clinical signs
Bearded dragons	Poor weight gain, dehydration, anorexia	*Coccidia* (*Eimeria* sp.) Pinworms (unknown if heavy infestation contributes) Atadenovirus Nutritional deficiencies UVB deficiency
Budgerigars	Diarrhea/pasted vents, emaciation, anorexia	*Macrorhabdus ornithogaster*, various gram-negative bacterial agents including *E.coli*

Supplemental vitamin C as well as an NSAID such as meloxicam (0.3-0.4 mg/kg PO q24h, Metacam, Boehinger Ingleheim, Ridgefield, CT, USA) can be given to any ill, stressed guinea pig. For individual animals with dermatophytosis, griseofulvin (50 mg/kg PO q24h × 3-4 wk, griseofulvin compounded to 50 mg/ml, Kusler's Compounding Pharmacy, Snohomish, WA, USA) has worked well.

Chinchillas

At present many of the chinchillas have been imported from Europe. Any with dermatophytosis are placed on griseofulvin (50 mg/kg PO q24h for a minimum of 14 d, griseofulvin compounded to 50 mg/ml, Kusler's Compounding Pharmacy, Snohomish, WA, USA). Lesions are topically treated with miconazole (1-2 drops each area q12h, Conofite 1% lotion, Merck Animal Health, Madison, NJ, USA). Cestodiasis can be treated with praziquantel (10 mg/kg PO or SC, repeat in 10 d, Droncit, Bayer, Whippany, NJ, USA).

Dwarf Hamsters

Many hamsters arrive at the facility severely infected with dermatophytes and with skin lesions and are euthanized. Dermatophytosis can be treated with griseofulvin (50 mg/kg PO q24h for a minimum of 14 d, griseofulvin compounded to 50 mg/ml, Kusler's Compounding Pharmacy, Snohomish, WA, USA). Lesions are topically treated with miconazole (1-2 drops each area q12h, Conofite 1% lotion, Merck Animal Health, Madison, NJ, USA).

Syrian Hamsters

There are three types of Syrian hamsters in the pet trade: teddy bear (long hair), fancy (short hair, various coat colors and patterns), and the wild-type golden (which is just called short-hair). Those being produced in the United States currently are problematic, succumbing to diarrhea wet-tail. Investigation of animals from several breeders has yielded many pathogens listed in Table 1. Derivation of pathogen-free animals has begun in the author's breeding program. Based on numerous cultures, necropsies, intestinal parasite analysis and response to treatments, progress has been made addressing the pathogens. Effective treatment for diarrhea is largely based on eliminating *Clostridium perfringens, E.coli* and *Klebsiella*. Ciprofloxacin (20 mg/kg PO q24h × 14-21 d, Cipro, Bayer, Whippany, NJ, USA compounded into solution) in combination with metronidazole benzoate (40 mg/kg PO q24h × 14-21d, Kusler's Compounding Pharmacy, Snohomish, WA, USA) delivered individually has been effective. Tapeworms, giardia, and Lawsonia have been eliminated from the colony. Diarrhea in pet store hamsters should be aggressively treated using a fluoroquinolone and metronidazole benzoate along with supportive care that includes fluid therapy and supplemental feeding. Prognosis is poor for any hamster with diarrhea that exhibits ataxia even with treatment.

Ferrets

It is fairly common for kits arriving at the distributor to have prolapsing rectal or anal tissue. Some may have some bleeding and abraded tissue. The stools are loose in all the kits largely due to the feeding of the diet mixed with water to the consistency of soup. Treatment consists of topical anti-hemorrhoidal ointment (applied to tissue q12h, Preparation-H, Pfizer, New York, NY, USA). Additional therapy may include amoxicillin (50 mg/kg PO q12h, BioMox, Lambert Veterinary, Fairbury, NE, USA). Most shipments contain ferrets with respiratory clinical signs including dyspnea, conjunctivitis, tonsillitis, and sneezing. Human influenza virus has been confirmed, and more cases are found in the months human influenza is more prevalent. Ferrets are isolated and may receive amoxicillin at the above regimen to control for secondary bacterial infection. Epistaxis and internal hemorrhaging and acute death are seen occasionally. As this appears to be ferret hemorrhagic syndrome, the author is continuing diagnostic investigation as cases arise. Coccidia have been found in some ferrets after being housed in pet stores. Ear mites usually are not found in the ferrets arriving directly from the breeder, but are found frequently after ferrets have been housed in pet stores.

Bearded Dragons

These animals are imported primarily from several sources. Those coming from tropical breeders are heavily infected with coccidia and pinworms. Atadenovirus has been found particularly in those that fail to thrive. At our facility we have pioneered an alternate housing that greatly decreases exposure to feces while maintaining optimal temperature and UVB exposure. At unpacking, bearded dragons are weighed and then dosed with fenbendazole (50 mg/kg PO once, Panacur, Merck Animal Health, Madison, NJ, USA) and ponazuril (30 mg/kg PO, repeated in 48h, Marquis, Boehinger Ingleheim, Ridgefield, CT, USA) and then placed in soaking trays to rehydrate. This treatment has significantly cut morbidity and mortality. In addition, bearded dragons are held for several weeks to confirm appetite and weight gain before being shipped to pet stores. Bearded dragons arriving from European breeders usually are free of intestinal parasites.

Budgerigars

Circovirus is present in budgerigars from several breeders in the United States, although it is unclear how pathogenic this virus strain is. Showering the birds after unpacking generally results in negative tests on PCR, while birds not showered have positive fecal swabs. The budgerigars are shipped in containers with white millet as the food and substrate along with orange slices. The birds often are dehydrated and many have stress droppings. Pasted vents, anorexia, and thin birds have had feces cultured, identifying various gram-negative bacteria. *Macrorhabdus ornithogaster* has been found on Gram stains. Severely ill birds are euthanized. Those less affected are placed in a hospital cage and treated with antibiotics as indicated based on culture and sensitivity testing. All birds are fed a pellet and seed mixture augmented with Avicakes (Lafeber Co, Cornell, IL).

Other Exotic Species

Sugar gliders are being called "sugar bears" and are being sold at malls and fair kiosks as well as over the Internet. Calling them sugar bears has resulted in confusion when the animal is presented ill at a veterinary hospital. Nutritional deficiencies and imbalances are still contributing to significant numbers of sugar gliders dying. Those being sold at malls, fairs and over the Internet often are barely weaned. There are a number of commercial diets being marketed that have not had proper feeding trials conducted.

Pet rats may carry *Streptobacillus moniliformes*, the agent of rat bite fever. Because of the zoonotic risk, many pet stores no longer carry live rats. This bacteria can be tested for and eliminated if breeders will do so. There will have to be certification of disease-free status before stores will resume selling rats. Developing a facility testing program can be based on laboratory animal programs.

Working with Breeders, Vendors, Distributors, and Pet Stores

We can offer expertise on diet, husbandry, breeding, and preventive health programs. This expertise will save the breeder, vendor, distributor or pet store money and produce a far healthier pet for the public. Convincing people of this is usually difficult as they look at the short-term money outlay. The first step in any facility is to identify what problems are present and that may take some time and diagnostics. USDA-inspected facilities should be satisfactory as to ventilation, environmental controls, and sanitation, but many are following a laboratory animal guideline that doesn't work when shipments are coming in and going out multiple times a week of multiple species. Close attention to disinfectants used, cage cleaning procedures, and animal handling often yield clues to the stresses being placed on these animals. Laboratory animal diets are available for many of the species and are actually the economical way for big facilities to provide good nutrition. Many in the pet trade however are not using these. Independent pet stores are often more willing to accept preventive health programs than large chain stores that have their own programs. Despite the Internet, good and accurate information about exotic pet species health and well-being is still not reaching owners. Most people believe the first sites that come up on a search and are not able to discern that most sites are posting information geared to sell animals rather than by those with medical and animal colony experience. We as a profession need to market exotic pet health and get that information to come up easily for the owners. Making the information available at the point of sale also would help.

Acknowledgments: Alex and Denice Orleans, NW Zoological Supply, LLC, Everett, WA, Dr. Drury Reavill, Zoo/Exotic Pathology Service, Carmichael, CA., Dr. Darrell Kraft, Kraft Mobile Services, Snohomish, WA, Dr. Robert Dahlhausen, Veterinary Molecular Diagnostics, Milford, OH.

References and Further Reading

1. Carpenter JW, Marion CJ. *Exotic Animal Formulary.* 5th ed. St. Louis, MO: Elsevier;2018.

2. Harkness JE, Turner PV, VandeWoude S, Wheler CL. *Harkness and Wagner's Biology and Medicine of Rabbits and Rodents.* 5th ed. Ames, IA: Wiley-Blackwell;2010.

3. Phalen D. Macrorhabdosis. In: Speer BL, ed. *Current Therapy in Avian Medicine and Surgery.* St. Louis, MO: Elsevier;2016:78-82.

4. Swisher S, Lennox AM. Disorders of the haemic, imunological and lymphatic systems. In: Johnson-Delaney CA, ed. *Ferret Medicine and Surgery.* Boca Raton, FL: CRC Press;2017:237-257.

5. Wellehan JFX, Johnson AF, Harrach B, et al. Detection and analysis of six lizard adenoviruses by consensus primer PCR provides further evidence of a reptilian origin for the atadenoviruses. *J Virol.* 2004;78: 13366-13369.

Animal Welfare in Wildlife Rehabilitation: Maintaining High Welfare Standards in a Private Practice Setting

Renée Schott, DVM, CWR

Session #2210

Affiliation: From the Wildlife Rehabilitation Center of Minnesota, 2530 Dale St. N, Roseville, MN 55113, USA.

Abstract: Animal welfare is an especially important consideration when rehabilitating wildlife, especially in a private practice setting. Wildlife in rehabilitation are always stressed and the negative effects of stress are well known. It is paramount that the rehabilitator and the veterinarian understand ways to decrease stress. This Masterclass will first briefly cover the basics of animal welfare, then cover the causes of stress in captive wildlife, what we can do to mitigate those stressors and finally the ethical implications of caring for wildlife in a rehabilitation setting. Special emphasis will be put on private practice scenarios and considerations. This Masterclass will simplify animal welfare concepts into tangible, easy to practice tools to take back to your clinic.

Introduction

Regardless of whether it is ethical to try and save all injured and ill wildlife,[1] people continue, and will continue, to rescue and bring injured and ill wildlife to wildlife rehabilitators. The need is present, and if a licensed wildlife rehabilitator isn't handy, the public often drops these animals off at the local veterinary clinic. Regardless of how the animal got there, now it is in your care and you have an obligation towards its welfare.

Animal welfare is a relatively new area of scientific inquiry, and there is very little published literature that considers the welfare of wildlife undergoing rehabilitation. Molony et al.[2] considered release the only positive welfare aspect of wildlife in rehabilitation, however, there is much more that can be done to improve the welfare of wild animals undergoing rehabilitation before release.

Surprisingly, there is no universal definition of animal welfare.[3] Veissier et al.[4] describes animal welfare as a multi-dimensional concept incorporating an animal's health, emotions, behavioral needs and psychologic needs among other things. Ng states welfare is net happiness; that is to mean enjoyment minus suffering.[5] The AVMA defines animal welfare as how an animal is coping with the conditions in which is lives.[6] These and most other definitions agree that an animal's welfare relates to the sensations the animal experiences.[7] External inputs (ie, from the environment) and internal inputs (ie, from the animal's body) are processed, integrated and interpreted by the animal's brain using past experiences, traits specifics to the species and to that individual. The result is the animal's current experience. All of the positive experiences and negative experiences are weighed against each other and this is often termed the animal's welfare status.[7] This status changes as the experience changes. This assumes that welfare only diminishes if the animal has a negative experience and only enhanced if the animal has a positive experience.[8] Often, the net balance of these states are weighed when assessing quality of life or whether a life is worth living.[9,10]

These states were first accepted in the Five Freedoms, which are: the freedom from thirst, hunger, and malnutrition; the freedom from discomfort and exposure; the freedom pain, injury and disease; the freedom from fear and distress; the freedom to express normal behavior.[11] The five freedoms are however limited because although they try and minimize negative states, they do nothing to encourage positive states. Currently, the concept of the Five Domains has replaced the Five Freedoms.[12] The Five Domains are divided into physical/functional domains (nutrition, environment, health, and behavior) and the affective experience domain (mental state); each have positive and negative aspects. For example, nutritional, environmental and health-related problems and problems with ability to express behaviors, can have physiologic and pathophysiologic effects. Once these are assessed, their anticipated affective consequences are assigned to the fifth domain, the mental state.[12] The mental state is made of the animal's experiences, which determines the animal's welfare.

Welfare is a continuum from very bad (all negative states and no positive states) to neutral (balanced bad and good states) to very good (all positive and no negative states). In assessing welfare, one isn't trying to only look for welfare compromise, but also welfare enhancement. The goal is not to completely eradicate the negative states (because that is impossible) but to minimize them, their effects and consequences.

Some examples of negative states include: breathlessness, thirst, hunger, nausea, pain, fear, dizziness, debility, weakness and sickness, anxiety, frustration, anger, helplessness, loneliness, boredom, panic. Some examples of positive states include: satiety, contentment, companionship, curiosity and playfulness. To simplify it, negative states create stress and all the negative physiologic consequences of stress, which include impaired growth in young,[13] impaired metabolism,[14] impaired immune system,[15] slowed wound healing[16] and sudden release of stress hormones such as cortisol and epinephrine[17] which can cause sudden death. Since the goal of wildlife rehabilitation is to get healthy animals back into the wild, stress is counterproductive. Alternatively, if welfare is improved enough to tip the scale and create a positive state, patients may heal quicker, fight disease faster and grow healthier, speeding release. If animals are released sooner, it frees up resources available to treat other patients. There are a finite amount of resources and an infinite number of wild animals that need rehabilitation services so prudent use of these resources is paramount.

Inherent Affective Experiences of Wildlife Undergoing Rehabilitation

Humans are predators to all wild animals. Every animal that enters the rehabilitation process feels like prey. Imagine if Godzilla picked you up, roared at you over and over, poked and prodded you, threw you in a cage and continued to repeat the process, you would be in a very negative state. It wouldn't matter if Godzilla was trying to soothe you or tell "It's ok, I'm trying to help" because you don't understand his language; it all sounds like "I want to eat you" to you. One can only assume wildlife in captivity feels the same way. Every day is the worst day of their life. They are scared, afraid of being eaten or hurt and always desperately trying to escape. Those that appear to calm down likely are simply finding a way to cope and survive, which shouldn't be confused with a positive experience. It is most likely impossible for humans to tip the welfare scale into the positive direction; however, it is possible to minimize the overall negative affective states by minimizing stress experiences.

Ways to Improve Welfare in Wildlife Undergoing Rehabilitation

When considering how to decrease stress in captive wildlife, it helps to consider the five senses and stressors associated with them.

Sight stressors

Humans: Humans are predators to all wild animals so limiting interactions with wildlife patients is paramount to decreasing their overall stress. Easy ways to limit interaction include: being efficient when you need to handle (eg, combine physical exams with medication administration, choose medications that need less frequent dosing); covering enclosure doors to prevent patients from seeing passersby; providing hide spots or hide boxes in enclosures so patients can choose to hide and not be seen; catching patients from a cage with the lights off; averting your eyes when approaching a patient can make you seem less like a predator; covering the animal's head while restraining or working with it. Above all else, no peeking!

Abnormal enclosure: everything about being in captivity is abnormal. Wild animals are used to being in natural habitats and able to choose wherever they want to go. Providing enclosures with as many natural items as possible will make it more familiar.

Other predators: wild animals have evolved to fear other wild predators other than humans. A rabbit seeing or smelling an opossum, cat, dog or ferret is extremely stressful, especially when the rabbit is trapped in a cage. Housing predators and prey animals in different rooms will decrease stress.

Normal day/night cycle: Being indoors and not seeing the day/night is abnormal; replicate it with as normal of a day/night cycle as possible with as close to a full-spectrum lighting as possible.

Sound stressors

Humans: Humans are predators to all wild animals and hearing them is incredibly stressful. Limit talking when working with or around wild animals.

Abnormal building sounds: Buildings are not natural for wild animals so doors slamming, dishes clashing or other sounds (eg, telephone ringing) is abnormal. Using background music has shown to decrease the startle response by drowning out abnormal noises.[18] This author recommends using a white noise machine or nature sounds.

Other predators: Prey species hearing predators is extremely distressing, so housing them in separate rooms is necessary.

Smell stressors

Humans: Perfumes and other body scents are abnormal. While we don't know exactly how they affect wild animals, it is best to stay away from strong odors.

Cleaners/disinfectants: Not only can these be caustic or physically harmful to certain species (ie, avian air sacs) but most are not found in nature and could be stressful.

Predators: Even if prey can't see predators, they can potentially smell them and be fearful.

Taste stressors

Abnormal food: Many nutritionally balanced rehabilitation diets are made of unnatural or processed products. Lack of a natural food source not only may cause hyporexia but can be mentally stressful. Offering as many natural food items as possible is good practice.

Touch stressors

Humans: Wild animals likely are in fear of their lives when being held or restrained by a human. It is paramount to limit all handling to the bare minimum.

Enclosure: Wild animals are used to being out in nature and choosing where to go. Being trapped in a cage with no escape is incredibly stressful. Ensure enclosures meet minimum standards[19] and allow normal behaviors, are supplied with natural materials, have a hiding spot and good footing.

Flooring: Slipping on traction-less surfaces is another very stressful experience. Ensure all floor and ground surfaces—including weight buckets, enclosures, exercise areas, physical exam areas—have a substrate that is not slippery.

Euthanasia

All wild animals in rehabilitation are either injured, ill, orphaned or a combination. By definition, they are suffering. With domestic animals, veterinary medicine aims to preserve life because even permanent disabled pets or pets in chronic pain often can have a decent quality of life. However, wildlife cannot be released if they are disabled or in chronic pain. Therefore, one of the best ways to decrease suffering, and thus increase welfare, in wildlife rehabilitation is appropriate triage of new admissions. If the new patient being admitted will not be releasable after being treated, and it will suffer the entire time it is captivity, then prompt euthanasia is the best welfare that can be offered.[20]

Transfer

Often, the veterinary clinic cannot offer a good welfare environment for wildlife undergoing rehabilitation. In this case, it is in the best interest of the animal to transfer to a rehabilitator with appropriate facilities. Rehabilitators are often looking for veterinarians to help with the veterinary care and usually welcome continued support.

Limits

It is important for the veterinarian to understand what appropriate treatment is for a wild animal. Some species can do well without a toe so a toe amputation maybe appropriate; however, can the species you are treating be released without a limb? Will the wing recover full function if you wrap it to the body? For how long? Do you have the skills to pin the fractured bone well enough so the animal can run and climb as well as its health counterpart? Sometimes knowing one's limits is more important than pulling out all the stops.

Conclusion

It is tempting to keep injured wildlife in a private practice setting for rehabilitation. Having such unique species is novel and exciting! The above stressors can make offering wildlife a good level of welfare in a clinic difficult. Consider what is best for the animal and how can its welfare be put before any humans wants. Keeping in mind that all wild animal patients are prey animals, can help veterinarians better interact and ultimately treat them. Working together with your local rehabilitator as a team will ensure each wild patient gets the best chance to be ultimately released.

References

1. Kirkwood JK, Sainsbury AW. Ethics of interventions for the welfare of free-living wild animals. *Anim Welf.* 1996;5(3):235-243.

2. Molony SE, Baker PJ, Garland L, Cuthill IC, Harris S. Factors that can be used to predict release rates for wildlife casualties. *Anim Welf.* 2007;16(3):361-367.

3. Hewson CJ. What is animal welfare? Common definitions and their practical consequences. *Can Vet J.* 2003;44(6):496-499.

4. Veissier I, Spoolder H, Rushen J, Mounier L. Editorial: Pigs crying, silent fish and other stories about animal welfare assessment. *Animal*. 2015;10(2):292-293. doi:10.1017/S1751731115002682.

5. Ng Y. How welfare biology and commonsense may help to reduce animal suffering. *Anim Sentience*. 2016:1-9.

6. American Veterinary Medical Association. 2018. Animal Welfare: What is it? Available at: https://www.avma.org/KB/Resources/Reference/AnimalWelfare/Pages/what-is-animal-welfare.aspx. Accesssed July 2, 2018.

7. Hemsworth PH, Mellor DJ, Cronin GM, Tilbrook AJ. Scientific assessment of animal welfare. *N Z Vet J.* 2015;63(1):24-30. doi:10.1080/00480169.2014.966167.

8. Robbins J, Franks B, von Keyserlingk MAG. More than a feeling: An empirical investigation of hedonistic accounts of animal welfare. *PLoS One*. 2018;13(3):1-11. doi:10.1371/journal.pone.0193864.

9. Yeates J. Quality of life and animal behaviour. *Appl Anim Behav Sci.* 2016;181:19-26. doi:10.1016/j.applanim.2016.04.018.

10. Mcmillan FD. Quality of life in animals. *J Am Vet Med Assoc.* 2000;216(12):1904-1910.

11. Mellor DJ. Updating animalwelfare thinking: Moving beyond the five freedoms towards a life worth living. *Animals*. 2016;6(3)21. doi:10.3390/ani6030021.

12. Mellor DJ, Beausoleil NJ. Extending the Five Domains model for animal welfare assessment to incorporate positive welfare states. *Anim Welf.* 2015;24(3):241-253. doi:10.7120/09627286.24.3.241.

13. Moberg GP. Biological response to stress. In: Moberg GP, Mench JA, eds. *The Biology of Animal Stress.* Wallingford, UK: CABI;2000:11-12.

14. Elsasser TH, Klasing KC, Filipov N, Thompson F. The metabolic consequences of stress. In: Moberg GP, Mench JA, eds. *The Biology of Animal Stress.* Wallingford, UK: CABI;2000:77-110.

15. Blecha F. Immune response to stress. In: Moberg GP, Mench JA, eds. *The Biology of Animal Stress.* Wallingford, UK: CABI Publishing;2000:111-121.

16. Guo S, Dipietro L a. Factors affecting wound healing. *J Dent Res.* 2010;89(3):219-229. doi:10.1177/0022034509359125.

17. Gaynor JS, Muir III WW. *Handbook of Veterinary Pain Management*. 3rd ed. St. Louis, MO: Elsevier;2015.

18. Goody A, Rachel F, Gelatos M, Yim C. Background music to reduce startle respnse in wild avian species. *J Wildl Rehabil*. 2013;33(1):13-18. http://theiwrc.org/wp-content/uploads/2010/07/JWR331FinalRev.pdf.

19. Miller E, ed. *Minimum Standards for Wildlife Rehabilitation*. 4th ed. St. Cloud, MN: National Wildlife Rehabilitators Association and International Wildlife Rehabilitation Council;2012.

20. Mullineaux E. Veterinary treatment and rehabilitation of indigenous wildlife. *J Small Anim Pract*. 2014;55(6):293-300. doi:10.1111/jsap.12213.

Section 2

Hybrid Technician Session

Pia Bartolini, CVT, VTS (CP-Exotics);
Ryan O'Shea CVT;
Liz Vetrano CVT, BS
Session Chairs

Avian and Exotic Companion Mammal Case Presentations

Cheryl B. Greenacre, DVM, DABVP (Avian Practice), DABVP (Exotic Companion Mammal Practice), Professor, Avian/Zoological Medicine

Session #2304

Affiliation: From the Department of Small Animal Clinical Sciences, College of Veterinary Medicine, University of Tennessee, Knoxville, TN 37996, USA.

Abstract: These case studies represent some of the most commonly encountered diseases in exotic animal medicine. Each case study provides a history and physical examination, and sometimes initial diagnostic findings. Your goal for each case study is to develop and commit to a thorough problem list, rule-out list, and diagnostic plan before looking at my lists just below. Then, after reviewing the findings of the diagnostic tests, develop and commit to a treatment plan before looking at my treatment plan just below. You will get the most out of this exercise if you commit to your own ideas before looking at my list. There are multiple perfectly good methods to work up the same case, so please keep in mind that my way is not necessarily the only way.

Case 1. Cockatiel

Review the information in Table 1 and formulate a problem list, rule-outs for each problem, and a diagnostic plan.

Table 1. Cockatiel case.

Factor	Description
Signalment	4-month-old cockatiel
Client complaint	1 week progressive anorexia, depression, lethargy, fluffed, this morning at the bottom of the cage
History	Purchased 2 weeks ago from a pet store; not polyoma vaccinated nor *Chlamydia* tested, nor PBFD tested; housed in a 2' × 2' × 2' white powder coated cage with newspaper on the bottom; stays indoors; only other pet in the household is a cat
Diet	Seed mix for cockatiels (millet, sunflower and safflower seeds), and millet sprigs
Physical examination	Moderate to severe depression, fluffed, eyes closed, moderately thin, nasal discharge, dyspnea, green urates, and decreased fecal output; auscultation wet crackles; weight 82 grams

Author's problems, rule-out lists, and diagnostic plans

The author's problems, rule-outs, and plans are presented in Table 2.

Table 2. Author's problems and rule-out lists, assessment plan, and diagnostic results for a cockatiel case.

Factor	Description
Problems	Nasal discharge, wet crackles ausculted/bacterial, chlamydial, fungal, viral, foreign body, neoplasia Green urates/liver disease (bacterial, viral, chlamydial, fungal, toxin, neoplasia) Anorexia, depression, lethargy, fluffed, bottom of cage, thin/many rule-outs for a very sick bird
Assessment	Upper and lower respiratory disease/f, and liver disease (chlamydiosis must be on rule-out list/fecal gram stain, CBC (differential), PCV, profile (AST, LDH, bile acids), radiograph, chlamydia test, ± PBFD test.
Results	Fecal Gram stain: 100% gram positives, 90% rods, 10% cocci, no yeast. Differential: estimated WBC 45,000 (leukocytosis), heterophils 70% (heterophilia), lymphocytes 15% (lymphopenia), monocytes 15% (monocytosis) PCV: 33% (anemia, regenerative) Profile: (limited due to size of bird, unless use new benchtop analyzers that can provide an entire chemistry profile on only 0.1 m; of blood) hypoproteinemia (1.8), hypoalbuminemia (1.0), elevated AST (375), LDH (450) and bile acids (80) Radiograph: hepatosplenomegaly

After reviewing information in Table 2, further formulate the diagnostic plan.

Chlamydia testing

Which test should be run? There are some new advances and it is best to perform a panel of the following 3 tests to obtain the most information, since there is no perfect test.

1) Serum antibody: Screening test, (+) = exposure, not necessarily disease

2) Blood PCR: Chlamydia DNA in WBC, suggests active infection

3) Fecal PCR: Means shedding chlamydia DNA in feces

An in-house fecal ELISA detects antigen in feces, but can obtain false positives and negatives.

The above cockatiel was negative on an in-house fecal ELISA, and the *Chlamydia* panel results will take 3-5 days to come back, so now formulate a treatment plan. Until get remainder of *Chlamydia* test results are received, what do you do?

Author's plan for avian patient care while awaiting *Chlamydia* test results

Several steps are outlined in Table 3, which provide the author's approach to caring for the avian patient while awaiting remaining test results.

Table 3. Author's plan for avian patient care while awaiting *Chlamydia* test results.

Step	Avian patient care
1	Hospitalize in 85-90 °F incubator with humidity
2	Fluids (LRS) 50 ml/kg/day maintenance, give half SQ or IO bolus, other half over next 24 hours
3	Broad spectrum, bactericidal antibiotic (enrofloxacin, 10 mg/kg IM q12h,would be a good choice, since it has some affect against chlamydia) in addition, administer doxycycline, 25 mg/kg PO q12h, (treatment of choice against chlamydia)
4	After fluids, tube feed small amount (5 cc) of hand-feeding formula
5	Lactulose 0.03 ml PO q12h
6	± Nystatin

Avian patient response

Twenty-four hours later, the bird is much improved! The bird is eating on its own, but not maintaining weight, so hand feeding is still required.

The diagnosis?

It is an enrofloxacin- and/or doxycycline-responsive disease, but is probably chlamydiosis, since the bird responded so swiftly to treatment.

Test results and diagnosis

Test results are received 48 hours later. The results include: Antibody (+), blood PCR (+), and fecal antigen (+).

Based on these results, the diagnosis is chlamydiosis. So now formulate a revised treatment plan.

Author's revised treatment plan

The revised treatment plan is described in Table 4. The plan includes treatment as well as zoonoses concerns and reporting requirements.

Table 4. Author's revised treatment plan.

Action	Description
Treatment	Doxycycline only, discontinue enrofloxacin, and continue administering lactulose; treatment must be for 45 days per the law; chlamydia is an intracellular organism
Zoonoses	If anyone who has come in contact with this bird is sick in any way ill or has flu-like symptoms in the next 6 months they should see their physician immediately
Reporting	Chlamydiosis is a reportable disease in most states (not Tennessee); call the state veterinarian or public health department immediately upon receiving the positive test results; legal cases require a culture of before treatment feces sent to one laboratory in Texas

Current addresses and phone numbers of laboratories that test for *Chlamydophila*, accepted by the American Veterinary Medical Association (AVMA) and the Association of Avian Veterinarians (AAV) can be found in the Compendium of Measures to Control *Chlamydophila psittaci* (formerly *Chlamydia psittaci*) Infection Among Humans (Psittacosis) and Pet Birds (Avian Chlamydiosis), 2004 by the National Association of State Public Health Veterinarians (NASPHV) at the following website: http://www.avma.org/pubhlth/psittacosis.asp

Miscellaneous items

What if antibody titer was positive only? Repeat antibody titer 2 weeks later, if it demonstrates a rising titer, then it is disease. Supposedly disease doesn't shed after the third day of treatment of doxycycline.

What about the cat? It is thought to be possible for a bird to transmit chlamydia to a cat. With regard to diet, add cuttle bone. Current diet recommendation for a cockatiel diet is half pellets and half seeds.

The word for green urates is biliverdinuria.

Taxonomy and nomenclature of *Chlamydophila*

The disease psittacosis goes by the names parrot fever, ornithosis, chlamydiosis, and chlamydophilosis. The term psittacosis refers to the disease in people originating from a parrot (a psittacine bird), whereas the term ornithosis refers to the disease in people originating from any species of bird. The term chlamydiosis is a more generic term and refers to infection in any animal or person with any organism in the *Chlamydia* or *Chlamydophila* genus. The term avian chlamydiosis is used to specify an infection with *Chlamydophila psittaci* in birds.

Note that *Chlamydophila psittaci* should not be confused with a related organism in people, *Chlamydia trachomatis*, a sexually transmitted disease of people or another related organism, *Chlamydophila pneumoniae*, a common mild respiratory pathogen of people.

In 1999 changes in nomenclàture occurred that reflected recent advances in DNA testing that revealed differences between organisms that were previously thought to be the same. What used to be just one genus *Chlamydia*, is now described as two, *Chlamydia* and *Chlamydophila*. A word of caution therefore, when reading any literature on chlamydial organisms prior to 1999, the reader may not be 100% certain as to which of the newly categorized organisms was being referred to in the document.

Epidemiology

The organism has been found in 130 species of birds worldwide and a variety of mammals, including humans, and is therefore a zoonotic disease. There are numerous potential avian species that may act as a source of infection for people. The most common source of infection (70% of all cases in the 1980's) is exposure to a recently acquired psittacine bird, most often a budgerigar, cockatiel or lovebird. Birds smuggled across the United States borders as well as those raised in captivity can potentially be infected with avian chlamydiosis. Other birds can be a potential source of infection, such as domestic or wild pigeons, passerines (soft-billed birds) or poultry. The organism *Chlamydophila psittaci* is relatively resistant, surviving in the soil for 3 months or within a bird dropping for up to 1 month. *Chlamydophila psittaci* is a dimorphic organism, meaning it exists in two different forms called phases, the infectious phase and the replicating phase. In general, the *C. psittaci* organism enters the host via inhalation or ingestion and replicates in the host's cells. The infectious form is small (0.2-0.6 microns) and called an elementary body. After the elementary body is inhaled or ingested it is endocytosed by a host cell where it transforms into a reticulate body. The reticulate body is larger (1.5 microns) than the elementary body and is the replicating, or metabolically active, form of the organism inside the host's cell. The reticulate body uses nutrients from the host cell and undergoes multiple rounds of binary division before releasing multiple elementary bodies from the cell when it ruptures. Two or more days pass between the time the host cell is infected and the elementary bodies are released. These infectious elementary bodies then infect other of the host's cells or are released into the environment via feces, nasal secretions, sputum, blood or infected tissues. The elementary body is metabolically inactive and resistant to environmental forces, so it can survive long enough to be inhaled or ingested by another host.

Is avian chlamydiosis considered a reportable disease by veterinarians?

Usually it is reportable, but it depends on the state. If a veterinarian diagnoses psittacosis in a bird, then most states require that that case be reported to the state veterinarian or public health department. Usually the state veterinarian's office works with the reporting veterinarian to either test or treat the other exposed birds involved with that case. If the bird came from a pet store or bird breeding facility where additional cases could spread to other people, the state veterinarian, will usually enforce a 45-day ban on movement of birds in and out of the facility during testing and treatment. Usually the public health department becomes involved if humans are affected.

Is human psittacosis considered a reportable disease by physicians?

No, human psittacosis is a nationally notifiable disease (as designated by the Council of State and Territorial Epidemiologists, in conjunction with CDC) and approximately 100-250 cases are reported to the CDC yearly. In 1992 it was a notifiable in 42 states, but in 1999 it became nationally notifiable. Understand that this does not mean reportable (ie, physicians are not required to report the disease in humans). The actual number of human cases is thought to be higher due to non-reporting or misdiagnosis. Psittacosis is deemed a Category B Bioterrorism Agent by the CDC, defined as a second highest priority agent and one that is moderately easy to disseminate, can result in moderate morbidity rates and low mortality rates and require specific enhancements of CDC's diagnostic capacity and enhanced disease surveillance.

To help identify multistate outbreaks of *C. psittaci*, local and state authorities should report suspected outbreaks to the Respiratory Diseases Branch, Division of Mycotic Diseases, National Center for Infectious Diseases, CDC, telephone (404) 639-2215.

Physicians are not required to report human cases of *Chlamydophila psittaci*, but reporting of *Chlamydia trachomatis* infections (sexually transmitted disease of humans) is required nationally.

Are imported birds tested for psittacosis?

No, imported birds are not tested for psittacosis. The Veterinary Services of the APHIS, USDA that regulates importation of pet birds ensures that exotic poultry diseases (such as Exotic Newcastle's Disease, aka END), and avian influenza (AI) are not brought into this country. While imported birds are quarantined they are only tested for END and AI and given chlortetracycline (CTC) pellets to prevent psittacosis shedding during quarantine, so as not to expose the people caring for the birds. The CTC is not appropriate for treating psittacosis, but it does prevent shedding of the organism.

Case 2. Rabbit

Review the information in Table 5 and formulate a problem list, rule-outs for each problem, and a diagnostic plan.

Table 5. Rabbit case.

Factor	Description
Signalment	5-year-old female, intact, French lop rabbit
Client complaint	2 week history of depression, anorexia, red urine, occasional soft stool
History	Owned since was a baby at the pet store. No previous problems. No other animals
Diet	Rabbit pellets, apples, carrots and seed treat daily
Housing	In a 6' × 6' × 4' wire mesh cage with newspaper under grate which is changed daily
Physical examination	Bright, alert, responsive, obese, well hydrated, 5 gut sounds/minute, 6 × 4 cm caudal abdominal mass, doughy, mammary glands enlarged and producing milk

Author's problems, rule-out lists, and diagnostic plans

The author's problems, rule-outs, and plans are presented in Table 6. Published reference ranges for complete blood count (CBC) and chemistry panel in rabbits are provided in Table 7.

Table 6. Author's problems and rule-out lists, assessment plan, and diagnostic results for a rabbit case.

Factor	Description
Problems	Red urine/porphyrinuria, hematuria, hemaglobinuria, myoglobinuria
	Occasional soft stool/low fiber diet, fruit in diet, seeds in diet, obesity making rabbit unable to reach cecatrophs to eat them, parasites, foreign body (carpet fibers), other GI disease
	Obese/low fiber diet, fruits and seeds, low amount of exercise
	Caudal abdominal mass/uterus, bladder, GI tract; neoplasia, abcess, grauloma
	Mammary gland enlargement with milk production/uterine tumor, pseudopregnancy, pregnancy
	Depression, could be anything, but indicates chronic rather than acute disease
	Anorexia, could be anything (GI including teeth problems, or non-GI), indicator of severe disease in rabbit (need to do something quick to get her eating again!)
	No hay or greens in diet, less than ideal fiber intake
Assessment	CBC, profile, radiograph, ultrasound, urinalysis (I did clean, free catch; you could also do cystocentesis during US) (save some in case this is all you can obtain for a culture later), oral examination, fecal float, direct and Gram's stain
	CBC: leukocytosis with a left shift (3% bands) indicates infection (uterus, bladder, GI?)
	Chemistry profile: hypercalcemia/diet, increased alkaline phosphatase/liver disease?
	Urinalysis: calciuria due to increased calcium in diet, hematuria could be due to blood from uterus, or blood from diapedesis through bladder wall or a cystitis, urolithiasis
	Radiograph/ultrasound uterine mass in an older rabbit is most likely uterine adenocarcinoma, cystic wall to uterus suggests concurrent cystic endometrial hyperplasia (common), fluid in uterus is probably mucus, but could also be pus
Results	Oral examination: within normal limits
	Fecal float, direct and Gram's stain: within normal limits
	Radiograph abdomen: 6 × 4 cm caudal abdominal mass, dorso cranial to bladder
	Ultrasound: mass is fluid filled with a cystic, thickened wall
	Urinalysis: calciuria, hematuria with WBC and bacteria

Table 7. Published complete blood count and chemistry reference ranges in rabbits.

Complete blood count	Value	Published reference range for rabbit
White blood cells	30,000	5,200-12,500
Neutrophils	82% (24,600)	20-75%
Bands	3% (900)	0%
Lymphocytes	11% (3300)	30-85%
Monocytes	4% (1200)	1-4%
Eosinophils	0%	1-4%
Basophils	0%	1-7%
Packed cell volume	30%	33-50%
Chemistry Panel		
Total protein	7.1 g/dl	5.4-8.3
Albumin	4.4 g/dl	2.4-4.6
Globulin	2.7 g/dl	1.5-2.8
Glucose	122 mg/dl	75-155
Blood urea nitrogen (BUN)	14 mg/dl	13-29
Creatinine	0.7 mg/dl	0.5-2.5
Alanine aminotransferase (ALT)	73 U/L	48-80
Aspartate aminotransferase (AST)	45 U/L	14-113
Alkaline phosphatase	36 U/L	4-16 (Most sick rabbits I test are higher than this)
Total bilirubin	0.1 mg/dl	0-0.7
Cholesterol	70 mg/dl	10-80
Sodium (Na)	147 mmol/L	131-155
Potassium (K)	4.4 mmol/L	3.6-6.9
Chloride (Cl)	108 mmol/L	92-112
Calcium (Ca)	20 mg/dl	5.6-12.5 (I hardly ever see it this low)
Phosphorus (P)	5.0 mg/dl	4.0-6.9
Lactate dehydrogenase (LDH)	62 U/L	34-129

Source: Quesenberry KE, Carpenter JW, eds. *Ferrets, Rabbits, and Rodents: Clinical Medicine and Surgery.* 3rd ed. St. Louis, MO: Elsevier Saunders;2012.

After reviewing information in Tables 6 and 7, further formulate the diagnostic plan.

Author's diagnostic and treatment plan

Do a thoracic radiograph before ovariohysterectomy (OHE) surgery to ensure there are no mets to the lungs, and look carefully at radiographs for any evidence of lytic bone lesions! When using antibiotics pre and long-term post-operatively, consider what is safer to use in a rabbit (enrofloxacin, trimethoprim-sulfa, metronidazozle, chloramphenicol). Pain medications pre-, intra-, and post-operatively include buprenorphine for the first 24 hours, then meloxicam. Note that tramadol lasts only 15 minutes in a rabbit, so do not use it. Supportive care includes a warm (75 °F) environment, fluids, greens, and hand feeding Critical Care for herbivores.

Rabbit response

The rabbit recovered well from surgery. Fluid that looked like pus within the uterus was cultured.

Test results and diagnosis

Histopathology showed ovarian and uterine adencarcinoma with concomitant cystic endometrial hyperplasia and pyometra. Both the urine and uterine cultures showed Pseudomonas sp. (resistant to enrofloxacin and sensitive to TMS)! So this rabbit had a bacterial pyometra and bacterial cystitis as well as cystic endometrial hyperplasia and uterine adenocarcinoma.

Author's revised treatment plan

The revised treatment plan is described in Table 8. The plan includes treatment and dietary recommendations.

Table 8. Author's revised treatment plan.

Action	Description
Medication	Continue long term (2-3 weeks post normal recheck urinalysis) antibiotic according to sensitivity, ability to use in rabbit and excreted in the urine
Dietary change	Increase roughage in diet, no more seeds at all! Add timothy or coastal Bermuda hay, fresh grass, change to a timothy hay based pellet (in order to decrease amount of excess calcium needing to be excreted by kidneys (yes, rabbits are like horses in that they lose excess calcium through the kidneys making them susceptible to calcium carbonate urolithiasis)

Uterine and ovarian adenocarcinoma

There is a high incidence of this tumor in rabbit, sometimes as high as 80% over 5 years of age (in the Silver Marten breed). If clinical signs are present, they include anorexia, depression, hematuria, dysuria or milk production. Diagnosis is based on abdominal radiography, palpation or ultrasound of an enlarged, possibly fluid filled uterus, sometimes with masses or cystic endometrial hyperplasia. Treatment is OHE, but a metastasis check should be done first.

Case 3. Guinea Pig

Review the information in Table 9 and formulate a problem list, rule-outs for each problem, and a diagnostic plan.

Table 9. Guinea pig case.

Factor	Description
Signalment	5-year-old, female, intact guinea pig
Client complaint	Partial anorexia, lethargy and depression for past week, urine dribbling for 2 days now
History	Owner obtained as a youngster about 4.5 years ago from a pet store
Diet	Eats guinea pig pellets, timothy hay, some greens and carrots, ad lib water from a bottle
Housing	In a 5' × 4' × 3' plastic sweater box with no top (can't jump out) with aspen shavings that are changed out weekly; has a hide box and uses this area to sleep and as a latrine
Physical examination	Moderately thin, possible caudal abdominal mass palpated but difficult to palpate awake, a cursory oral examination shows no evidence of oral disease, there is urine on perineum and the smell of urine on the hair of the rear

Author's problems, rule-out lists, and diagnostic plans

The author's problems, rule-outs, and plans are presented in Table 10.

Table 10. Author's problems and rule-out lists, assessment plan, and diagnostic results for a guinea pig case.

Factor	Description
Problems	Anorexia, depression, lethargy/anything Urine dribbling/cystitis, renal disease or cystic calculi
Assessment	Differential or CBC, urinalysis with C/S (aerobic and anaerobic), chemistry profile, radiograph, ultrasound
Results	CBC: heterophilic leukocytosis Profile: within normal limits US: +4 bacteria, and WBC Radiograph: urinary cystic calculi, 1 × 1 cm Ultrasound: stone with mildly thickened bladder wall

After reviewing information in Table 10, formulate the treatment plan.

Author's diagnostic and treatment plan

The plan was to perform surgery to remove the stone from the bladder, but an examination just prior to anesthesia (actually when we gave the pre-operative butorphanol) we noticed a swollen vulva. It was the stone. With mild digital pressure and isoflurane sedation, the stone was digitally manipulated out and surgery was cancelled! Stone analysis showed it to be 100% calcium carbonate. The culture showed an aerobic as well as an anaerobic infection, therefore treatment consisted of not only fluids and enrofloxacin, but also metronidazole for the anaerobic bacteria. Antibiotics were continued for at least 3 weeks past a normal urinalysis.

Teaching Parrots Alternatives to Biting and Lunging: Angry Birds

Melody Hennigh, RVT, KPA-CTP

Session #2387

Affiliation: From the Medical Center for Birds and Busy Beaks Academy, 3805 Main Street, Oakley, CA 94561, USA.

Abstract: Biting and lunging are common behavioral problems of companion parrots that may lead to reasons for relinquishment. In the absence of functional assessment, behavior-change strategies are more likely to be forceful, less ethical, and can result in an augmentation of the problem behavior(s). Using the principles of functional assessment, this discussion demonstrates how a scientifically based behavior-change plan can be formulated and implemented in a practice setting. Enrichment of maintenance behaviors and target training are emphasized.

Introduction

When describing any behavior it is important to describe observable actions (ie, what we see and what can be seen easily by others). These types of descriptors often have different meanings to different people. For example, an aggressive bird may look like a bird slamming its toys around the cage to one person, but to another person aggressive may look like a lunging, biting bird. Another example is of a depressed bird, which needs to be petted constantly to be happy. Happy and depressed can have many different meanings to different people, yet they are very commonly used as explanations for why a bird behaves as it does. For example, what is the difference between an obnoxious or rude bird and a bird that screams constantly when the owner leaves for the day? One seeks to describe the bird's personality (the owner might say "That's just how the bird is!"), while the latter describes an actual behavior, something the bird is doing, that can be measured, observed, repeated, and therefore is subject to behavioral change. Without taking time to paint a clear picture of a bird's behavior, we can be a greater hindrance than help to our patients and clients.[1-3]

Enrichment

Foraging is a necessary form of enrichment for most companion birds. Working for food by means of searching through toys or puzzles, and shredding or chewing through paper or cardboard are great ways to keep birds busy, improve cognition, engagement and daily maintenance activities. Many birds will need to be taught these natural behaviors since they are not innate, and are often times have been very limited in the lives we offer for our caged birds, especially when some natural chewing behaviors are considered a nuisance to the pet owner and discouraged. Beginning foraging behaviors and shredding can be taught by demonstration in the exam room and continued at home. These food acquisition activities can range from very simple hidden treats in half covered bowls to very complex puzzles and time consuming boxes to untie or open. By increasing foraging in daily activities often times we can eliminate or lessen inappropriate behaviors.[4]

Target Training

We can teach birds numerous behaviors using target training. By this method, we teach the bird that if it targets (ie, touches with a body part such as its beak) an object (such as a chop stick or closed hand), it will receive a favorable outcome (ie, positive reinforcement).[5] Target training is used commonly to train a large number of animal species. Target training can be used to train birds to do anything from playing basketball, allowing intra-muscular injections, to stepping up onto the owners hand or exiting the cage. Using small approximations and the individual strength of different reinforcers we can create new behaviors. A reinforcer is any consequence that increases the strength or frequency of the behavior.[1-4] Target training is a tool used to communicate with our birds what we would like them to do, step by step and allows the human animal bond to be strengthened through positive interaction and reinforcement opportunities.

There are several applications of target training for behavior-change strategies with aggressive parrots, often which have resulted from broken human-animal bonds. Owners can engage their birds with target training while the bird is in the cage. This allows the owner to interact with their companion while it is still in the cage, making the owner feel safe from being bitten and preventing the bird from practicing biting. Treats for positive reinforcement can be dropped into a bowl inside the cage, given by hand, or delivered on a spoon or from a syringe. Target training is a great starting point for stationing behaviors, where the bird is reinforced and taught to stand on a perch or platform. Standing on a designated perch and lunging a person are incompatible behaviors. By reinforcing stationing and not reinforcing lunging or biting behaviors, the aggressive behaviors become less valuable to the bird and they chose to perform these behaviors less often.

Conclusions

Behavior problems are common reasons for parrot owners to visit their veterinarian and even re-home or surrender their birds. When properly identifying the problem, using functional assessment to find an acceptable replacement behavior and adding enrichment activities to their daily activities, difficult behavior problems can be managed.

References

1. Friedman SG, Edling TM, Cheney CD. Concepts in behavior: Section 1. The natural science of behavior. In: Harrison G, Lightfoot T: *Clinical Avian Medicine.* Vol 1. Palm Beach, FL: Spix Publishing;2006:46-49.

2. Friedman SG, Haug L. From parrots to pigs to pythons: Universal principles and procedures of learning. In: Tynes VV, ed. *The Behavior of Exotic Pets.* Ames, IA: Blackwell Publishing. In press.

3. Fridman SG, Martin S, Brinker B. Behavior analysis and parrot learning. In: Luescher AU, *Manual of Parrot Behavior.* Ames, IA: Blackwell Publishing;2006:147-164.

4. Echols S. *Captive Foraging. The Next Best Thing to Being Free.* Lake Worth, FL: Zoological Education Network;2009.

5. Heidenrich B. Learning to fly. *Good Bird Magazine.* 2006;3-1:30-37.

Examples of Complementary Medicine in Exotic and Wildlife Patients

Ian Kanda, RVT

Session #2168

Affiliation: From the Center for Veterinary Health Sciences, Oklahoma State University, Oklahoma State University, Stillwater, OK 74078, USA.

A 1-year-old Ayam Cemani rooster presented for not using its right leg for 2 days. The bird was diagnosed with septic arthritis of the knee joint. Following surgical removal of caseous debris from the joint, physical therapy was initiated with passive range of motion and placement of ice packs over the affected area twice daily. Six days after surgery, acupuncture with electrostimulus, and laser therapy with a class 4 therapeutic laser was administered. A modified flotation device was fit to allow for the use of an underwater treadmill to encourage the use of the affected leg. Significant improvement in leg use was observed. Prior to and immediately after surgery the patient held its right leg up and did not attempt to use it. Two weeks post-operation, the patient was bearing partial weight on the right limb, was moving the leg in a walking motion, and was better balanced. Additional examples of these methods in other species will be given.

Avian Behavior and Implementing Natural Positive Reinforcement Techniques at Home

Liz Vetrano, BS, CVT

Session #2016

Affiliation: From Mount Laurel Animal Hospital, Mount Laurel, NJ 08054, USA.

Abstract: Educating clients on the foundation and etiology of avian behavior is essential when beginning to implement training protocols in the home. Wild bird behavior can assist in providing the basis of understanding the development of certain behaviors at home and attempting modifications. It also can assist in the simplicity of standard husbandry and setting your client up for success. Implementing positive reinforcement techniques that are commonly used in zoological settings (in addition to the proper set-up) have proven to be most beneficial and effective. Training clients to execute these techniques can be challenging, but using step-by-step customized training protocols can facilitate behavioral transformations.

Where to Start

Considering the standard education process with a client, the best place to begin is setting clients and their companions up for success. Doing this will prevent future behavioral issues and preventing is always the most effective and easiest route when it comes to behavior. What exactly does "set up for success" mean? It means getting off on the right foot and providing the patient with the proper care and environment before attempting behavioral transformations. It is essential for clients to understand that these birds are still considered wild animals. Allowing them to utilize their innate instincts and expend that energy towards useful actions in our homes will allow for easier behavioral modifications.

What does a wild bird's day consist of? Many studies have been done, but it's been noted that these wild animals can be quite elusive and difficult to study in detail. What we do know is that their standard day is made up of these common behaviors: foraging for food, socializing, preening, flying, and vocalizing. Shaping the bird's environment to mimic some of these actions can prevent that energy from being used towards unwanted behaviors. Making sure your owner understands how to accomplish this is a whole other challenge. Initially setting the bird up for the success is a crucial part of the training process, as it will prevent most unwanted behaviors. The following tips are all part of the training process and setting the animal up for success by starting with the proper husbandry, activities, and socializations.

Foraging, Foraging, and More Foraging

When we have our birds' food so readily available in their bowls, we are actually doing them a disservice. Birds' innate instincts include working for food (Fig 1), yet we provide it for them, so what can they do with that extra time? Screaming? Self-mutilation? Working towards 100% foraging techniques for feeding has proven to be beneficial for the bird and occupy a majority of his/her time alone. There is a plethora of foraging toys available, but owners can also make their own (eg, tissue boxes, shoe boxes, wrapped in newspaper balls). Begin by having the bird search for something really high in value (eg, nut, fruit) and then they can graduate little by little with their pellets.

Figure 1. Innate instincts include working for food.

Appropriate Interactions and Relationships

One of the biggest challenges is advising an owner that their beloved companion shouldn't be cuddled, stroked, or touched in certain places. Understanding what this means is one of the hardest obstacles considering most bird stores tend to advertise by saying," She loves to cuddle!" Undeniably the bird does because it's stimulating, but not in the way most owners are perceiving it, unfortunately. Stimulating the bird to perform masturbatory behaviors can not only lead to physical ailments such as chronic egg-laying and heart disease, but is usually the number one offender for unwanted hormonal behaviors. These can be the most difficult to manage despite the advancements of hormonal therapies such as leuprolide acetate and deslorelin implants (which can also be very costly for the owner).

Developing customized training exercises for the owner, can motivate them to appropriately interact with their companion and solidify that relationship. For example, training the bird to fly "A to B" exercises, move toys, talk, etc. can eliminate the cuddling and improper behaviors on both ends of the bond between bird and human. This is where teaching your client how to utilize positive reinforcement techniques (ie, reinforce the good, ignore the bad). They can also learn how to shape behaviors and give bonuses when they want to capture a certain behavior. It can be confusing for the average bird owner at first, but developing a customized protocol just for them and providing them with handouts containing step-by-step instructions prove to be advantageous.

Positive Reinforcement: Tell Them What to Do, Not What Not to Do

This saying goes for both the clients and their birds. As mentioned above, these birds have innate instincts that allow them to expend their energy towards activities including foraging, socialization, flying, etc. Whether they are given these opportunities or not in the home, they will find a way to expend this energy. Normally, this produces unwanted behaviors that we typically see such as plucking, screaming, biting, self-mutilating, and/or masturbating. By directing the clients and tell them what to do, we make the instructions clear and easy for them to follow. A simple way to do this could be customizing training protocols. Instructing them when to reinforce their bird, how, and when not to. This goes for the training, as well. Telling a bird what to do by implementing positive reinforcement techniques, is more likely to lead the bird to success then using negative reinforcement or punishment.

You cannot force a bird to do anything. All you can do is ask. Try not to push them past their threshold (ie, stop training session if they start to exhibit unwanted behaviors such as excitement, screaming, or biting).

Behavior examples

- A bird performs a behavior you like, which results in reinforcement (eg, the bird says hello or whistles instead of screams).
- The bird steps up nicely after being asked.
- Clients can also shape behaviors. Owner catches bird singing or twirling in a circle. Reinforce that behavior.

Reinforcement examples

- Break up pieces of nuts (eg, almonds, peanuts, brazil nuts); use smaller pieces so birds do not get a full crop in a short period of time.
- Attention
- Toys

Bonus reinforcements: If the bird does something for the first time, or something the owner really likes, reinforce with a larger bonus treat such as a whole nut.

Consistency can be another hurdle to overcome. A house full of family members can prove to be beneficial yet also cumbersome when it comes to training protocols. Making sure that everyone is on the same page and has the same interactions with bird, is crucial for any behavioral modifications. It only takes one person and one action to break down a behavior that took months to develop.

Our practice offers customized training protocols along with mobile home visits for behavioral consultations. This way, we can visualize the birds' environment and daily activities to be able to accurately assess the situation. We can then evaluate the options and develop an appropriate protocol for that specific bird.

Finally, Allow the Bird to Be a Bird

Some of the most successfully kept birds in captivity belong to free-flight bird shows. Considering the facts that they get to act like real birds and fly, work for their food, and interact with other birds, this shouldn't come as a surprise. Consider keeping the bird flighted and completing a very minimal wing trim. Not only does this help expend energy, but it also has physical benefits (ie, decreasing chances of heart disease, increasing cardio health). Increasing the likelihood of a healthy patient is also a huge contributor to desirable behaviors (which also includes diet).

Educating the client on the origins of bird behavior, in the beginning stages of their relationship is absolutely crucial to their success with any behavioral modifications. Our practice has developed a plethora of handouts for bird owners to take home and have on hand as a resource. This includes specific instructions on behavior and training methods. Without providing the above changes: foraging, flying, and appropriate interactions, behavioral changes will ultimately become very difficult with the development of undesirable actions. Customized training protocols in conjunction with setting the bird and owner up for success, can allow for a healthy relationship and patient long term.

Basics of Clinical Pathology in Small Mammals

Thomas M. Donnelly, DVM, DACLAM, DABVP (Exotic Companion Mammal Practice), DipECZM (Small Mammal)

Session #2501

Affiliation: From the University Veterinary Hospital Center of Alfort, French National Veterinary School at Alfort, 7 Avenue du General de Gaulle, 94704 Maisons-Alfort Cedex, France.

Abstract: Important clinical pathology tests such as fecal examination for parasites and complete blood counts with total protein can be performed by technicians. As these tests involve microscopic examination of body fluids, they may seem intimidating – but actually they are not. There will always be a few challenging examinations, but generally these samples and slides will be referred to veterinary specialists in pathology or parasitology, as they are also challenging for the veterinary practitioner. Systematic preparation and microscopic examination of slides allows quantification and qualification of material on the slide. This presentation will present methods to perform coprological and hematological examinations on exotic pets that can be easily performed in busy veterinary practice.

Fecal Examination

Examine fecal specimens using the 10X objective. Begin the fecal examination at one corner of the slide and end at the opposite corner, moving over the slide in a systematic pattern (Fig 1).

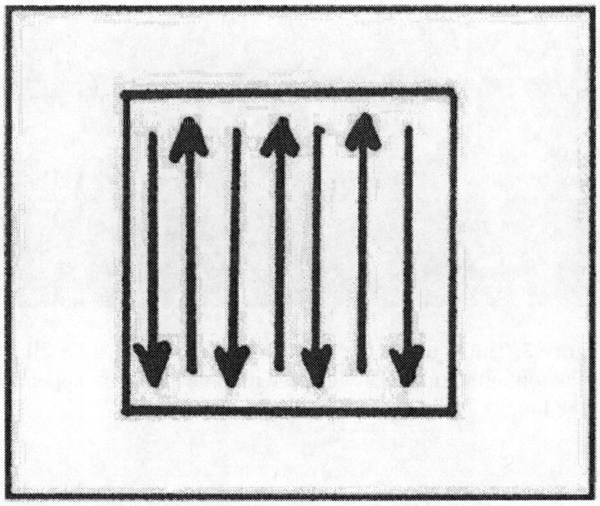

Figure 1. Direction of slide under microscopic for fecal examination.

Direct fecal smear

The direct smear is made by breaking up a very small particle of feces in a drop of saline. It is simple and quick. You can smear the feces sticking to the rectal thermometer directly on a microscope slide. Use a coverslip as it improves the optics, subdues swirls and currents of the saline, and helps prevent dirtying the objective lens of the microscope. Direct smears of fresh fecal material also allow us to observe the motility of amoebae, flagellates and nematode larvae.

Fecal smears are useful for guinea pigs with diarrhea or soft feces. Often you see numerous protozoa. Identification of the protozoa is not necessary; what is important is that there are significantly increased numbers of protozoa.

Fecal flotation

The fecal flotation is a concentration method for fecal examination. This technique uses a test tube or vial, in which feces and flotation solution are mixed (Fig 2). A coverslip or microscope slide is placed on top of the test tube, and the unit is allowed to sit undisturbed. Any parasite eggs in the feces float to the top and adhere to the underside of the coverslip or microscope slide. The coverslip or slide is then removed and microscopically examined for parasitic material.

Figure 2. The liquid in the fecal flotation tube must be filled to the top of the tube. The liquid will have a convex appearance at the top.

Fecal flotation methods are based on the specific gravity of parasite eggs and fecal debris. The specific gravity of most parasite eggs is between 1.100 and 1.200 g/ml, whereas the specific gravity of water is 1.000.

To allow for flotation of parasite eggs the flotation solution must have a higher specific gravity than that of the parasite eggs. Several salt and sugar solutions work well for flotation (Table 1). Most have a specific gravity of 1.200 to 1.250. In this range, heavy fecal debris sinks to the bottom of the container, and parasitic material rises to the top of the solution.

Table 1. Fecal flotation solutions.

Solution	Ratio (Chemical/H_2O)	Advantages	Disadvantages
Sugar solution	454 g:355 ml	Inexpensive Does not distort eggs	Floats few eggs Messy
Sodium nitrate	315 g:1 L		Forms crystals Distorts eggs Expensive
Sodium chloride	Add salt until no longer dissolves in boiling water	Cheap Easy to melt Readily available	Corrodes lab equipment Forms crystals Distorts eggs
Zinc sulfate	386 g:1 L	Inexpensive Easy to make	Forms crystals

A solution commonly used for flotation is saturated sugar solution. Sugar solution is inexpensive and does not crystallize or distort eggs. Sugar solution may be made anywhere and has a long shelf life. Although sticky to work with, spilled sugar solution may be removed with warm, soapy water.

Sodium nitrate solution is the most common fecal flotation solution used in veterinary hospitals today. This solution is efficient for floating parasite eggs, oocysts, and larvae. It may be purchased with commercial diagnostic test kits or in individual aliquots. The major disadvantage of using sodium nitrate is the expense. Sodium nitrate also forms crystals and distorts eggs if allowed to sit longer than 20 minutes.

The least desirable solution used is saturated sodium chloride solution. This solution corrodes laboratory equipment, forms crystals, and severely distorts parasite eggs. Saturated sodium chloride solution is also a poor flotation medium because the maximum specific gravity obtainable is 1.200, allowing heavier eggs to remain submerged.

Zinc sulfate solution is more commonly used in diagnostic laboratories. Zinc sulfate floats protozoal organisms with the least amount of distortion. It generally is used in combination with one of the previously mentioned solutions.

Fecal flotation is useful for identifying coccidia oocysts. Coccidia are an important cause of diarrhea in young rabbits. Nematodes of pet rabbits and rodents are not important disease-causing organisms as they are in cats and dogs. The presence of nematode eggs generally indicates poor husbandry.

Commercial flotation kits

Commercial flotation kits use the same principle as the standard flotation technique. These kits contain a vial with a filter; some also include prepared flotation solution (Fig 3). Examples of commercial flotation kits include Fecalyzer (EVSCO Pharmaceuticals, Buena, N.J.), Ovassay Plus (Synbiotics, San Diego), and Ovatector (BGS Medical Products, Venice, Fla.). These kits are simple to use, but expensive when compared with the simple flotation technique.

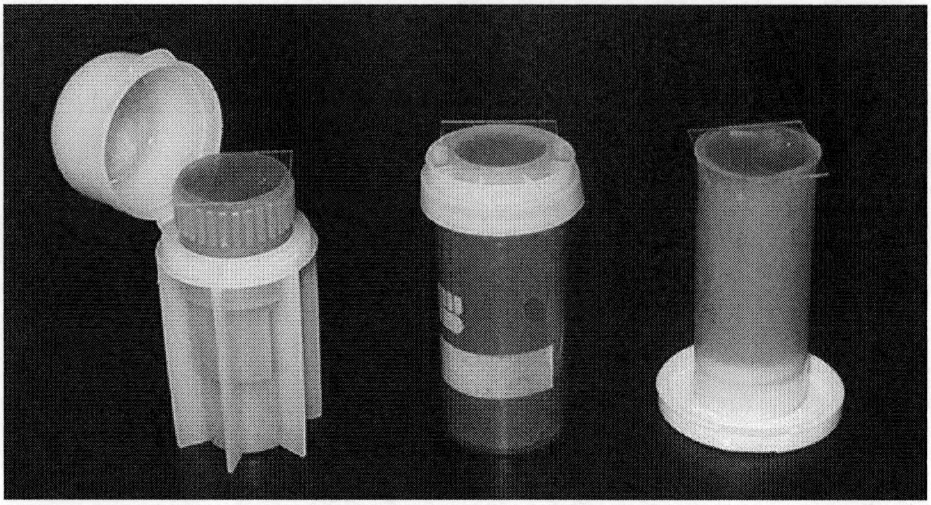

Figure 3. Commercial fecal flotation kits. From left to right, Fecalyzer, Ovassay, and Ovatector.

Detecting parasite antigens in feces

The detection of antigens in feces (or coproantigens) by antigen-capture immunoassays is becoming more and more routine. Methods have existed for some time for the laboratory detection of parasite antigens in feces, particularly those of *Giardia* and *Cryptosporidium*. Now the method exists for the routine in-house detection of the cyst wall antigen of *Giardia* using the IDEXX SNAP *Giardia* Test.

The enzyme-linked immunosorbent assay (ELISA) is one of the most sensitive and specific diagnostic technologies available. ELISA is the method of choice in reference laboratories worldwide because of its unparalleled accuracy and ease of use. The SNAP assay device (Fig 4) contains reagents needed to conduct the steps associated with the ELISA in-clinic with minimal hands-on time.[1] The SNAP assay was developed by scientists at IDEXX Laboratories to overcome limitations of other clinic-based assays, which included first generation ELISA-, colloidal gold-, and latex particle–based tests. The SNAP product combines the simplicity and immediate results of an in-office test with the accuracy of a reference laboratory-format ELISA.

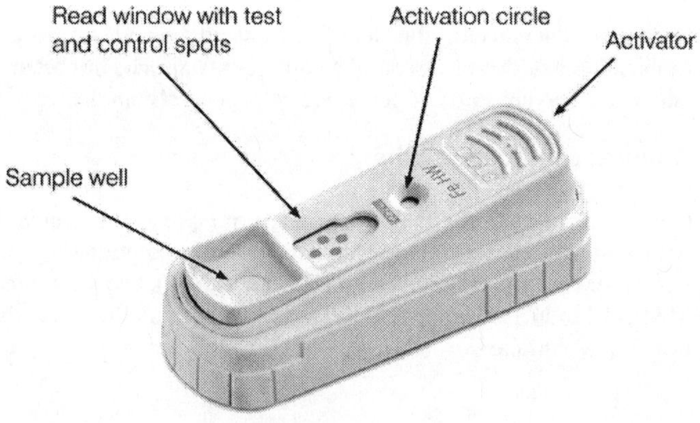

Figure 4. The SNAP assay device.

Group-housed chinchillas in fur ranches and research colonies had a high prevalence of giardiasis; however, the role of *Giardia* in causing disease in chinchillas is difficult to establish. *Giardia* are found in both healthy and sick captive-bred animals. In a Belgian survey, eighty pet chinchillas from both pet owners and breeders were screened and 66% excreted *Giardia* cysts. Recently weaned juvenile chinchillas, immunosuppressed animals and those living in crowded environments appear to be at highest risk of showing clinical disease from *Giardia*.

Factors influencing the reliability of a fecal examination[2]

1. Fecal samples must be of adequate size. For companion animals, 3-5 grams of material is ideal, while 1 gram is the minimal amount.

2. Fresh fecal samples are required because the stage of the parasite (eg, cyst, single-cell stage, morula stage, larval stage) in fresh feces is characteristic: a thick-shelled egg containing a single cell is characteristic of ascarid nematodes (roundworms), while thin-shelled morula stage eggs are characteristic of hookworms; however, as soon as eggs and cysts are passed in feces, they begin to develop (eg, single-celled eggs become morulated, morulated eggs become larvated and larvated eggs can hatch) and these more mature parasite stages can be difficult to identify. Parasite eggs desiccate and cannot be identified in old fecal samples.

3. Routine fecal examinations are usually done with a flotation procedure. All flotation procedures rely on the same basic principle: a salt (usually zinc sulfate or sodium nitrate) or sugar solution is used to float eggs to the surface, which occurs because eggs are lighter than the specific gravity of these solutions; however, not all flotation methods are equal, with some procedures having better recoveries than others do. For example, those procedures that rely on gravity to do the separation between the parasitic stages and the debris ("standing" flotations) have poorer recoveries than those using centrifugation. Thus, centrifugal fecal flotations (performed in a laboratory) are the gold standard when it comes to routine fecal examination.

4. The importance of well-trained personnel to conduct and read the tests cannot be overstated. Many fecal artifacts resemble parasites and can confuse the untrained technician.[3]

Blood Examination

An important role of veterinary technicians is collecting blood samples. If blood is not collected correctly and samples are not appropriately prepared preanalytical errors occur. Small mistakes can give rise to test results that are not consistent with preconceived expectations for a given case. An outstanding review by Gilor and Gilor is recommended.[4] This review is most useful when read in full before collecting a sample.

Hematologic procedures are performed on anticoagulated whole blood. The preferred anticoagulant is ethylenediaminetetraacetic acid (EDTA) because it does not interfere with blood cell morphology and staining. EDTA is commercially available in purple-top Vacutainer tubes in a variety of sizes. In exotic pet medicine because of the small amount we might obtain, heparin tubes are often used for hematology and biochemistry.

Packed cell volume (hematocrit)

The packed cell volume (PCV) or hematocrit tells us if the patient is anemic and tells us the hydration status of the patient.

Equipment to perform a PCV. Blood sample in EDTA or heparin tube, capillary tube (plain), microhematocrit reader (Fig 5), microhematocrit centrifuge, soft plasticine or Cristoseal.

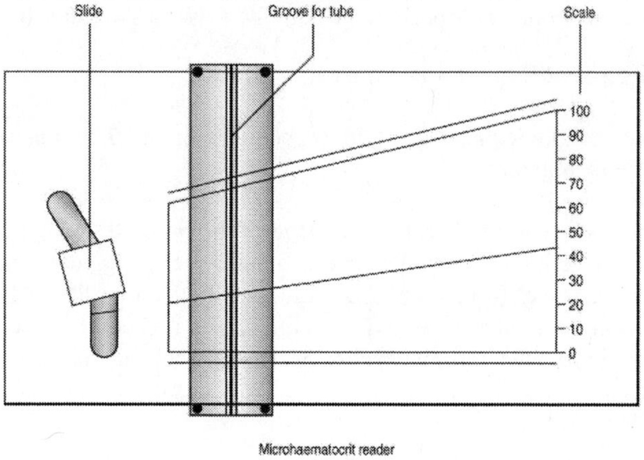

Figure 5. A microhematocrit reader.

1. Collect a blood sample in an EDTA or heparin tube. The anticoagulant and prevents the blood from clotting.

2. Gently agitate the sample and place the end of a capillary tube in the blood.

3. Hold the tube at a 45° angle and allow the tube to fill until it is three-quarters full. Blood will be drawn up the tube by capillary action.

4. Place the opposite end of the tube into the plasticine pad. If you use the end containing the blood, you will contaminate the plasticine.

5. Twist the capillary tube two or three times and take it out of the plasticine. This creates a plug that will prevent blood running out of the tube.

6. Wipe the tube with soft tissue.

7. Place the tube in one groove of the centrifuge with the plasticine plug facing outwards against the outer rim. Centrifugal force causes the fluids to spin outwards. The sealed end prevents blood escaping as it is forced towards the perimeter.

8. Place a similar tube on the opposite side of the centrifuge. This balances the tube under test and reduces vibration and subsequent damage to the centrifuge.

9. Screw the safety plate down over the tubes and close the lid. The safety plate prevents the tubes spinning out of position. Both plate and lid are essential for safety.

10. Centrifuge for 5 minutes at 10 000 rpm.

11. When the machine stops, remove the lid, safety plate and tube.

12. Place the tube into the groove on the microhematocrit reader (Fig 5). The blood will have separated into three layers (Fig 6); from the top downwards:
    - plasma
    - buffy coat, white blood cells
    - red blood cells

13. Line up the top of the plasticine plug with the line on the bottom of the reader (Fig 6). The top of the plug marks the lowest point of the blood column.

14. Line up the top of the plasma with the diagonal line at the top of the reader (Fig 6). The bottom of the plasma meniscus is used as the measuring point.

15. Move the slide so that the middle line is level with the top of the red cells.

16. Read the measurement from the scale on the right side of the reader. The scale is marked 1 to 100 and the reading can be expressed as a percentage (ie, 45 becomes 45%).

17. The PCV is the percentage of whole blood that consists of red blood cells.

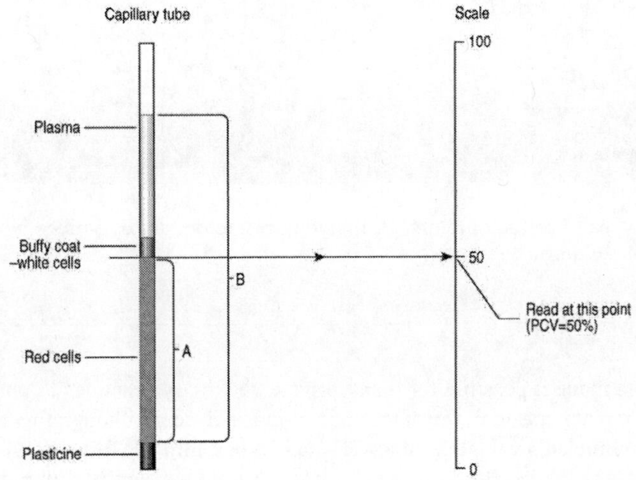

Figure 6. Measuring packed cell volume. A is the length of the tube occupied by red cells; B is the total length of the column of blood. Using a microhematocrit reader card measure A; the top of the red cells as read on the card is the PCV.

Plasma protein determination

The plasma protein tells us if the patient is hypoproteinemic (associated with kidney, liver or gastrointestinal disease) or hyperproteinemic (associated with certain tumors). With the PCV, plasma protein tells us the hydration status of the patient.

The determination of the plasma protein concentration is another standard component of the routine CBC. After plasma color and turbidity have been noted, the capillary tube used for measuring the PCV is broken at a point slightly above the buffy coat (cream-colored layer of WBCs and platelets just above the RBCs), and the plasma is allowed to run through the unbroken end onto the prism of a refractometer by capillary action. Plasma protein values obtained with a refractometer (Fig 7) are accurate as long as the plasma is clear. If the plasma is lipemic, hemolyzed, or cloudy, the refractive index will be increased and provide a false high protein measurement.

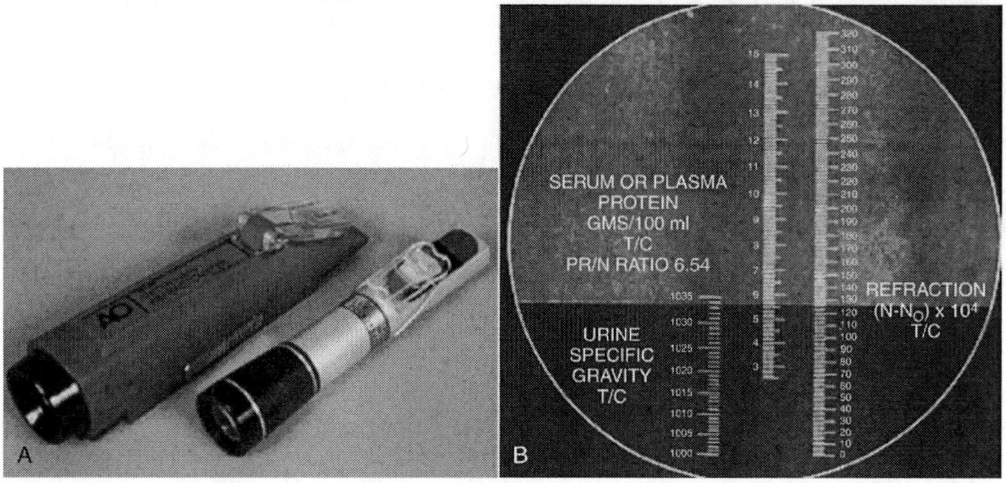

Figure 7. A. Two types of refractometers. **B.** Image in refractometer eye-piece when a transparent fluid is placed on the sample window.

Blood film evaluation

Technical advances have made it possible for many private veterinary practices to purchase reasonably priced automated hematology instruments to perform in-clinic blood analyses. Although these instruments can quickly provide numbers to the clinician, evaluation of a well-made blood film can often provide information critical to the interpretation of those numbers. Blood film review is essential to identify important abnormalities such as neutrophilic left shifts and toxic change, neoplastic cells, hemoparasites, and erythrocyte morphologic changes that may suggest the cause of an anemia. In addition, the blood film provides an important quality control measure for the automated hematology results.

Examination of the morphology (appearance) of the blood cells on a blood film smear is one of the most valuable parts of the complete blood count (Fig 8). Excellent reviews are available on exotic pet blood film evaluation.[5-7] Although pathologists or expert-trained hematology technicians usually perform a differential count (they count 100 cells and provide the numbers of each type of cell as a percentage), an untrained individual can quickly assess the blood smear.

Cells that appear too big or too small, are indications of disease. Too many cells also often indicate disease. The morphology of cells is best examined by spreading a very thin layer of blood over a glass microscope slide. The feather edge or tail of the smear, where the cells are spread at their thinnest, is often the best location for examination of the blood smear.

Manual staining methodologies with traditional Wright-Giemsa stains are labor intensive and require frequent filtering of the stains. More equipment, a source of distilled water, and critical timing are necessary. Several modifications of the traditional Wright's stain are available for the suitable staining of blood films for veterinary practices.

The most common is Diff-Quik (Baxter S/P). This stain is economical, less time consuming, and technically easy to use and maintain. The different colored liquids can be kept in individual Coplin jars. The lids of the Coplin jars should always be securely replaced after use, and the stains should be freshened or refilled as necessary. Stains should be replaced on a regular basis to prevent bacterial contamination and the accumulation of debris with repeated usage.

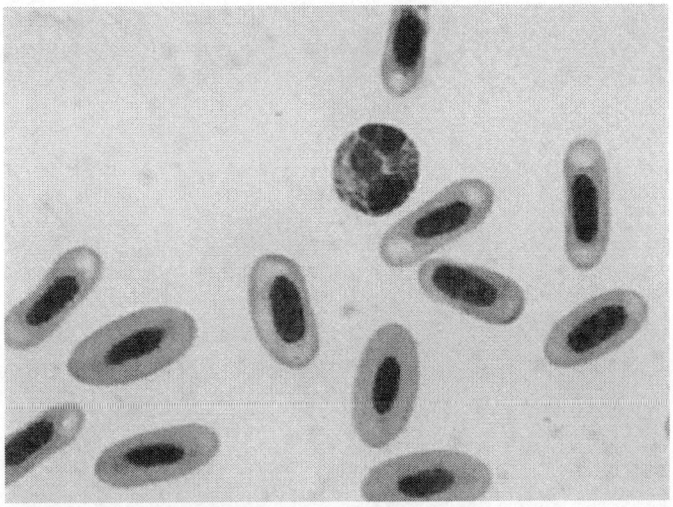

Figure 8. Nucleated red blood cells. Unlike mammals, birds and reptiles have nucleated red blood cells.

References

1. O'Connor TP. SNAP Assay Technology. *Top Companion Anim Med.* 2015;30:132-138.

2. Ballweber LR, Beugnet F, Marchiondo AA, et al. American Association of Veterinary Parasitologists' review of veterinary fecal flotation methods and factors influencing their accuracy and use--is there really one best technique? *Vet Parasitol.* 2014;204:73-80.

3. Gates MC, Nolan TJ. Comparison of passive fecal flotation run by veterinary students to zinc-sulfate centrifugation flotation run in a diagnostic parasitology laboratory. *J Parasitol.* 2009;95:1213-1214.

4. Gilor S, Gilor C. Common laboratory artifacts caused by inappropriate sample collection and transport: how to get the most out of a sample. *Top Companion Anim Med.* 2011;26:109-118.

5. Campbell TW. Evaluation of the blood film. *Vet Clin North Am Exot Anim Pract.* 2015;18:117-135.

6. Lindstrom NM, Moore DM, Zimmerman K, et al. Hematologic assessment in pet rats, mice, hamsters, and gerbils: blood sample collection and blood cell identification. *Vet Clin North Am Exot Anim Pract.* 2015;18:21-32.

7. Moore DM, Zimmerman K, Smith SA. Hematological assessment in pet rabbits: Blood sample collection and blood cell identification. *Vet Clin North Am Exot Anim Pract.* 2015;18:9-19.

3D Printing: Art and Engineering in the Clinic

Ryan O'Shea, CVT

Session #2004

Affiliation: From ZooTampa at Lowry Park, 1101 W Sligh Ave, Tampa, FL 33604, USA.

The development and application of 3D printing has grown exponentially in the last decade. Many of today's consumer-level printers are based on the fused deposition modeling (FDM) technique patented in 1992 by Scott Crump of Stratasys Inc.[1] FDM printing heats and extrudes a thermoplastic material to a build plate in successive vertical layers until the final height of the model is reached. One of the earliest reported uses of 3D printing in veterinary medicine occurred at Cornell University College of Veterinary Medicine in 2009. Using computed tomographic images, university staff were able to replicate a 3D model of a patient's limb to adapt a surgical approach prior to taking the patient to surgery.[2] Since then, 3D printing has been used in multiple cases to either evaluate an anatomic region or even replace damaged anatomy via the use of prosthetics.[3] In addition to 3D printing's role of enhancing surgical techniques, the veterinary staff of ZooTampa at Lowry Park have utilized 3D printing to create various implements, supplies, and tools that are used on a regular basis. The utilization of a consumer-grade FDM printer has simplified tasks of the veterinary support staff and animal care staff at the zoo, and new concepts are being developed every day. The development of the 3D printed models is not without a steep learning curve and interested parties should consider their ability to design objects in a three-dimensional atmosphere and trouble shoot printer quirks that regularly occur prior to investing in such a device.

References

1. Sculpteo. The History of 3D Printing: 3D Printing Technologies from the 80s to Today. Available at: https://www.sculpteo.com/blog/2016/12/14/the-history-of-3d-printing-3d-printing-technologies-from-the-80s-to-today/ Accessed April 4, 2018.

2. JAVMA News. 3D Printing makes its way to veterinary medicine. Available at: https://www.avma.org/News/JAVMANews/Pages/140701j.aspx/ Accessed April 4, 2018.

3. 3D Printing Industry. 10 Animals Who Got a Second Chance at Life with 3D Printing. Available at: https://3dprintingindustry.com/news/10-animals-who-got-a-2nd-chance-in-life-with-3d-printing-51221/ Accessed April 4, 2018.

Streamlined Approaches and Increased Efficiency in Exotic Emergency and Trauma Triage Protocols

Simon Chavez, MS, MVM, CVT

Session #2041

Affiliation: From SCV Veterinary Management LLC, Chicago IL, USA, Animal Internal Medicine Specialty Services, 1333 9th Street, San Francisco, CA 94122, USA, Ocean State Veterinary Specialist, 1480 S Country Trail, East Greenwich, RI 02818 USA.

Abstract: In the past few years veterinary emergency protocols have introduced point-of-care ultrasound (POCUS) as a tool for rapid, real time diagnosis of life threating conditions in dogs and cats. Frequently used techniques such as cage-side organ assessment with sonography for trauma (COAST), triage and tracking, focused ultrasound assessment to the abdominal cavity (AFAST), and the thoracic cavity (TFAST), focused organ evaluation and veterinary bedside lung ultrasound exam (Vet BLUE) have been beneficial to other species outside of felines and canines like small mammals (eg, rabbits, rats, ferrets, chinchillas and birds). A full discussion of POCUS techniques is beyond the scope of this manuscript, but the mostly common techniques used in exotic patients will be discussed.

Introduction

The use of point-of-care ultrasound (POCUS) is feasible after minimal training in the inexperienced sonographer. This resource should be considered to assist in addition to the initial triage and physical examination findings, while minimizing patient stress and handling. All techniques do not require any shaving of fur from the patient. It should be considered during the examination and through the hospitalization of the patient for tracking changes in the patient in addition to clinical sings or laboratory changes.

Focused Ultrasound Assessment

Frequently used triage techniques in exotic companion mammals are the following: focused ultrasound assessment targeted to the abdominal cavity (AFAST), the thoracic cavity (TFAST), and veterinary bedside lung ultrasound exam (Vet BLUE) and when all three techniques together are used it is commonly referred to as global FAST (GFAST). This combination carries greater potential to guide the team and improve the patient outcome by detecting condition or complications otherwise missed on traditional physical examination and other diagnostics.

In the small mammal patient POCUS has been used as aid to determine the median values for the left atrium to aorta ratio (LA;Ao). It can be used to assess the evidence of cardiac chamber dilation, pleural effusion, pulmonary edema vs pulmonary contusion, ascites, neoplastic effusion, chylothorax, pyothorax in the during the Focus assessment in the thoracic cavity.

In ferrets, POCUS can be used for the assessment of congestive heart failure (CHF) in the severe respiratory distress presentation. POCUS of the abdominal cavity can be used for assessment of hemoabdomen, neoplasia and foreign body.

In birds, POCUS can be used for the assessment of cardiac abnormalities. If fluid is present in the coelom, it can be aspirated with ultrasound guide. This can be done prior of taking radiographs especially in the dyspneic avian patient.

References

1. Boysen SR, Lisciandro GR. The use of ultrasound for dogs and cats in the emergency room: AFAST and TFAST. *Vet Clin North Am Small Anim Pract.* 2013;43(3): 773-797.

2. McMurray J, Boysen S, Chalhoub S. Focused assessment with sonography in nontraumatized dogs and cats in the emergency and critical care setting. *J Vet Emerg Crit Care.* 2016;26(1):64-73.

3. Olin JM, Smith TJ, Talcott MR. Evaluation of noninvasive monitoring techniques in domestic ferrets. *Am J Vet Res.* 1997:58(10):1065-1069.

4. McLaughling A, Strunk A. Common emergencies in small rodents, hedgehogs, and sugar gliders. *Vet Clin North Am Exot Anim Pract.* 2016;19(2):465-499.

5. Graham J, Basseches J, Liver lobe torsion in the poet rabbits: Clinical Consequences, diagnosis, and treatment. *Vet Clin North Am Exot Anim Pract.* 2014;17(2):195-202.

6. Lisciandro GR, Fosgate GT, Fulton RM. Frequency and number of ultrasound lung rockets (B-lines) using a regionally based lung ultrasound examination named Vet BLUE (veterinary bedside lung ultrasound exam) in dogs with radiographically normal lung findings. *Vet Radiol Ultrasound.* 2014;55(3):315-322.

Toxicity in Amphibians

Erica Mede, CVT

Session #2520

Affiliation: From Friends of Scales Reptile Rescue, PO Box 553, Wheeling, IL 60090, USA.

Amphibians are vulnerable to various toxicities. Amphibian toxicities are unique because unlike mammals, birds, and reptiles, when an amphibian encounters a toxin it is systemically affected due to permeability of the skin. In captivity, amphibians are increasingly vulnerable to toxins encountered in their environment (eg, cleaning products, contamination with pesticides and herbicides from outdoor plants and branches) and by ingestion (eg, inappropriate supplementation). These toxic encounters present as various disease processes and some can appear similar to disease processes from other etiologies.

Amphibian Nutrition and Gastrointestinal Parasites

Erica Mede, CVT

Session #2521

Affiliation: From Friends of Scales Reptile Rescue, PO Box 553, Wheeling, IL 60090, USA.

Amphibians offer a unique set of challenges when it comes to captive nutritional requirements. Several common issues seen by veterinary professionals can be linked to nutritional issues. Deficiencies are the most common form of nutritional issues along with over supplementation of vitamins and minerals. Inappropriate protein to fat ratios lead to unhealthy patients with the potential for reduced longevity and/or reproductive rates; however, dietary issues are not always the cause of patient gastrointestinal disease. Parasitism is a common issue in amphibians, even those in captivity.

Clinical Application of Local and Regional Anesthesia in Reptilian Species

Pia Bartolini, CVT, VTS (CP Exotics)

Session #2522

Affiliation: From Terrestrial & Arboreal, LLC, 5831 Lady Bug Lane, Melrose, FL 32666 USA.

Abstract: The use of local and regional anesthesia as a multimodal approach to analgesia historically has been overlooked in non-domestic species such as reptiles.[1] Clinically, the use of local blocks could assist with a reduction of injectable and inhalant anesthetics with providing a smoother and possibly shorter anesthetic and recovery period;[2] however, clinical studies on the efficacy of local and regional anesthesia in non-domestic species are limited.[1] The use of nerve stimulators to locate nerves can be extrapolated from the known reptilian anatomic landmarks.[1,3]

Introduction

The large number of reptilian species coupled with an insufficient amount of information on reptilian neuroanatomy makes identification of appropriate sites for local blocks challenging.[1] Use of a nerve locator provides both a means for appropriate placement of local anesthetic drugs and an objective measure of their efficacy.[1] Local anesthetics have the advantage of producing minimal systemic effect at appropriate dosages.[1] Peripheral nerve locators use a pulsatile electric current passed through an insulated needle in contact with the motor nerve that innervates and thus stimulates an identifiable muscle.[2] As in small animals, local blocks are used to decrease or eliminate pain pre-, intra- and postoperatively and can preemptively control the wind-up phase.[4]

Types of Blocks and Uses

Splash block

a) Blocking glottis prior to intubation

b) Surgical wound debridement

Topical block

a) Chelonian phallus[5]

Local and incisional block

a) Endoscopic surgery[6]

b) Surgical placement of telemetry devices

c) Surgical wound management

Dental block[1]

a) Infraorbital

b) Maxillary

c) Mental

d) Mandibular

e) Inferior alveolar

Radial-ulnar-median nerve block

a) Surgical digit amputation

Brachial plexus block

a) Forelimb procedure or biopsy

Femoral and sciatic block

Epidural block

a) Intrathecal "epidural"[7]

Intra-articular block

Discussion

Reducing injectable and inhalant anesthesia in reptilian species would in turn reduce the anesthetic recovery period.[1] General anesthesia blocks the perception of pain by inducing unconsciousness; whereas, local and regional anesthesia completely blocks the transmission of noxious impulses from the blocked location. By blocking the nerve locally, less general anesthesia may be needed.

There is a certain amount of specialization of the nervous system of various reptilian species such as a snake versus a chelonian. There are more similarities among reptilian taxa than there are differences.[8]

The use of a peripheral nerve stimulator to accurately locate the nerve prior to blocking will result in the use of less local anesthetic drugs. There are species-specific considerations as well as considerations of patient size. In addition, it is not recommended to dilute more than 50%. There is a preferred optimum temperature zone (POTZ). The pH of the tissue can affect the efficacy of the block.[2] Remember the total dose being used per patient is not to exceed the toxic dose in specific species:

- Lidocaine 1-2 mg/kg
- Bupivacaine 1-2 mg/kg
- Mepivicaine 1-2 mg/kg

Limited pharmacokinetic/pharmacodynamic (PK/PD) studies are available in reptilian species. More research needs to be conducted to ensure efficacy in all species. This technique can be one of many options that will need to be considered in the event of an anesthetic plan in reptiles.

References

1. Wellehan JXF, Gunkel CI, Kledzik D, et al. Use of a nerve locator to facilitate administration of mandibular nerve blocks in crocodilians. *J Zoo Wildl Med.* 2006;37:405-408.

2. West G, Heard D, Caulkett N, eds. *Zoo Animal and Wildlife Immobilization and Anesthesia.* Ames, IA: Blackwell Publishing;2008.

3. Wyneken J. Reptilian neurology: Anatomy and function. *Vet Clin North Am Exot Anim Pract.* 2007;10: 837-853.

4. Grimm K, Tranquilli W, Lamont L. *Essentials of Small Animal Anesthesia and Analgesia.* Hoboken, NJ, IA: Wiley-Blackwell;2011.

5. Rivera SS, Divers SJ, Knafo SE, et al. Sterilization of hybrid Galapagos tortoises (*Geochelone nigra*) for island restoration. Part 2: Phallectomy of males under intrathecal anesthesia with lidocaine. *Vet Rec.* 2011;168:78.

6. Hernandez-Divers SJ, Stahl SJ, Ferrell R. An endoscopic method for identifying sex of hatching Chinese box turtles and comparison of general versus local anesthesia for coelioscopy. *J Am Vet Med Assoc.* 2009;234:800-804.

7. Mans C, Steagall PVM, Lahner LL, et al. Efficacy of intrathecal lidocaine, bupivacaine, and morphine for spinal anesthesia and analgesia in red-eared slider turtles (*Trachemys scripta elegans*). *Proc Annu Conf Am Assoc Zoo Vet.* 2011;135.

8. Wellehan JXF, Gunkel CI. Emergent diseases in reptiles. *Semin Avian Exot Pet Med.* 2004;13(3):160-174.

Nidovirus Case Report: A Fatal Outbreak of Respiratory Virus in a Green Tree Python (*Morelia viridis*)

Pia Bartolini, CVT, VTS (CP Exotics)

Session #2523

Affiliation: From Terrestrial & Arboreal, LLC, 5831 Lady Bug Lane, Melrose, FL 32666, USA.

Abstract: Nidovirus has been linked to fatal respiratory disease in reptiles. Pythons are commonly kept and bred, and strategies must be used for managing infectious disease outbreaks. Discovery and diagnosis of nidovirus has led to many answers of a causative agent of fatal respiratory disease in python species; however, there are more questions than answers. More research is needed on transmission, treatment, and prognosis.

Introduction

An adult male green tree python (*Morelia viridis*) was part of a newly acquired collection of approximately 25 green tree pythons. The previous owner also had a large group of approximately 350 ball pythons (*Python regius*) that were sold to other buyers during that time. The adult male and the other green tree pythons were placed into quarantine July 2016 and subsequently started showing respiratory signs of wheezing, slight open mouth breathing, oral discharge and bubbles, lethargy, and an overall waxy appearance 3 weeks into the 90-day quarantine. The first animal to show clinical signs was an 11-year-old female green tree python that showed respiratory signs of oral discharge and wheezing on August 4, 2016 and while accommodations were made to be seen by a veterinarian, it was found dead within 7 days of first clinical signs. A necropsy was performed and revealed pneumonia with a viral etiology, an electron microscopy (EM), nidovirus and an IBD PCR was performed post mortum. A second green tree python that died 2 days later also had a necropsy that showed similar pneumonia and was tested for nidovirus and IBD. The EM did not show viral particles; however, both snakes were positive for nidovirus. Twelve snakes were showing clinical signs of respiratory infection and 4 of the 12 received sterile lung washes for culture and sensitivity and were started on a multimodal combination antibiotic therapy and supportive care. Twelve days later 2 more snakes died, a fifth died 2 days later, a sixth snake 2 days later, and a seventh 3 days later. By the September 15, 2016 a choanal swab was submitted for nidovirus PCR to Dr Mark Stenglein's lab at Colorado State University (CSU) of the remainder of the snakes. Dr Stenglein was continuing his research on nidovirus and accepted the samples for further research and sequencing. Nidovirus is a newly discovered coronavirus in reptile species with first publication in 2011 with more publication in 2014.[1-5] Nidovirus was first discovered in ball pythons with fatal respiratory disease. Symptoms include respiratory signs such as wheezing, oral discharge, lethargy, and death. Common necropsy findings include proliferative interstitial pneumonia, hyperplasia of the pulmonary tissue. A recent paper from 2017 was the first published report of nidovirus in green tree pythons.[4] In this paper next-generation sequencing revealed that the nidovirus found in green tree pythons was < 85% identical to the previously identified ball python nidovirus. There have been a few published papers of the ball python nidovirus.[2,3]

Clinical Report

Physical examination findings of the male green tree python showed oral exudate in the choana, wheeze upon exhalation, dull waxy appearance, 7-10% dehydrated assessed by tacky mucous membrane, good body condition with a body condition score of 2/5, and no other significant findings. A sterile lung wash was performed on multiple green tree pythons including the male, which revealed heavy growth of *Pseudomonas* and moderate growth of *Escherichia coli* and *Citrobacter* that was sensitive to enrofloxacin. Treatment of enrofloxacin 10 mg/kg SC q48h (diluted 1:1) and ceftazidime 20 mg/kg IM q72h, Lactated Ringer's Solution 20 ml/kg SC along with supportive care of increased thermal gradient from an approximate range 75-82 °F to 85-92 °F. A sterile choanal swab was submitted for a nidovirus q-PCR to a research lab at CSU. Blood work showed a packed cell volume of 24%, total solids 8.1 g/dl, leukocytopenia at 3K. The male died after approximately 1 month after treatment was started. The necropsy revealed a significant proliferative interstitial pneumonia and anemia. The necropsy reported "the functional airways of the snake were compromised by the increase inflammatory cells and proliferation of the mucosal epithelium. The cellular population along with hypertrophy/hyperplasia are indicative of a chronic inflammatory condition. Chronic inflammation can lead to decreased red blood cell production (erythropoiesis) due to several mechanisms including but not limited to: iron sequestration in tissue, blunted response to erythropoietin, and targeting of red blood cells for destruction by the immune system. It is hard to say whether the ultimate cause of death was the anemia, the pneumonia, or the combination of the two. Both of these conditions can readily lead to decreased oxygenation of vital organs, multi-organ failure, and subsequent death."

Nidovirus has been identified on necropsy in ball pythons with proliferative interstitial pneumonia. Currently, a long-term study is being conducted with the remainder of the nidovirus positive snakes. Serial q-PCR are being conducted quarterly to evaluate progression or regression of the virus. As of the last q-PCR submitted in October 2017, of the 22 snakes remaining, which includes *Morelia* species previously owned, 7 animals have had 2 negative PCR after being positive. More research needs to be conducted to evaluate whether the animals have recovered, have an immunity, or are intermittently shedding. Multiple of snakes have had positive PCR without showing any clinical signs.

Chronic viral infections can lead to weakened immune system and conditions like anemia, chronic disease, or overgrowth of normal bacterial flora as seen in the male green tree python and other green tree pythons that were affected. Commonly, in python species such as ball python, green tree pythons, and carpet pythons with respiratory signs, normal bacterial flora will grow on culture and sensitivity, and inflammatory response will be seen on cytologic evaluation of lung washes. These viral respiratory diseases are not responsive to antibiotic therapy. Common presenting complaints would be waxing and waning respiratory symptoms that return and are non-responsive to antibiotic therapy.

Discussion

Nidovirus has been linked to fatal respiratory disease in reptiles. With these pythons being commonly kept and bred, together with the data on outbreaks, strategies must be used for managing infectious disease outbreak. These outbreaks can be compared to respiratory or gastrointestinal virus of dogs and cats in a shelter environment.[6] It is important to diagnose and isolate the pathogen causing the fatality. Then identify and manage the exposed animals to avoid further spread of the outbreak. Once the pathogen is identified, decontamination of the environment can proceed. If possible, stop the acquisition of new animals to minimize risk of transmission. Lastly, documentation and communication is the key to prevent and minimize other outbreaks.

Conclusions

Discovery and diagnosis of nidovirus has led to many answers of a causative agent of fatal respiratory disease in python species; however, there are more questions than answers. More research must be done on transmission, treatment, and prognosis.

References

1. Bodewes R, Lempp C, Schürch AC, et al. Novel divergent nidovirus in a python with pneumonia. *J Gen Virol.* 2014;95:2480-2485.

2. Stenglein MD, Jacobson ER, Wozniak EJ, et al. Ball python nidovirus: A candidate etiologic agent for severe respiratory disease in *Python regius*. *MBio.* 2014;5:e01484-414.

3. Uccellini L, Ossiboff RJ, de Matos RE, et al. Identification of a novel nidovirus in an outbreak of fatal respiratory disease in ball pythons (*Python regius*). *Virol J.* 2014;11:144.

4. Dervas D, Hepojoki J, Keller S, Hetzel U. Nidovirus infection of green tree pythons: Identification and morphological features of a fatal respiratory disease in *Morelia viridis*. *J Comp Pathol.* 2017;156doi:10.1016/j.jcpa.2016.11.014.

5. Dervas E, Hepojoki J, Laimbacher A, et al. Nidovirus-associated proliferative pneumonia in green tree pythons (*Morelia viridis*). *J Comp Pathol.* 2017;91(21):e00718-17.

6. Litster A. Top 5 Strategies for Managing Shelter Infectious Disease Outbreaks. Available at: https://www.cliniciansbrief.com/article/top-5-strategies-managing-shelter-infectious-disease-outbreaks. Accessed July 1, 2018.

Section 3

Plenary Session

Lauren V. Powers, DVM, DABVP (Avian Practice),
DABVP (Exotic Companion Mammal Practice);
Tim Tristan, DVM, DABVP (Reptile and Amphibian Practice);
Charly Pignon, DVM, DABVP (Exotic Companion Mammal Practice)
Session Chairs

Avian Bornavirus: A New Vaccine and the Pathogenesis of Proventricular Dilatation Disease

Samer S. Hameed, BVMS, MSc, PhD,
Ian Tizard, DVM, PhD, DSc, Dipl ACVM,
Susan Payne, PhD

Session #2253

Affiliation: From Schubot Exotic Bird Health Center, Texas A&M University, College of Veterinary Medicine and Biomedical Sciences, College station, TX 77843, USA (Hameed, Tizard, Payne), Department of Pathology and Avian Pathology, University of Baghdad, College of Veterinary Medicine, Baghdad, Iraq (Hameed).

Abstract: Proventricular dilatation disease (PDD) may be a fatal viral disease that affects mainly psittacine birds. In order to investigate methods of protection, we developed and tested inactivated whole virus and recombinant N protein vaccines in cockatiels as a part of our efforts to protect against PDD in parrots followed by challenge with virulent parrot bornavirus-2. The present study suggests for the first time that the recombinant N protein vaccine can protect against the development of clinical PDD and its pathology and mortality as a result of a switch from a type 1 cell-mediated immune response to a type 2 humoral immune response.

Introduction

Proventricular dilatation disease (PDD) may be a fatal viral disease that affects mainly psittacine birds, although some non-psittacine species can be affected. Gastrointestinal and neurologic signs can be seen in PDD-affected birds. These signs are likely a result of non-suppurative inflammation of the central, peripheral and autonomic nervous systems. Parrot bornavirus (PaBV) was discovered as a causative agent of PDD in 2008. The prognosis of PDD may be poor once clinical signs have developed and no specific treatments (except symptomatic treatments) or commercial vaccines are available to date.

In order to investigate methods of protection, we have developed and tested inactivated whole virus and recombinant N protein vaccines in cockatiels as a part of our efforts to protect against PDD in parrots followed by challenge with virulent PaBV-2.

Methods

In the first experiment, inactivated PaBV-2-infected cells were administrated via intramuscular (IM) inoculation, three times, followed by one IM injection of recombinant N protein vaccine. In a second experiment, the recombinant N protein vaccine was administered alone, two times at three weeks interval, via the IM route. Over 28 weeks post challenge, cloacal swabs were collected and body weights and any clinical signs were noted. Using tissues collected at necropsy, histopathologic examination was used to detect the presence of characteristic microscopic lesions. Quantitative real-time PCR (qRT-PCR) and immunohistochemistry (IHC) were used to detect PaBV-RNA and antigen, respectively. Using Western blots assays, we were not able to detect a humoral response after vaccination with the inactivated vaccine. Conversely, the recombinant N protein vaccine stimulated production of anti-N antibodies.

Results

The results of both vaccine studies indicated that the recombinant N protein vaccine was able to protect subsequently challenged birds from macroscopic and microscopic lesions associated with PDD. Further, the vaccine protected birds from PDD-related morbidity and mortality; however, the vaccine did not protect "efficiently" against infection, as PaBV- RNA and antigen were detected in organs and cloacal swabs of vaccinated and unvaccinated birds. Moreover, there was no evidence that vaccination, either before (vaccine groups) or after challenge (vaccine post challenge group), increased the severity of the infection in comparison with the control group. We hypothesized that this protective response was a result of a switch from a type 1 cell-mediated immune response to a type 2 humoral immune response. This hypothesis was supported by a third study which showed that treatment of cockatiels with cyclosporine A, an immunosuppressant, at the time of challenge with virulent PaBV-2, also conferred complete protection against PDD at the expense of increasing the viral load. This experiment showed that PDD might be an immunologically mediated disease, similar to Borna disease in mammals.

Discussion

In all, the present study suggests for the first time that the recombinant N protein vaccine can protect against the development of clinical PDD and its pathology and mortality as a result of a switch from a type 1 cell-mediated immune response to a type 2 humoral immune response.

Comparison of Dexmedetomidine-Ketamine-Midazolam and Isoflurane for Anesthesia of Black-Tailed Prairie Dogs (*Cynomys ludovicianus*)

David Eshar, DVM, DABVP (Exotic Companion Mammal Practice), DipECZM (Small Mammal) DipECZM (Zoo Health Management),
Geoffrey R. Browning, DVM, MS,
Hugues Beaufrere, DVM, PhD, Dipl ACZM, DABVP (Avian Practice), DipECZM (Avian)

Session #2034

Winner of 2017 AEMV Research Grant

Affiliation: From the Department of Clinical Sciences, College of Veterinary Medicine, Manhattan, KS 66502 USA (Browning & Eshar) and the Department of Clinical Studies, Ontario Veterinary College, University of Guelph, Guelph, ON N1G Canada (Beaufrere).

Nine, 6-month-old intact male black-tailed prairie dogs were anesthetized with a combination of dexmedetomidine (0.25 mg/kg; Dexdomitor, Zoetis, New York, NY, USA), ketamine (40 mg/kg; Ketaset, Zoetis, Spain), and midazolam (1.5 mg/kg; Versed, West-ward, Eatontown, NJ) (DKM) administered intramuscularly in a randomized, crossover study design. For reversal, atipamezole (0.15 mg/kg; Antisedan, Zoetis, Spain) and flumazenil (0.05 mg/kg; Hikma Farmaceutica, Portugal) were administered 45 minutes after induction. For comparison, isoflurane was administered at 5% in 100% oxygen at 5 L/min in an induction chamber, and maintained with 2% in 2 L/min oxygen via facemask for 45 minutes. Induction and recovery time, respiratory rate, heart rate, body temperature, SpO_2, indirect blood pressure, and reflexes were monitored every 5 minutes during the anesthetic period. Blood samples for venous blood gas, PCV, and total solids were obtained at 5 and 45 minutes. Both protocols appeared to achieve safe and effective anesthesia. Significant differences were observed in all anesthetic parameters between the two treatments except for blood pressure. Isoflurane anesthesia resulted in a slightly longer induction and lower respiratory rate and temperature but greater odds of absent reflexes. The DKM anesthesia resulted in a faster induction and less hypothermia but also showed prolonged recovery, and lower heart rate and SpO_2 readings. In conclusion, isoflurane provides a more stable and consistent anesthetic plane, while DKM anesthesia appears as a suitable alternative for short procedures that require fast induction and also analgesia; however, clinicians should be aware of differences in the physiologic effects of each protocol.

Peri-Testicular Cyst in a Green Sea Turtle: Medical and Surgical Management

Terry M. Norton, DVM, Dipl ACZM

Session #2033

Affiliation: From the Georgia Sea Turtle Center, Jekyll Island Authority, 214 Stable Road, Jekyll Island, GA 31527, USA.

A green sea turtle presented to the Georgia Sea Turtle for cold stunning. The turtle was thin, low plasma albumin, glucose, and calcium and elevated uric acid and white blood cell count. Initial treatment consisted of intravenous crystalline and colloidal fluids, glucose, calcium, broad spectrum antimicrobials, and slow rewarming. The turtle improved over the next month and was being considered for release. At week 8, the packed cell volume dropped suddenly to 8%. Blood culture was taken and broad-spectrum antimicrobials instituted. Dexamethasone sodium phosphate was instituted 0.25 mg/kg SQ q24h for 10 days and then tapered. During this time, the packed cell volume (PCV) increased to 14% and continued to rise after steroid therapy discontinued. Severe hyperglycemia (329 g/dl) developed while the turtle was on steroids. The left side of the plastron was enlarged now. Radiographs, ultrasound and computed tomography revealed a large mass encompassing over half the coelom. Aspiration of the mass revealed a large quantity of clear fluid which cytologically appeared to be a transudate. The PCV dropped again 2 weeks later. A similar approach was taken. Once the PCV had stabilized, a laparoscopy was performed which revealed a large cystic mass. It was determined that surgical removal of the cystic mass was needed. Anesthesia and surgery were performed. A combined laparoscopic and surgical approached was used to remove the mass. A post-surgical anemia developed but responded to supportive care. The mass was diagnosed as a peri-testicular cyst and did not regrow. The turtle was released 1 year after initial presentation.

Plasma Concentrations of Voriconazole Delivered in Poloxamer 407 in Select Aquatic Bird Species

Keven Flammer, DVM DABVP (Avian Practice),
Gigi S. Davidson, BSPH, DICVP,
Wendy Massey,
Caroline Meek,
J. Gregory Massey, DVM

Session #2320

Affiliation: From the Department of Clinical Sciences (Flammer, J Massey, W Massey, Meek) and NC State Veterinary Hospital (Davidson), College of Veterinary Medicine, North Carolina State University, 1060 William Moore Dr, Raleigh, NC 27606, USA.

Aspergillosis is a common disease in rehabilitated oiled sea birds and can be treated with voriconazole. The reported voriconazole MIC_{90} for avian *Aspergillus* strains is ≤ 0.5 µg/ml[1,2] and therapeutic regimens target maintaining plasma drug concentrations near 1 µg/ml. In separate dose-ranging experiments with small groups of birds, voriconazole (VFEND, Pfizer, New York, NY USA) was compounded with a thermoreversible biocompatible gel (25% poloxamer 407) and injected subcutaneously into healthy brown pelicans (*Pelecanus occidentalis*) at 10 mg/kg, double-crested cormorants (*Phalacrocorax auritus*) at 10 and 20 mg/kg and ring-billed gulls (*Larus delawarensis*) at 30, 60, and 90 mg/kg. The poloxamer 407 was used as a vehicle to potentially slow drug absorption and prolong drug activity. There was marked individual variation in plasma voriconazole concentrations in all experiments. Therapeutic plasma voriconazole concentrations were maintained in individual birds as follows: 1) 3-5 days in pelicans receiving 10 mg/kg voriconazole/P407, 2) 0-3 days in cormorants receiving 10 mg/kg voriconazole/P407 (some birds failed to achieve therapeutic concentrations) 3) 4-6.5 days in cormorants receiving 20 mg/kg voriconazole/P407, and 4) 10-16 hours in gulls receiving 90 mg/kg voriconazole/P407. These are preliminary results from experiments with small numbers of birds. Additional studies are needed to determine if these doses are safe and effective, and to determine if adding voriconazole to P407 has substantial benefit.

Acknowledgments: Funded by the Oiled Wildlife Care Network, Wildlife Health Center, School of Veterinary Medicine, Davis, CA.

References

1. Silvanoze CD, Bailey TA, Di Somma A. Susceptibility of fungi isolated from the respiratory tract of falcons to amphotericin B. *Vet Rec.* 2006;159:282-284.

2. Beernaert LA, Pasmans F, Van Waeyenberghe L, et al. Avian *Aspergillus fumigatus* strains resistant to both itraconzole and voriconazole. *Antimicrob Agents Chemother.* 2009;53:2199-2201.

Longitudinal Evaluation of Adrenal Gland Volume in Chemically versus Surgically Neutered Ferrets

Nico Schoemaker, DVM, PhD, DipECZM (Small Mammal) DipECZM (Avian), DABVP (Avian Practice),
Evianne Drijver, MSc,
Joost Bandsma, MSc,
Erik Groeneveld, BSc,
Marja Kik, DVM, PhD, DipECZM (Herpetology),
Sara Galac, DVM, PhD,
Yvonne van Zeeland, DVM, MVR, PhD, DipECZM (Avian), DipECZM (Small Mammal), CPBC

Session #2285

Affiliation: From the Department of Clinical Sciences of Companion Animals, Faculty of Veterinary Medicine, University of Utrecht, Utrecht, Yalelaan 108, 3584 CM, The Netherlands.

Surgical neutering has been associated with the development of hyperadrenocorticism in ferrets.[1] A slow-release gonadotropin releasing hormone (GnRH) analogue (deslorelin) containing implant (GnRH A-SRI; Suprelorin®, Virbac, France) is considered an effective alternative for surgical neutering;[2] however, it remains uncertain whether chemical castration with a GnRH A-SRI indeed reduces the incidence of hyperadrenocorticism. The purpose of this study was to perform a longitudinal evaluation comparing the adrenal volumes, plasma hormone concentrations and post-mortem evaluation of the adrenal glands in chemically (annual placement of a GnRH A-SRI containing 4.7 mg deslorelin; n = 18) versus surgically (n = 18) neutered ferrets. Computed tomographic scans were performed annually to assess the adrenal volume, in combination with blood collection. Adrenal volumes were assessed by determining the surface areas of the left and right adrenal glands in each slice, followed by multiplying the sum of the surface areas with the slice thickness (ie, 1 mm). Plasma hormone (androstenedione, 17-OH-progesterone and estradiol) analysis was performed from samples obtained in the first year after neutering and prior to death/euthanasia. Histologic evaluation took place post mortem. In the surgically neutered ferrets, adrenal volumes increased significantly more ($P = 0.29$); adrenal adenomas were found in more animals (62% versus 29%); and plasma hormone concentrations increased in more animals (75% versus 33%); compared to those that received the GnRH A-SRI. These results indicate that, similar to what has been demonstrated in ferrets receiving a GnRH-vaccine,[3] the incidence of adrenal tumors is decreased, but not fully prevented when neutering is performed by using a deslorelin implant.

References

1. Schoemaker NJ, Lumeij JT, Rijnberk A. Current and future alternatives to surgical neutering in ferrets to prevent hyperadrenocorticism. *Vet Med.* 2005;100:484-496.

2. van Zeeland YRA, Pabon M, Roest J, Schoemaker NJ. Use of a GnRH agonist implant as alternative for surgical neutering in pet ferrets. *Vet Rec.* 2014;175:66.

3. Miller LA, Fagerstone KA, Wagner RA, Finkler M. Use of a GnRH vaccine, GonaConTM, for prevention and treatment of adrenocortical disease (ACD) in domestic ferrets. *Vaccine.* 2013;31:4619-4623.

Evaluation of the Feasibility of Spinal Anesthesia in Bearded Dragons (*Pogona viticeps*)

Christoph Mans, Dr med vet, Dipl ACZM, DipECZM (Zoo Health Management), and
Tatiana H. Ferreira, DVM, MSc, PhD, DACVAA

Session #2240

Affiliation: From the School of Veterinary Medicine, University of Wisconsin-Madison, 2015 Linden Drive, Madison, WI 53706, USA.

A prospective, randomized, controlled, blinded, complete crossover study was performed to assess the feasibility of spinal anesthesia following intrathecal injection of lidocaine in 8 adult bearded dragons. After sedation was induced with alfaxalone (15 mg/kg SC, Jurox Inc, Kansas City, MO), intrathecal injections were performed at the sacrococcygeal area with 2 mg/kg of preservative-free lidocaine (AuroMedics Pharma LLC, Dayton, NJ) 1% or 0.2 ml/kg of 0.9% sodium chloride in a complete cross-over design with a 7-day washout period. The injections were repeated if the initial intrathecal injection was not considered successful. Measurements consisted of bilateral mechanical stimulation of hind limbs, front limbs and 25%, 50%, 75% of the trunk's length and assessment of cloacal tone. Animals were monitored for any untoward side effects by monitoring heart rate, respiratory rate, righting reflex and head position. Spinal anesthesia was induced following the first injection of lidocaine in 50% (4/8 animals) and increased to 75% (6/8 animals) following a second injection. Onset of motor/sensory block occurred within 5 minutes and duration and extent was variable with most bearded dragons presenting motor/sensory block of cloaca (6/6), hind limbs (6/6) and 25% of the trunk (5/6) with mean ± SD (range) duration of 49 ± 28 (25-100), 48 ± 25 (25-90) and 40 ± 14 (25-50) minutes, respectively. Intrathecal lidocaine administration caused a significant increase in depth and duration of sedation as well as an increase in heart rate compared to intrathecal saline injection.

Section 4

Hybrid Aquatics Session

Mike Corcoran, DVM, CertAqV, DABVP (Reptile and Amphibian Practice)
Session Chair

Magnetic Resonance Imaging of the Fish Brain: Feasibility, Technique and Interspecies Anatomic Differences

Jennifer L. Ruby, DVM,
Jessica Comolli, DVM,
Alvin C. Camus, DVM, PhD,
Stephen J. Divers, BVetMed, DZooMed, DipECZM (Herpetology),
DipECZM (Zoo Health Management), Dipl ACZM, FRCVS,
Karine P. Gendron, DVM, Dipl ECVDI, Dipl ACVR

Session #2269

Affiliation: From the Department of Veterinary Biosciences and Diagnostic Imaging (Ruby, Gendron), the Department of Small Animal Medicine and Surgery (Comolli, Divers) and the Department of Pathology (Camus), College of Veterinary Medicine, University of Georgia, 2200 College Station Road, Athens GA 30602, USA.

Recent advances in aquatic animal medicine and continued growth of the fish hobbyist and aquaculture communities have led to a developing interest in antemortem diagnostic imaging of aquatic species. To the authors' knowledge, normal imaging anatomy of the fish brain remains unknown. The aim of the present study was to determine whether advanced neuroimaging can be safely achieved in the live fish, to optimize imaging parameters, and to develop a comparative magnetic resonance imaging (MRI)-histology atlas of a few fish species of economic or research value. Two each of channel catfish (*Ictalurus punctatus*), koi (*Cyprinus rubrofuscus*), and grass carp (*Ctenopharyngodon idella*) of at least 12 inches in length were individually anesthetized for MRI evaluation of the brain. After imaging, the fish were recovered from anesthesia, humanely euthanized, and immediately processed for brain histopathology. Necropsy confirmed sex and health status of each fish. All fish achieved an adequate anesthetic level for prolonged immobilization during imaging, and were successfully recovered from anesthesia prior to euthanasia. Diagnostic quality images were obtained on all subjects; however, spatial resolution was maximized with larger fish. Imaging protocols were optimized for standard neuroimaging sequences (T2 and T1-weighted, fat saturation). In addition, inversion times for fluid-attenuation inversion recovery (FLAIR) sequences were adapted to the naturally high protein content of fish cerebrospinal fluid. Quantification of brain transverse relaxation times was determined. A comparative MRI and histology atlas was created of these species' brain.

Clinical Use of Cross-Sectional Imaging (Computed Tomography and Magnetic Resonance Imaging) in Two Koi Carp (*Cyprinus carpio*)

Jessica R. Comolli, DVM,
Karine Gendren, DMW, Dipl ECVDI, Dipl ACVR,
Jörg Mayer, DVM, MS, DABVP (Exotic Companion Mammal Practice),
Jennifer Ruby, DVM, DipECZM (Small Mammal), Dipl ACZM,
Spencer Kehoe, DVM,
Stephen Divers, BVetMed, DZooMed, DipECZM (Herpetology),
DipECZM (Zoo Health Management), Dipl ACZM

Session #2178

Affiliation: From the Department of Small Animal Medicine and Surgery (Zoological Medicine) (Comolli, Mayer, Kehoe, Divers), Department of Veterinary Biosciences and Diagnostic Imaging (Gendren, Ruby), College of Veterinary Medicine, University of Georgia, 2200 College Station Rd, Athens, GA 30602, USA.

Reported are clinical cases of two koi carps, in which computed tomography (CT) and magnetic resonance imaging (MRI) were utilized. The first koi was presented to the University of Georgia Zoological Medicine Service for evaluation of a left caudodorsal mass. CT was performed and a large fat-dense intramuscular mass was identified. After biopsy and surgical removal, a repeat CT scan was performed indicating a resolved lipoma with no evidence of mass regrowth or regional osteomyelitis. A second koi carp was presented for evaluation of erratic and neurologic behaviors. On CT, numerous vertebral anomalies and mild scoliosis were present. High-field MRI (3T) of the brain and spinal cord demonstrated again scoliosis and vertebral morphology abnormalities, as well as multifocal areas of spinal cord compression within the cranial spine and an irregular subarachnoid column and spinal cord compression within the caudal spine. Necropsy revealed severe myelomalacia with vertebral subluxation. Anesthesia was maintained during imaging using an out-of-water technique for CT and full submersion for MRI. Results indicate that advanced cross-sectional imaging is a useful modality for evaluation of clinical disease in koi carp.

Effect of Temperature on Clinical and Biochemical Relationships of MS-222 Anesthesia in Goldfish

Alyssa M. Scagnelli, DVM,
Jeannette Cremer, DVM, Dr med vet, DACVAA, CVA,
Kevin Kleinow, DVM, PhD

Session #2216

Affiliation: From the School of Veterinary Medicine, Louisiana State University, Skip Bertman Drive, Baton Rouge, LA 70810, USA, present address Avian & Exotic Animal Clinic of Arizona, 1911 S. Lindsay Rd, Mesa, AZ 85204, USA (Scagnelli), School of Veterinary Medicine, Louisiana State University, Skip Bertman Drive, Baton Rouge, LA 70810, USA (Cremer), Department of Comparative Biomedical Sciences, Baton Rouge, LA, USA, Louisiana State University, Skip Bertman Drive, Baton Rouge, LA 70810, USA (Kleinow).

Our objective was to evaluate if acclimation temperature and acute temperature divergence influenced MS-222 anesthesia in goldfish (*Carassius auratus*). Ninety-six goldfish were used in this study and divided into a 9-cell treatment matrix consisting of 3 separate acclimation temperature groups (AC10, AC20, and AC30 °C) versus 3 acute temperature treatment groups (A10, A20, and A30 °C). Following a 30-day temperature acclimation period (± 1 °C), individual fish were exposed to un-dosed water (2 L) at the desired acute temperature for 50 minutes, followed by scototaxic stress evaluation at the same temperature for 10 minutes. Using the acute temperature and a buffered 80 mg L -1 MS-222 dose (2 L) fish were anesthetized for a maximum of 30 minutes and the following times were recorded: time to sedation, partial loss of equilibrium, complete loss of equilibrium, and surgical plane of anesthesia. Data were analyzed using a metric emulating survival analysis with Bonferroni post hoc comparisons ($P < 0.05$). Fish at AC10:A10 and AC20:A20 °C exhibited statistically indistinguishable times to surgical depth anesthesia (median 768 seconds [420-1792 seconds] vs 1067 seconds [601-1320 seconds]) while AC30:A30 °C never attained surgical depth. The AC10:A30 °C treatment reached surgical plane anesthesia rapidly (median 65 seconds [12-152 seconds]) while the AC30:A10 °C group performed similarly, albeit without use of MS-222. These results indicate that acclimation and divergent water temperatures significantly influence MS-222 anesthesia as a function of complex adaptive changes in the goldfish.

Section 5

Hybrid Mixed Bag Session

Mike Corcoran, DVM, CertAqV, DABVP (Reptile and Amphibian Practice)
Session Chair

A New Inexpensive Pressure-Cycle Ventilator for Use in Exotic Animal Medicine

Stephen J. Divers, BVetMed, DZooMed, DipECZM (Herpetology), DipECZM (Zoo Health Management), Dipl ACZM, FRCVS

Session #2052

Affiliation: From Department of Small Animal Medicine and Surgery (Zoological Medicine), College of Veterinary Medicine, University of Georgia, 2200 College Station Road, Athens, GA 30602, USA.

Abstract: Intermittent positive pressure ventilation is well established in human and veterinary anesthesia. It is essential for all reptiles and offers advantages over spontaneous respiration in birds and mammals. A new pressure cycle ventilator is available and has proven invaluable for the maintenance of intubated exotic species. A review of ventilation and the specific value and practical use of this relatively new and inexpensive ventilator is provided.

Introduction

Mechanical ventilation (also called positive pressure ventilation) is well established in human and veterinary anesthesia.[1] It is routinely required for all reptile anesthetic events but also has important benefits for mammals and birds.[2,3]

Ventilation improves lung expansion, respiratory gas exchange, and can overcome the depressive effects of opiates, and the physical restrictions of obesity and weight (eg, keel mass). Mechanical ventilation can fully or partially replace spontaneous breathing. It is indicated to overcome respiratory inadequacy, which is defined as insufficient oxygenation, insufficient alveolar-faveolar-airsac ventilation, or both. Ventilation is essential for thoracoscopy-thoracotomy and laparoscopy-coelioscopy procedures that involve insufflation.

Following an inspiratory trigger, a predetermined mixture of oxygen and anesthetic gases is forced into the airway through the endotracheal tube. As the lungs inflate, the intra-alveolar (mammals), airsac (birds) or intrapulmonary (reptiles) pressure increases. A termination signal eventually causes the ventilator to stop forcing air into the airway by redirecting through a low pressure exhaust outlet. Expiration follows passively, with gases flowing from the higher pressure lung and airsac spaces to the lower pressure central airway.

Ventilation

Methods of ventilation

Intubation is considered essential, although it may be possible to ventilate using some supraglottic airway devices. Ventilation can be done manually but extreme care must be taken to avoid over/under-inflation. Over-inflation can result in lung damage, but the most common disadvantage of manual bagging is the variability in frequency and pressure, which in turn creates greater instability in blood and CNS inhalant drug levels. Volume cycle ventilators offer the advantage that they can usually overcome a blocked endotracheal tube; however, the margin of error in volume delivery is small. Consequently, pressure-cycle ventilators are often preferred for small exotic species because they are generally safer, and the zoological medicine service currently uses the SAV4 Small Animal Ventilator (Vetronics Services, Units 1 & 2 Henley's Business Park, Manor Road, Abbotskerswell, Devon TQ12 5NF, UK; http://www.vetronic.co.uk/index.php/S=0/product/sav04-small-animal-ventilator).

Pressure-cycle ventilation using the SAV4

In order to use ventilation, the patient must be intubated using the largest possible tube to reduce leakage. In addition, airway resistance increases dramatically as tube radius decreases:

$$Airway\ resistance = \frac{8nl}{\pi r^4}$$

$$n = dynamic\ viscosity;\ l = length\ of\ airway; r = radius\ of\ endotracheal\ tube$$

Mainstream capnography should be connected between the endotracheal tube connector and the ventilator. Pediatric connectors should be used to minimize dead space. Sampling volumes from sidestream units, if large compared to the size of the patient, may result in dilution and artificially depress $ETCO_2$ values, and again, pediatric settings should be used.

The endotracheal tube is connected to the valve mechanism, along with the incoming oxygen line and exhaust (with or without a bag). The initial gas flow is set to 2-3x minute volume and, with the ventilation (IPPV) switch set to off, the ventilator acts as a simple non-rebreather circuit.

Minute volume is estimated:

$$MV = \frac{V_T\ x\ RR}{1000}$$

MV = minute volume (L/min); V_T = tidal volume (ml); RR = respiratory rate (/min)

To initiate ventilation, the expiration length is set to 1 second and the initial trigger pressure set to 2 cm water pressure. The IPPV function is switched on. The animal is observed and the trigger volume is adjusted to match pre-anesthesia, conscious respiratory excursions. Once pressures are appropriate the expiratory length is adjusted to achieve pre-anesthesia, conscious respiratory rates. Finally, the oxygen flow rate is adjusted to achieve normal inspiratory function. Decreasing oxygen flow will extend the duration of the inspiration phase, while increasing flow decreases it. Both ventilation frequency (expiration length) and depth (pressure) should be fine-tuned to maintain appropriate $ETCO_2$ values during anesthesia.

Conclusion

The author has used various models of the Vetronics ventilator since 1997, and the University of Georgia has purchased 5 devices since 2001 for research and clinical use. These units have consistently performed well with birds and mammals up to about 6-8 kg (because of the high gas flows required for non-rebreather circuits), and with reptiles up to 30-40 kg (because of the greatly reduced flow requirements even in large animals). Subjectively, the use of this ventilator has helped maintain a more consistent and controllable anesthetic plane. The most current model (SAV4) retails for approx. £1200 in the UK ($1700, €1400), which is considerably lower than other ventilators on the market. There are several distributers in Europe and the device can be ordered and shipped to the United States.

Disclaimer: The author has no financial or conflicts of interest associated with this product or the company.

References

1. Luchetti M. Mechanical ventilation during general anesthesia. In: Hertzog M, Kuhn Z, eds. *General Anesthesia Research Developments.* Hauppauge, NY: Nova Science Publishsers;2009.

2. West G, Heard D, Caulkett N. *Zoo Animal and Wildlife Immobilization and Anesthesia.* Hoboken, NJ: John Wiley & Sons;2014.

3. Hernandez-Divers SM, Schumacher J, Stahl S, et al. Comparison of isoflurane and sevoflurane anesthesia after premedication with butorphanol in the green iguana (*Iguana iguana*). *J Zoo Wildl Med.* 2005;36: 169-175.

Plasticosis from Intact and Micronizing Plastic in Birds, Reptiles, and Mammals Including Humans

Evan White, PhD,
Christopher R. Gregory, DVM, PhD,
Jason Locklin, PhD,
Branson W. Ritchie, DVM, MS, PhD, DABVP (Avian Practice), DipECZM (Avian)

Session #2404

Affiliation: From New Material Institute, University of Georgia, Athens, GA 30602, USA (White, Locklin, Ritchie), Infectious Diseases Laboratory, University of Georgia, Athens, GA 30602, USA (Gregory).

Petroleum-based plastics and their toxins are increasingly contaminating our rivers, oceans and soil. Recent estimates suggest that 5.2 trillion pieces of plastic already contaminate the world's oceans. As plastic degrades it micronizes placing smaller fauna and juvenile animals (particularly endangered species) at risk of gastrointestinal impaction. The clinical implications of our team's research to document the health effects of plastic ingestion and exposure and our progress in replacing petroleum-based plastics with biodegradable alternatives will be discussed.

Vital Components of an Animal Vendor Standards and Audit Program

Whitney L. Miller, DVM, MBA, Dipl ACVPM

Session #2312

Affiliation: From PETCO Animal Supplies, Inc, San Antonio, TX 78245, USA.

Abstract: Retail pet stores are a common source for many companion pets including reptiles, small mammals, and birds. Pet stores must ensure breeders and distributors of animals they work with are adhering to proper standards for the care and health of animals in their facilities.

US Department of Agriculture and State Inspection and Licensing

All breeders and distributors of animals must be licensed by the US Department of Agriculture and the state when applicable. Pet stores should verify active licensure and require notification when inspections occur, any citations issued, and corrective actions taken.

Animal Care Standards and Protocols

Standards should be in place within any facility breeding or distributing animals for retail pet stores. Pet stores should review standards and protocols and ensure they align with expectations. Appropriate expertise, including veterinarians should be involved in creation and regular review of animal care protocols and processes.

Standards should address:

- Facility requirements
- Biosecurity
- Cleaning and sanitation
- Storage of food, medicine, and bedding
- Feeding and watering
- Habitat requirements
- Veterinary care and other health standards
- Disease prevention and management
- Transportation of animals in and out of the facility

- Emergency and disaster plans
- Recordkeeping
- Training of staff

Validation of Proper Care and Standards

Pet stores should seek to work with breeders and distributors that not only have appropriate care standards and protocols in place, but also have a system of verification of adherence to those standards. Verification should go beyond regular USDA and/or state licensing inspections. Facilities should have an internal validation process in place, as well as consider third-party validation. Large national pet store chains currently have auditing processes in place.

Conclusions

Sourcing animals from quality breeders and distributors is critical to ensuring their health and welfare, and should be a top priority for any retail pet store. Veterinarians should be prepared to advise pet stores, animal breeders and distributors on creating, adhering to, and validating standards at their facilities.

References

1. Pet Industry Joint Advisory Council Small Animal Care Standards. Available at: https://pijac.org/sites/default/files/SmallAnimalCareStandards2017.pdf. Accessed July 1, 2018.

Building Blocks in a Successful Retail Pet Animal Care and Welfare Program

Whitney L. Miller, DVM, MBA, Dipl ACVPM

Session #2308

Affiliation: From PETCO Animal Supplies, Inc, San Antonio, TX 78245, USA.

Abstract: Retail pet stores are a common source for many companion animals including reptiles, small mammals, and birds. All retail pet stores, whether independent, a franchise, or part of a larger corporation should have a strong animal care program to ensure the wellbeing of animals in their care and to provide appropriate veterinary care when animals are sick or injured. Partnering with retail pet stores is an opportunity for veterinarians to work directly with a supplier of pets to potential clients and to assist and support the proper care of animals while in a pet store environment.

Sourcing of Animals

Retail pet stores that sell animals must have appropriately licensed breeders or distributors to source animals. If stores work with multiple breeders and/or distributors of animals they must have appropriate systems and tracking in place for traceability and to address any health or welfare concerns with animals.

Receiving of Animals

Animals can be transported from breeders and distributors to pet stores by different means, but regardless of method, procedures should be in place to minimize stress and ensure safety. Animals should be evaluated by pet store staff upon arrival for any signs of injury or illness, damage to shipping containers, or any other concerns. If any concerns are identified they should be communicated back to the breeder or distributor to address with their staff and delivery service.

Due to stress of transport to a pet store a consideration to support health of the animals is for any pet store to quarantine away from the sales floor so animals can acclimate to the new environment, any differences in care, and other changes. This is not always achievable, but can be, at minimum, discussed as a best practice.

In-Store Care Standards

Pet stores must have formal care instructions for any species housed in their store. These care standards should include: maximum number of animals allowed per habitat, size of habitat per species if not standard, type of food and water source provided and how often, temperature and humidity ranges, substrate provided, décor, lighting, and any supplements, treats, or fresh food should be provided. Certain husbandry parameters can be automated based on the habitat types in the store (eg, lighting cycle and type), but verification by store staff should be ongoing.

Direction and schedules, including tracking of completion, should be created for maintenance of habitats, cleaning and disinfection, and any other tasks important for care and health of the animals. In creation of recommended cleaning and disinfection schedules consideration should be given to how often new animals are entering the store to minimize any potential disease transmission. Onboarding and training must also be provided to all caretakers in the store so they can provide appropriate care and be held to store standards.

Pet stores must also be aware and stay current on any regulatory requirements in their state and locality. This includes if they must abide by certain minimum care standards, if there are certain species that are illegal to sell, requirements around inspection and providing veterinary care, guarantee periods, and any other applicable regulations.

Veterinary Care for Illness, Injury, and Preventive Needs

All pet stores must have a relationship with a veterinarian to provide care to animals when ill, injured, or for preventive care needs. If veterinarians are not open or accessible all hours the store is open, the store should also partner with an emergency facility when necessary. The pet store must provide appropriate training and guidance for staff on when and how an animal must be provided care by a veterinarian. There should never be a situation where an animal in need of veterinary care does not receive care within a reasonable timeframe. A recommended standard used in the industry is care must be provided within 24 hours of a noticed ailment, unless deemed to be an emergency in which case care should be provided as soon as possible.

Regular Health Checks

While it can be assumed that animals in pet stores will be regularly monitored for health, as they are usually open regular hours, it should be recommended to any store that there be a formal system in place to check on animals on regular and ongoing basis. Training and guidelines should be given on what signs would indicate a potential health concern for species carried in the store.

Education of Customers

Pet store staff should be discussing proper care with any customer interested in purchasing an animal. Focus should not only be on proper care and husbandry based on species, but also that animal is a good fit to be in a forever home with that customer. Printed or online care sheets/guides should be provided to reinforce proper care for the customer. These are required by regulations in certain jurisdictions. ARAV has companion care brochures available for some reptile and amphibian species.

Conclusions

While working with retail pet stores can have its challenges and frustrations, it is critical for veterinarians to be engaged not only in the care of sick and injured animals, but also the regular care provided in the store to healthy animals. While meant only as a temporary housing situation for these various species, the highest standards of husbandry are paramount to ensure a healthy pet makes it from the retail environment into an owner's home.

Section 6

Hybrid Master Classes

Anna Osofsky, DVM, DABVP (Avian Practice);
Michelle Hawkins, VMD, DABVP (Avian Practice)
Session Chairs

Putting it all Together: Clinical Cases Correlated with Clinical and Anatomic Pathology

Anneliese Strunk, DVM, DABVP (Avian Practice),
Kendal E. Harr, DVM, MS, DACVP,
Drury Reavill, DVM, DABVP (Avian Practice), DABVP (Reptile and Amphibian Practice), DACVP

Session #2360

Affiliation: From The Center for Bird and Exotic Animal Medicine, 11401 NE 195th St, Bothell, WA 98011, USA (Strunk), Urika, LLC, 8712 53rd Place West, Mukilteo, WA 98275, USA (Harr), Zoo/Exotic Pathology Service, 7647 Wachtel Way, Citrus Heights, CA 95610, USA (Reavill).

Abstract: Real clinical cases (covering avian, small mammal, and reptilian species) will be presented in an interactive format with the audience. The speakers will discuss differential diagnoses with likely pathogenesis using evidence-based medicine, develop diagnostic plans, interpret data, correlate histopathologic findings, and discuss case management.

Introduction

Laboratory information is vital to understanding and monitoring clinical cases; however, not every case demonstrates the classically expected findings for different disease processes. Working as a team with a clinical pathologist and anatomic pathologist lead to a deeper understanding and more thoughtful approach to cases, even when the definitive diagnosis is elusive. Discussion of relevant disease processes follow case presentations.

Case 1. Macaw

Information in Table 1 describes a clinical case involving a macaw.

Table 1. Macaw case.

Factor	Description
Signalment	21-year-old male severe macaw (*Ara severus*)
Presenting complaint	Owner noted regurgitation the day before
Physical examination	Quiet per owner, but no significant findings were noted on physical examination; the regurgitation had stopped by the time of presentation

Diagnostics

Tables 2 and 3 provide results of hematology and biochemistry analyses. Reference intervals also are provided.

Table 2. Selected hematologic values from a male severe macaw with recent regurgitation.

Analytes	April 11, 2014	Dec 15, 2014	May 31, 2017	June 19, 2017	Dec 27, 2017	Feb 28, 2018	Unit	Reference Interval
Hematocrit	57	50	39	42	48	47	%	39-54
Total solids	5.70	4	6.5	6.2	3.8	3.8		
White blood cells	43,200	10,800	18,000	14,400	20,000	17,500	/µl	5000-20,000
Bands	3024	0	720	0	0	0	/µl	0-0
Heterophils	28,944	5184	10,620	8640	10,400	9625	/µl	1100-15,000
Lymphocytes	5184	3996	4500	3744	3200	3675	/µl	700-8700
Monocytes	6048	1512	1980	2016	6200	4200	/µl	0-900
Eosinophils	0	0	0	0	0	0	/µl	0-100
Basophils	0	0	180	0	0	0	/µl	0-400

Table 3. Selected biochemistry values from a male severe macaw with recent regurgitation.

Analytes	May 23, 2017	June 6, 2017	June 19, 2017	July 24, 2017	Sept 20, 2017	Sept 27, 2017	March 10, 2018	Unit	Reference Interval
CA	8.5	8.4	8.1	7.8	8.6	9.6	10	mg/dl	7.5-11.9
PHOS	2.9	2.4	2.9	2.5	5.3	1.8	1.8	mg/dl	1.4-6.7
TP	6.8	4.9	5.6	6	4.4	4.5	4.7	g/dl	2.1-5.2
ALB	0.4	0.6	0.7	0.7	1.1	1.9	2	g/dl	0.7-1.9
GLO	6.4	4.3	4.9	5.3	3.3	2.6	2.7	g/dl	1.4-3.5
GGT	3	4	5	6	7	11	9	U/L	0-22
AST	515	518	338	242	289	188	199	U/L	70-490
CK	390	430	263	390	1005	223	244	U/L	98-814
CHOL	304	339	412	252	184	293	264	mg/dl	122-441
Uric acid	16.2	8.9	10.5	7.9	14.5	5.8	5.6	mg/dl	2.3-12.7
Bile acids	300	548	473	352	28	18	30	umol/L	5-103
Triglycerides	136	145	260	154	105	204	144	mg/dl	46-564

Additional diagnostic tests and case information included radiographs, ultrasound, celioscopy, and histopathology. Widened hepatic silhouette and splenomegaly were noted on radiographs. Ultrasound revealed rounded liver margins. Celioscopy for liver and kidney biopsy was performed. The liver had a very thickened capsule, to the degree that it was difficult to get to the liver parenchyma. Histopathology of the liver revealed fibrosis with severe lymphoplasmacytic cholangiohepatitis and rare heterophilic cholangiohepatitis. There was diffuse moderate membranoglomerulopathy noted in the kidney.

Treatment

The initial treatment plan included subcutaneous fluid therapy with antibiotics (enrofloxacin (25 mg/kg PO q24h; Baytril, Shawnee Mission, KS, USA), amoxicillin/clavulanic acid (125 mg/kg PO q12h; Clavamox, Zoetis, Kalamazoo, MI, USA), meloxicam (1 mg/kg PO q24h; Metacam, Boehringer Ingelheim Vetmedica, Inc, St. Joseph, MO, USA), colchicine (0.2 mg/kg PO q12h; compounded at Woodinville Pharmacy, Woodinville, WA, USA), and liver supplement (silymarin (Hepasyl, USANA, Salt Lake City, UT, USA) and SAM-e (Denosyl,

Nutramax, Lancaster, SC, USA) suspended in lactulose (Generlac, Morton Grove Pharmaceuticals, Morton Grove, IL, USA)). The bird was clinically normal within a week of initial presentation. Subcutaneous fluids were discontinued and other medications maintained long term.

A follow-up biopsy was performed 4 months after the initial biopsy. The liver appeared more normal during the procedure (approach was from ventral midline, allowing visualization of both liver lobes). Two biopsies were obtained from each lobe. Histopathology revealed mild acute focal cholangitis.

Case 2. Red-Lored Amazon

Information in Table 4 describes a clinical case involving a red-lored Amazon parrot.

Table 4. Red-lored Amazon case.

Factor	Description
Signalment	28-year-old male red-lored Amazon (*Amazona autumnalis*)
Presenting complaint	1-week history of labored breathing and nasal discharge
Physical examination	Weakness, dehydration, emaciation (BCS 1/9), dull plumage with stress bars, intermittent, irregularly irregular arrhythmia, wheezing auscultable in caudal airsacs, and a crusty brown discharge from both nares

Diagnostics

An extensive diagnostic plan was initiated, starting with complete blood count (CBC) and chemistry panel. Results are provided in Table 5.

Table 5. Selected hematologic values from a male red-lored Amazon with labored breathing.

Analytes	May 6, 2013	June 6, 2013	Aug 23, 2013	Dec 6, 2013	March 14, 2014	Oct 1, 2014	Jan 9, 2015	Aug 13, 2015	Unit	Reference Interval
Hematocrit	44	46	58	49	54	53	55	56	%	39-54
White blood cells	50,000	80,000	32,000	40,000	20,000	20,000	15,000	14,400	/µl	5000-15,000
Bands	2,000	16,000	320	800	800	800	150	144	/µl	0-0
Heterophils	38,000	71,200	24,960	30,400	14,200	11,400	6,900	7,776	/µl	1500-11,000
Lymphocytes	3,000	0	1,280	2,800	1,000	6,200	3.150	4,032	/µl	1,000-10,000
Monocytes	6,500	7,200	5,760	6,000	3,800	1,600	4,800	2,016	/µl	0-900
Eosinophils	500	0	0	0	200	0	0	432	/µl	0-400
Basophils	0	0	0	0	0	0	0	0	/µl	0-300

Additional diagnostic tests and case information included diagnostic imaging, Gram's stain, culture, and other testing. Radiographs showed severe increased coelomic soft tissue with widened hepatic silhouette. A barium study showed dilated proventriculus, isthmus, ventriculus with overall slow gastrointestinal (GI) motility. Crop Gram's stain revealed > 90% gram negative rods and crop culture identified multi-drug resistant *Klebsiella pneumonia*, *Enterobacter*, and *Enterococcus.* Bornavirus PCR from a cloacal swab was positive. Other diagnostics that were negative or within normal limits: EPH; choanal Gram stain; fecal direct, float, acid fast, and gram stain; blood culture, *Aspergillus* panel, PCR for *Mycobacterium*, *Chlamydia psittaci*, *Macrorhabdus ornithogaster*, *Helicobacter* genus, heavy metal levels, and electrocardiogram.

Treatment

Initial treatment included antibiotics (piperacillin/tazobactam (200 mg/kg IM q12h; Piperacillin and Tazobactam for Injection, USP, Hospital Worldwide, Lake Forest, IL, USA) and amikacin (15 mg/kg IM q24h; Amikacin 50 mg/ml injectable, compounded by MedsforVets, Sandy, UT, USA), beneficial bacteria gel (1 gm daily; Bene-Bac Plus Bird and Reptile Gel, Pet-Ag, Hampshire, IL, USA), celecoxib (10 mg/kg PO q24h; compounded at Kusler's Compunding Pharmacy, Snohomish, WA, USA)). Moderate clinical improvement was observed, but CBC changes worsened.

Crop biopsy and celiotomy for biopsy of ventriculus and air sac were performed. No lesion aside from severe thin walled/dilated proventriculus/ventriculus was noted. The procedure was truncated due to patient instability. Histopathology revealed no lesions in any section.

Ongoing treatment included enrofloxacin (10 mg/kg PO q12h; Baytril, Shawnee Mission, KS, USA), metoclopramide (0.5 mg/kg PO q12h; metoclopramide oral solution, USP, Alpharma, Baltimore, MD, USA), celecoxib (10 mg/kg PO q24h; compounded at Kusler's Compunding Pharmacy, Snohomish, WA, USA), and liver supplement (silymarin (Hepasyl, USANA, Salt Lake City, UT, USA) and SAM-e (Denosyl, Nutramax, Lancaster, SC, USA) suspended in lactulose (Generlac, Morton Grove Pharmaceuticals, Morton Grove, IL, USA)). Trial with azithromycin (50 mg/ml PO q24h; Azithromycin for oral suspension, Greenstone LLC, Peapack, NJ, USA) for 30 days did not cause improvement in clinical or laboratory picture, so a therapeutic trial with itraconazole (10 mg/kg PO q24h; Sporanox oral solution, Janssen Pharmaceutica NV, Beerse, Belgium) was instituted. The patient was clinically improved within 1 week, and back to normal 1 month later. The long-term treatment plan included itraconazole, celecoxib, and the liver supplement at doses noted above. Signs recrudesced 34 months after initial presentation.

Case 3. Guinea Pigs

Information in Table 6 describes a clinical case involving two guinea pigs.

Table 6. Guinea pig cases.

Factor	Description
Signalment	5-year-old spayed female and 4-year-old male guinea pigs (*Cavia porcellus*)
Presenting complaint	Annual examination, male had mild inappetance and lethargy
Physical examination	No significant findings on examination, male had mild lingual elongation of rostral lower cheek teeth and decreased GI sounds

Diagnostics

The initial plan recommended for the female guinea pig included routine annual diagnostics, including a CBC and chemistry panel. A previous CBC and chemistry panel were performed at the time of her spay 1 year prior, noted to be within normal limits. For the male, a full workup was recommended, but the owner elected only a CBC and mini chemistry panel. Results are presented in Table 7.

Table 7. Hematologic values from a clinically normal female guinea pig and a male guinea pig with mild dental disease and signs of gastrointestinal stasis.

Analytes	Female guinea pig	Male guinea pig	Unit
Hematocrit	40	34	%
Total solids	4.4	5.4	
White blood cells	15,000	90,000	/µl
Bands	0	0	/µl
PMN	1800	4500	/µl
Lymphocytes	10,800	84,600	/µl
Monocytes	2250	900	/µl
Eosinophils	150	0	/µl
Basophils	0	0	/µl

The male's CBC is > 75,000 cells/µl. This is most likely a small cell lymphocytic leukemia of T cell origin. Immunostains are required for definitive identification of the cell type. A grade 5 lymphoma cannot be differentiated from a disseminated lymphocytic leukemia. Prognosis is guarded to poor.

Additional diagnostic tests and case information for the female guinea pig included a followup CBC that confirmed continued leukocytosis with lymphocytosis. Three months after the lymphocytosis was noted, a 2-mm diameter pedunculated pink conjunctival mass was noted at the medial canthus of the left eye. Histopathologic examination revealed lymphoid hyperplasia. The patient has had one episode of a urinary tract infection (UTI) that resolved with antibiotics, and otherwise remains clinically normal on examination, with persistent leukocytosis with predominant lymphocytes on serial CBCs.

Treatment

Additional case information for the male included the patient being sent home on metacam pending results, in preparation for an occlusal adjustment within the following week. Once the CBC results were back, the owner elected euthanasia.

Case 4. Rabbit

Information in Table 8 describes a clinical case involving a rabbit.

Table 8. Rabbit case.

Factor	Description
Signalment	9-year-old castrated male rabbit (*Oryctolagus cuniculi*)
Presenting complaint	Seen by referring veterinarian for prolapsed third eyelids
Physical examination	Prolapsed third eyelids, noncompressible thorax

Diagnostics

Radiographs revealed a thoracic mass. The CBC and biochemistry panel were submitted. Results are provided in Table 9.

Table 9. Complete blood count and selected biochemistry values from a rabbit with a thoracic mass.

Analytes	Jan 21, 2017	April 8, 2017	Unit
Hematocrit	43	36	%
Total solids	8.8	54	
White blood cells	10,800	5000	/µl
Bands	0	0	/µl
Heterophils	5940	2550	/µl
Lymphocytes	3780	2150	/µl
Monocytes	972	200	/µl
Eosinophils	0	100	/µl
Basophils	108	0	/µl
ALP	24	24	IU
AST	30	NR	IU
GGT	4	NR	IU
ALT	44	55	IU
Creatinine	1.6	1.3	mg/dl
Albumin	4.3	2.3	g/dl
Total protein	7	5.5	g/dl
Globulins	2.7	3.2	g/dl
Urea	22	13	mg/dl
Cholesterol	56	NR	mg/dl
Glucose	130	187	mg/dl
Calcium	14.9	14	mg/dl
Phosphorus	3.3	2.1	mg/dl

In results of the CBC on 4/8/17, there was morphologic change. While there is no strong evidence of a neoplastic lymphocytic population in the sample, it was interesting to note the apoptotic and lytic large lymphocytes in the sample. These likely originate in the irradiated thymoma.

Additional diagnostic tests and case information included a computed tomography (CT) scan that revealed a large cranial mediastinal mass that was partially cavitated. Thymoma was suspected. A cranial mediastinal lymphadenopathy, thymic lymphoma, granulomatous lymphadenopathy, or ectopic thyroid carcinoma were considered less likely. There was no evidence of pulmonary metastasis. Cardiomegaly was noted.

Ultrasound guided fine needle aspirate (FNA) was performed. Cytologic evaluation was consistent with a probable thymoma.

Respiratory signs developed and progressed quickly, requiring oxygen support for transport to a tertiary facility for radiation therapy. The rabbit recovered well and was considered in remission 3 months later. He developed cataracts after his radiation therapy and was noted to be *E. cuniculi* positive on testing 1 month after remission.

An *E. cuniculi* panel was IgG titer 1:128, considered moderate positive; IgM titer < 1:64, considered not supportive of active acute infections; C reactive protein < 0.1mg/dl, with a reference range 0-20 mg/dl.

Treatment

Respiratory signs recurred 9 months after remission, and treatment with chemotherapy and another course of radiation was elected. Remission was achieved 3 months later, with an expectation of at least 6 months before signs recur.

Case 5. Bearded Dragon

Information in Table 10 describes a clinical case involving a bearded dragon.

Table 10. Bearded dragon case.

Factor	Description
Signalment	1-year-old male bearded dragon (*Pogona vitticeps*)
Presenting complaint	Lesion on tip of tail
Physical examination	Black discoloration of the tail tip with bone exposure

Diagnostics

A CBC was performed. Results are provided in Table 11.

Table 11. Complete blood count values in a bearded dragon with tail tip necrosis.

Analytes	Value	Unit
Hematocrit	18	%
Total solids	9.3	
White blood cells	18,000	/µl
Bands	0	/µl
Heterophils	5040	/µl
Lymphocytes	11,520	/µl
Monocytes	1440	/µl
Eosinophils	0	/µl
Basophils	0	/µl

Treatment

Treatment included antibiotics (ceftazidime) and meloxicam. The tail was amputated and submitted for histopathology. Histopathology revealed focally extensive severe erosive to ulcerative subacute dermatitis with inflammatory exudates and intralesional fungus (CANV-like). No inflammation or skin changes were noted at the amputation site. Atadenovirus PCR was positive.

The patient was started on a liver supplement (silymarin (Hepasyl, USANA, Salt Lake City, UT, USA) and alcohol free dandelion extract (Herb Pharm, Williams, OR, USA) suspended in lactulose (Generlac, Morton Grove Pharmaceuticals, Morton Grove, IL, USA)) for life. The tail healed well. The owner has declined further recommendations for monitoring the animal's condition with biochemistry panels and coelomic evaluation via ultrasound.

Case 6. Bearded Dragon

Information in Table 12 describes a clinical case involving a bearded dragon.

Table 12. Bearded dragon case.

Factor	Description
Signalment	1.5-year-old male bearded dragon (*Pogona vitticeps*)
Presenting complaint	Lethargy, decreased appetite/activity with progressive swelling of proximal right pelvic limb
Physical examination	Bright, alert, responsive, mild dehydration, thin body condition (body weight of 144 g), severe swelling of tissue surrounding the right femur, to the point where dramatic separation of scales was noted; no crepitus or mobility was identified, but pain was elicited on palpation

Diagnostics

A CBC, biochemistry panel, ionized calcium, and radiographs were performed. Results are provided in Table 13.

Table 13. CBC, selected biochemistry values, and ionized calcium values in a bearded dragon with right proximal pelvic limb swelling.

Analytes	Value	Unit
Hematocrit	20	%
White blood cells	23400	/µl
Bands	0	/µl
Heterophils	6800	/µl
Lymphocytes	10300	/µl
Monocytes	6300	/µl
Eosinophils	0	/µl
Basophils	0	/µl
AST	1	IU
CK	441	IU
Cholesterol	302	mg/dl
Glucose	142	mg/dl
Calcium	10.1	mg/dl
Phosphorus	5.7	mg/dl
Uric acid	2.5	mg/dl
Triglycerides	66	mg/dl
iCa	0.83	mmol/L

Radiographs revealed irregular bony proliferation of the right mid femur with severe soft tissue swelling. There was also an enlarged cardiac silhouette. Ultrasound showed mild pericardial effusion. In-house cytologic evaluation of leg swelling FNA demonstrated inflammation with intracellular structures.

Treatment

The patient was given subcutaneous fluids and sent home with nutritional support (5 ml daily, Oxbow critical care for Herbivores mixed 2:1 with Oxbow critical care for carnivores, Oxbow Animal Health, Murdock, NE, USA) and antibiotics (ceftazidime (20 mg/kg IM q72h; Ceftazidime for injection, USP, WG Critical Care, LLC, Paramus, NJ, USA)).

Referred cytologic evaluation showed granulomatous inflammation with possible intracytoplasmic protozoa (1-2 µm), acid fast negative. Biopsy was recommended, but the patient passed away within 2 weeks of receiving the results.

Histopathologic necropsy findings included focally extensive severe subacute to granulomatous cellulitis and rhabdomyositis with intralesional protozoa within the right proximal pelvic limb. There was lymphoplasmacytic infiltration in the myocardium, lung, intestines, brain (lymphocytic/histiocytic), with numerous granulomas containing intracytoplasmic coccoid bacteria and suspected microsporidia in the trachea and testis.

Culture of both leg and pericardial effusion indicated *Salmonella* sp. Frozen leg tissues were submitted to University of Florida for PCR. It was noted to be microsporidia positive and identified as a unique *Encephalitozoon* sp.

Case Discussions

Case 1. Macaw

Cholangitis and cholangiohepatitis:[1-3] Cholangitis generally suggests an ascending inflammatory process through the biliary tree from the gastrointestinal tract. Usually concurrent lesions can be identified in the gastric or intestinal tract, as well as the pancreas. The common infectious disease agents are bacterial (including *Chlamydia*), bacterial endotoxins, and less commonly metazoan parasites. Severe cases, resulting in necrosis, can progress to hepatic fibrosis. Cholangiohepatitis indicates that the inflammatory lesion has extended beyond the limiting plate and into the adjacent periportal region.

Renal membranoglomerulopathy: Membranous glomerulopathy is a non-specific lesion of the kidney. The glomerular lesion can suggest a systemic inflammatory disease process. In mammals, the lesions are commonly seen post-infections such as from bacteria, protozoa, and heartworm or as a component of immune-mediated diseases. In psittacine species at this time, membranous glomerulopathy has been associated with avian polyomavirus infection and represents antigen/antibody complexes, although other causes are also likely.

Case 2. Red-lored Amazon

Systemic fungal infection:[4-6] The specific etiologic cause of this patient's fungal infection was not determined either antemortem or postmortem, but based on the primarily GI-related symptoms and positive response to treatment with itraconazole, either candidiasis or aspergillosis would seem most likely. *Aspergillus* sp. is more commonly reported as a pathogen of the respiratory tract of birds, but can infect any organ system. Commonly reported avian fungal infections include candidiasis, cryptococcosis, aspergillosis, zygomycosis, and avian gastric yeast (*Macrorhabdus ornithogaster*). There are several reports of concurrent fungal infections in birds affecting multiple body systems.

Gastric carcinoma:[6-9] Gastric carcinomas produce a thickening and irregularity of the proventricular and/or ventricular wall with a variable amount of hemorrhage and necrosis. There may be ulceration and perforation. Gastric carcinomas can penetrate through the muscular tunics to adjacent organs as well as metastasize. Sites of metastasis include the pancreas and lungs. The clinical signs reported include weight loss, weakness, inability to perch, head tilt, melena, passing of undigested seeds in the feces, beak overgrowth, and polyuria. Although this tumor is rarely diagnosed before death, when tested, the feces tested positive for occult blood. Ventricular carcinoma is a common neoplastic disorder of older Amazon parrots, and has been reported in a variety of both Old World and New World psittacine species. Only rarely has antemortem diagnosis been made with exploratory surgery/biopsies. There are no reported attempts of medical or surgical treatment of this neoplasm. Although it is certainly possible that the patient's primary disease process was ventricular carcinoma and that the positive response to treatment was due solely to management of secondary bacterial and fungal infections, the longevity of the patient following initial presentation is unique to other reported cases of avian ventricular carcinoma (average lifespan < 1 month). An alternative hypothesis is that the gastric carcinoma could have developed secondary to a fungal ventriculitis, since chronic inflammation has been linked to development of neoplastic disease in other species.

Case 3. Guinea pigs

Lymphocytic leukemia: Lymphoid neoplasia is infrequently reported in guinea pigs relative to other veterinary species; this may due in part to an enzyme with L-asparaginase activity discovered in guinea pig serum;[1] however, disseminated/multicentric,[11-15] ocular,[11,15] and epitheliotropic[16-18] lymphoma, as well as lymphocytic leukemia[13,19-21] have been well-described in guinea pigs. Guinea pigs possess an endogenous type C retrovirus[22] which has been associated with leukemia and exogenous transmission in an in-bred strain of guinea pigs,[23,24] but spontaneous viral-induced leukemia has not been well-established in a natural setting. Two reports have identified epitheliotropic lymphoma as a T-cell phenotype based on characteristic histopathological features as well as immunohistochemistry for CD3,[16,17] but validation of these antibodies for use in guinea pigs was not performed. Lymphoma has also been experimentally-induced in guinea pigs, secondary to chronic antigen stimulation.[25]

Lymphoid hyperplasia of the conjunctiva: The lymphoid hyperplasia is suggestive of chronic antigenic stimulation to the eye. In other animal species *Chlamydia* sp. or *Mycoplasma* sp. would be considered as etiologies. A hypersensitivity reaction should also be considered (to environmental allergens). There is also an acute lesion, which may be secondary to the severe proliferative response of the conjunctiva. This is most likely secondary to mechanical irritation.

Case 4. Rabbit

Thymoma:[26-30] Neoplasms in adult rabbits are uncommon with a prevalence of disease in typical pet populations of < 3%. Thymoma, neoplasm of the thymus, have been reported in numerous species. Paraneoplastic syndromes that have been associated in people include cytopenia, polymyositis, hypercalcemia, and commonly acquired myasthenia gravis, which is found in almost one third of people with thymoma. They can be classified based on the cells undergoing neoplastic transformation (epithelial and/or lymphocytic). These anterior mediastinal masses are slow growing, encapsulated and metastases have not been reported in rabbits. Clinical signs associated with the thymomas include hyperpnea, open-mouth breathing, swelling of the head, persistenly prolapsed third eyelids and exophthalmos. Generally, the rabbits are middle-aged to older (6-8 years). Surgical removal has been described and this approach reduces the risk of recurrence; however, because of its proximity to the heart and large blood vessels, this option carries significant risks.

Radiation therapy is effective in reducing the size of the mass. Side effects, including heart fibrosis, pneumonia and the creation of mass-to-heart adhesions have been described. A success rate approaching 80% is reported with an often very rapid response after a few treatments.

Palliative treatments with chemotherapy protocols are described, but the response to the latter is mixed.

In general, consultation with a veterinary oncologist is advised in order to develop an individualized treatment plan for each case.

Case 5. Bearded dragon

Nannizziopsi guarroi (CANV):[31-36] Fungal isolates formerly referred to as "*Chrysosporium* anamorph of *Nannizziopsis vriesii* (CANV) complex" have recently been assigned to 16 species, either within *Nannizziopsis* or within the new genera *Paranannizziopsis* and *Ophidiomyces*. *Nannizziopsi guarroi* (formerly *Chrysosporium guarroi*) is the main causal agent of deep granulomatous dermatomycosis (commonly referred to as yellow fungus disease) in several lizard species, including captive bearded dragons (*Pogona vitticeps*). Lesions initially appear as patchy yellowing of the skin that turns dark and becomes necrotic and thickened. Infection often progresses to involve the deeper dermis. In bearded dragons, infection is often fatal and may disseminate and involve organs such as the liver.

Atadenovirus:[37-39] Agamid adenovirus 1 has become established in certain lizard breeding groups of the genus *Pogona* (bearded dragons) in United States and other countries. The common clinical signs in young dragons are a progressive weakness and anorexia, but severe enteritis, hepatitis, and sudden death may be seen. The adenovirus infections has been reported to be associated with a second virus, Dependovirus. This virus is defective and appears to need the adenovirus to replicate. Dependovirus is not associated with a disease condition. Co-infection with protozoa and *Encephalitozoon* is also described.

Case 6. Bearded dragon

Microsporidiosis:[40-43] Microsporidia, currently classified as fungi, include *Encephalitozoon* sp., *Pleistophora* sp., *Nosema* sp. and other species. A microsporidium species specific to bearded dragons has been identified as *Encephalitozoon pogonae*. Most reports in reptiles are of a systemic granulomatous fatal infection. Treatment is extrapolated from human studies (benzimidazole derivatives including albendazole).

References and Suggested Reading

1. Reavill DR, Lennox A. Comparison of serum biochemistry values and liver biopsy histopathology results in psittacines. *Proc Annu Conf Eur Assoc Avian Vet.* 2007;327-342.

2. Harrison JL, Turek BJ, Brown DC, et al. Cholangitis and cholangiohepatitis in dogs: A descriptive study of 54 cases based on histopathologic diagnosis (2004-2014). *J Vet Intern Med.* 2018;32:172-180.

3. Callahan Clark JE, Haddad JL, Brown DC, et al. Feline cholangitis: A necropsy study of 44 cats (1986-2008). *J Feline Med Surg.* 2011;13:570-576.

4. Beernaert LA, Pasmans F, Van Waeyenberghe L, et al. Aspergillus infections in birds: A review. *Avian Pathol.* 2010;39:325-331.

5. Hubbard GB, Schmidt RE, Eisenbrandt DL, et al. Fungal infections of ventriculi in captive birds. *J Wildl Dis.* 1985;21:25-28.

6. Schmidt RE. Pathology of gastrointestinal diseases of psittacine birds. *Semin Avian Exot Pet Med.* 1999;8: 75-82.

7. Campbell TW, Turner O. Carcinoma of the ventriculus with metastasis to the lungs in a sulphur-crested cockatoo (*Cacatua galerita*). *J Avian Med Surg.* 1999;13:265-268.

8. Zehnder A, Graham J, Reavill DR, et al. Neoplastic diseases in avian species. In: Speer BL, ed. *Current Therapy in Avian Medicine and Surgery.* St. Louis, MO: Elsevier Saunders;2016:107-141.

9. Reavill DR. Tumors of pet birds. *Vet Clin North Am Exot Anim Pract.* 2004;7:537-560.

10. Schalk AM, Nguyen H-A, Rigouin C, et al. Identification and structural analysis of an L-asparaginase enzyme from guinea pig with putative tumor cell killing properties. *J Biol Chem.* 2014;289:33175-33186.

11. Steinberg H. Disseminated T-cell lymphoma in a guinea pig with bilateral ocular involvement. *J Vet Diagn Invest.* 2000;12:459-462.

12. Debout C, Caillez D, Izard J. A spontaneous lymphoblastic lymphoma in a guinea-pig. *Pathol Biol.* 1987;35:1249-1252.

13. Hong CC, Liu PI, Poon KC. Naturally occurring lymphoblastic leukemia in guinea pigs. *Lab Anim Sci.* 1980;30:222-226.

14. Kołodziejska A, Depta A, Szweda M. Lymphoma in a domestic guinea pig. *J Comp Pathol.* 2010;143:319.

15. Allgoewer CP, Pfleghaar S. Lymphosarcoma with conjunctival manifestation in a guinea pig. *Vet Ophthalmol.* 1999;2:117-119.

16. Martorell J, Such R, Fondevila D, Bardagi M. Cutaneous epitheliotropic T-cell lymphoma with systemic spread in a guinea pig (*Cavia porcellus*). *J Exot Pet Med.* 2011;20:313-317.

17. Koebrich S, Grest P, Favrot C, et al. Epitheliotropic T-cell lymphoma in a guinea pig. *Vet Dermatol.* 2010;22:215-219.

18. Heuer L, Stötter M, Eydner M, et al. Metastasizing epitheliotropic T-cell lymphoma (*Mycosis fungoides*) in two guinea pigs (*Cavia porcellus*). *Berl Munch Tierarztl Wochenschr.* 2014;127:333-336.

19. Ediger RD, Rabstein MM. Spontaneous leukemia in a Hartley strain guinea pig. *J Am Vet Med Assoc.* 1968;153:954-956.

20. Congdon CC, Lorenz E. Leukemia in guinea-pigs. *Am J Pathol.* 1954;30:337-359.

21. Day MJ, Briggs EK. CD3-SmIg-lymphoblastic leukaemia in an outbred domestic guinea pig. *Aust Vet J.* 1997;75:217-218.

22. Nayak DP, Murray PR. Induction of type C viruses in cultured guinea pig cells. *J Virol.* 1973;12:177-187.

23. Opler SR. Pathology of cavian viral leukemia. *Am J Pathol.* 1967;51:1135-1151.

24. Gross L, Feldman DG. Comparative study of the morphology and distribution of virus particles causing leukemia and lymphosarcoma in mice, rats, cats, and guinea pigs. *Arch Geschwulstforsch.* 1970;36:1-9.

25. Ziegler V, Ziegler B. Experimentally induced malignant lymphoma due to chronic antigen stimulation. *Contact Dermatitis.* 1994;30:77-79.

26. Orr JN. A malignant tumor of the thymus of a rabbit. *Am J Cancer.* 1939;35:269-274.

27. Kostolich M, Panciera RJ. Thymoma in a domestic rabbit. *Cornell Vet.* 1992;82:125-129.

28. Klebanow ER. Thymoma and acquired myasthenia gravis in the dog: A case report and review of 13 additional cases. *J Am Anim Hosp Assoc.* 1992;28:63-69.

29. Van Zeeland Y. Rabbit oncology: Diseases, diagnostics, and therapeutics. *Vet Clin North Am Exot Anim Pract.* 2017;20:135-82.

30. Andres K, Kent M, Siedlecki C, et al. The use of megavoltage radiation therapy in the treatment of thymomas in rabbits: 19 cases. *Vet Comp Oncol.* 2012;10(2):82-94.

31. Paré JA et al. An overview of reptile fungal pathogens in the genera *Nannizziopsis*, *Paranannizziopsis*, and *Ophidiomyces*. *J Herpetol Med Surg.* 2016;26:46-53.

32. Bowman MR, Pare JA, Sigler L, et al. Deep fungal dermatitis in three inland bearded dragons (*Pogona vitticeps*) caused by the *Chrysosporium* anamorph of *Nannizziopsis vriesii*. *Med Mycol.* 2007;45:371-376.

33. Schmidt-Ukaj S, Loncaric I, Klang A, et al. Infection with *Devriesea agamarum* and *Chrysosporium guarroi* in an inland bearded dragon (*Pogona vitticeps*). *Vet Dermatol.* 2014;25:555-558.

34. Schneider J et al. Characterization of *Nannizziopsis guarroi* with genomic and proteomic analysis in three lizard species. *Med Mycol.* 2018;56:610-620.

35. Le Donne V, Crossland N, Brandão J, et al. *Nannizziopsis guarroi* infection in 2 inland bearded dragons (*Pogona vitticeps*): clinical, cytologic, histologic, and ultrastructural aspects. *Vet Clin Pathol.* 2016;45: 368-375.

36. Sigler L, Hambleton S, Paré JA. Molecular characterization of reptile pathogens currently known as members of the Chrysosporium anamorph of *Nannizziopsis vriesii* complex and relationship with some human-associated isolates. *J Clin Microbiol.* 2013;51:3338-3357.

37. Doneley RJ, Buckle KN, Hulse L. Adenoviral infection in a collection of juvenile inland bearded dragons (*Pogona vitticeps*). *Aust Vet J.* 2014;92:41-45.

38. Kim DY et al. An outbreak of adenoviral infection in inland bearded dragons (*Pogona vitticeps*) coinfected with dependovirus and coccidial protozoa (*Isospora* sp.). *J Vet Diagn Invest.* 2002;14(4):332-334.

39. Schilliger L et al. Triple infection with agamid adenovirus 1, *Encephaliton cuniculi*-like microsporidium and enteric coccidia in a bearded dragon (*Pogona vitticeps*). *Tierarztl Prax Ausg K Kleintiere Heimtiere.* 2016;44(5):355-358.

40. Jacobson ER, Green DE, Undeen AH, et al. Systemic microsporidiosis in inland bearded dragons (*Pogona vitticeps*). *J Zoo Wildl Med.* 1998;29:315-323.

41. Richter B, Csokai J, Graner I, et al. Encephalitozoonosis in two inland bearded dragons (*Pogona vitticeps*). *J Comp Pathol.* 2013;148:278-282.

42. Sokolova YY, Sakaguchi K, Paulsen DB. Establishing a new species *Encephalitozoon pogonae* for the microsporidian parasite of inland bearded dragon *Pogona vitticeps* ahl 1927 (reptilia, squamata, agamidae). *J Eukaryot Microbiol.* 2016;63:524-535.

43. Strunk A, Reavill DR. Intracellular parasite in the pelvic limb of a bearded dragon, *Pogona vitticeps*. *Proc Annu Conf Assoc Reptilian Amphib Vet.* 2011;152.

Therapeutic Contraindications in Pet Exotic Species

Sue Chen, DVM, DABVP (Avian Practice), and
Olivia A. Petritz, DVM, Dipl ACZM

Session #2311

Affiliation: From Gulf Coast Avian and Exotics – Gulf Coast Veterinary Specialists, 3800 Southwest Freeway; Houston, TX 77027, USA (Chen), Department of Clinical Sciences, North Carolina State University, College of Veterinary Medicine, 1060 William Moore Drive, Raleigh, NC 27607, USA (Petritz).

Abstract: Just as certain drugs are contraindicated in certain breeds of dogs or in cats, there are specific drugs where their use is contraindicated in commonly seen exotic pets. The selection and dosing of medications for exotic pets is already challenging as most drugs are used in an extra-label manner without pharmacokinetic, pharmacodynamic, efficacy, or safety studies in the species of interest. Doses often are extrapolated from common domestic animals; however, the bioavailability and therapeutic levels are different for each species, and what may be a safe and commonly used medication in one species can be deadly in another. Drugs with documented contraindications in certain exotic pet species are outlined in this review and the pathophysiology, clinical signs, and treatment options are described when applicable.

Antiparasitic Medications

Ivermectin

Ivermectin is a macrocyclic lactone that targets the ivermectin-sensitive glutamate-gated chloride channel receptors, only found in invertebrates, and the gamma amino butyric acid (GABA) receptors.[1] Ivermectin does not cross the blood brain barrier in many vertebrate species; however, in certain species, neurologic signs can occur following ivermectin administration, even at recommended doses.

Clinical signs

Paresis or flaccid paralysis occurred with a single dose as low as 0.05 mg/kg in red-footed tortoises (*Geochelone carbonaria*). The leopard tortoise (*Geochelone pardalis*) is even more susceptible with paresis developing with a dosage of as low as 0.025 mg/kg, and death with dosages as low as 0.3 mg/kg. Ivermectin was previously evaluated in 5 red-footed tortoises (*Geochelone carbonaria*); each received a single IM injection of ivermectin (0.4 mg/kg), and subsequently developed paresis or flaccid paralysis.[2] Other species of chelonians susceptible to ivermectin toxicosis at dosages of 0.1 mg/kg or less. Based on this and other published data, the use of ivermectin in any chelonian species is not recommended.

Other reptilian species that have been reported of having suspected cases of ivermectin toxicosis include a chameleon (*Chameleo sengalensis*), prehensile-tailed skinks (*Corucia zebrata*), crocodilians, and indigo snakes, though some individual sensitivities may be possible.[3-5] There are anecdotal reports of ivermectin toxicity in finches.[3,6] There are also many non-veterinary references alleging ivermectin-induced morbidity and mortality of hedgehogs though there are no scientific studies that substantiate this claim.

Treatment

Treatment of ivermectin toxicity in reptiles is largely supportive. Respiratory support must be provided for the duration of action of ivermectin at the neurotransmitter site (7 days).[7] Anecdotal reports of full recoveries after 4-6 weeks of supportive care have been reported in tortoises.[8]

Fipronil

Fipronil is a phenylpyrazole insecticide used in dogs and cats for the treatment of various ectoparasites including fleas, ticks, lice, and mites.[9] Fipronil prevents chloride ion uptake thereby blocking GABA receptors in the central nervous system which results in excessive central nervous system stimulation. Because fipronil has a greater binding affinity for insect GABA receptors than mammalian receptors, most mammals are not adversely affected. Rabbits appear to be more sensitive to fipronil than other mammals. Though dermal absorption is only 0.07%, if the rabbit licks off product during grooming, absorption can reach upwards of 30-50%.[10] The narrow safety margin in rabbits makes them susceptible to the neurologic effects of this drug.

Clinical signs

Rabbits suffering from fipronil toxicity present with neurologic symptoms ranging from tremors to seizures that develop anytime between a few hours up to a few weeks after exposure. Other reported clinical signs include anorexia, lethargy, ptyalism, gastrointestinal stasis, diarrhea, emaciation, hypothermia, and sudden death.

Treatment

If exposure was within 48 hours, the rabbits should be bathed with a mild puppy/kitten shampoo in warm water to remove all toxin from the surface of the skin.[11] Activated charcoal may be administered to limit additional absorption of the drug. Midazolam (1-2 mg/kg IM, IV) and levetiracetam (20 mg/kg PO q8h) can be administered to treat seizures.[12] The prognosis is guarded for rabbits that have seizures. Parental fluids and nutritional support should be administered rabbits that are anorectic or exhibiting other signs of gastrointestinal stasis.

Benzimidazoles

Benzimidazoles, are anthelmintics used in dogs, cats, birds and rabbits. In exotic animal medicine, they have been used commonly for the treatment of *Encephalitozoon cuniculi*. These drugs inhibit microtubule formation which interferes with the parasite cytoskeleton development as well as bind to parasitic intestinal cells, which decreases glucose uptake leading to starvation of the parasite.[13] Though their binding affinity is greater to parasitic tubulin, vertebrate tubulin can also be affected, especially in rapidly dividing cells such as the bone marrow and cells lining the intestinal tract. The resulting bone marrow dysplasia and gastrointestinal ulceration can result in peracute bacterial or fungal infections and sepsis.[14,15] Consequently, this medication can result in radiomimetic lesions such as depletion of the myeloid, erythroid, and megakaryocytic cell lines.[16] This pancytopenia can lead to severe immunosuppression and subsequent bacterial and/or fungal infections, which may be fatal. Additionally, gastrointestinal erosion and crypt necrosis can contribute to generalized septicemia. Benzimidazoles undergo extensive hepatic metabolism (by cytochrome P450 and others) following oral administration, therefore if hepatic disease is present, a patient may be at greater risk for toxicosis. Benzimidazole-associated toxicities have been reported in avian,[14,17-20] reptilian,[21] and mammalian species.[16,22]

Clinical signs

Non-specific clinical signs are typically seen within 48 hours of drug administration. Rabbits have presented for signs of anorexia, lethargy, diarrhea, pale mucous membranes, epistaxis, petechial to ecchymotic hemorrhage, abdominal hemorrhage, GI stasis, fever, and acute death.[16] Feather abnormities have been noted in birds treated with fenbendazole during active molt. Reported biochemical changes include transient hypoglycemia, hyperuricemia, and hyperphosphatemia, in Hermann's tortoises (*Testudo hermanni*).[21]

Treatment

Immediate cessation of benzimidazoles is recommended if lethargy or significant decreases in the complete blood count are noted. Sucralfate and acid-reducers, such as famotidine and omeprazole, may be considered for the treatment and prevention of gastrointestinal ulceration. Antibiotic and parental fluid therapy are recommended if septicemia is suspected. Blood transfusions should be performed in patients that are clinical for a severe anemia and may need to be repeated until the bone marrow is functional. A biochemical profile is used to screen for hepatic disease, as decreased hepatic function may predispose a patient to toxicosis. Complete blood counts should be performed prior to treatment as well as weekly throughout treatment to assess for anemia, thrombocytopenia, and/or leukopenia. If considering the use of a benzimidazole, it is highly advisable to have confirmed diagnosis of a parasitic infection as well as pre-treatment biochemical profiles and complete blood count performed.

Based on published studies and anecdotal reports of toxicity in various zoo and exotic pet species, clinicians should carefully consider the risk of morbidity and mortality from a parasitic infection compared with the risk of potential septicemia following damage to hematopoietic and gastrointestinal systems from fenbendazole therapy.

Antimicrobials

Antibiotics: Beta-lactams, cephalosporins, macrolides

Although any antibiotic can potentially cause disruption of the gastrointestinal flora of any species, special care must be taken in those that employ hindgut formation as part of their digestive strategy, such as rabbits, guinea pigs, and chinchillas. Antibiotics that target primarily gram-positive microbial flora typically found in the cecum of these species can lead to fatal dysbiosis. Antibiotics such as penicillins, cephalosporins, clindamycin, lincomycin, erythromycin, and tylosin leads to the eradication of normal gastrointestinal flora, thereby allowing for the overgrowth of pathogenic bacteria. One such bacteria, *Clostridium difficile*, produces exotoxins that can result in secretory diarrhea, mucosal damage, and hemorrhagic typhlitis.[23] The adverse effects most commonly are associated with oral administration of these antibiotics, but lethal hemorrhagic typhlitis associated with parental penicillin has been reported in guinea pigs.[24] In contrast, parental penicillin does not appear to affect the rabbit gastrointestinal tract and can be used successfully for penicillin-sensitive bacterial infections or abscesses.

Clinical signs

Profuse and watery diarrhea after the oral administration of the offending antibiotic is usually noted. Varying levels of dehydration is common, and hypovolemic shock may be present in severely affected animals. Generalized signs of gastrointestinal stasis such as anorexia, lethargy, abdominal distension, and diminished fecal production are sometimes noted prior to the development of diarrhea. Signs of abdominal pain include a hunched posture, reluctance to move, and resistance to abdominal palpation.

Treatment

The antibiotic should be discontinued immediately in any small herbivorous mammal presenting with diarrhea. If the patient is still bright and alert, subcutaneous fluids may be adequate to treat dehydration and electrolyte imbalances; however, intravenous fluids are recommended if patient is lethargic, hypothermic, and showing signs of hypovolemic shock. A complete blood count and biochemical profile should be performed to evaluate for underlying metabolic and systemic disorders that may or may not be associated with the gastrointestinal tract. Pain medication should be administered to any patients exhibiting signs of pain (hunched behavior, bruxism, reluctance to move). Cholestyramine, an ion exchange resin that binds to clostridial iota toxins, has been reported to be effective in preventing deaths in rabbits challenged with acute clostridial enterotoxemia when given at 2 g/20 ml water by mouth once daily for 18-21 days;[25] however, anecdotal experience with this medication has been equivocal. Gram stains of the diarrhea may reveal large numbers of spore-forming bacteria. Metronidazole (20 mg/kg PO, IV q12h) has been reported to reduce the number of mortalities associated with enterotoxemia from clostridial overgrowth.[26]

Itraconazole

Itraconazole is a synthetic triazole antifungal agent used for the treatment of various cutaneous and systemic mycoses. It is thought to disrupt fungal cellular membranes by inhibiting ergosterol synthesis by blocking cytochrome P450. The pharmacokinetics of itraconazole has been studied in a variety of avian species, with significant interspecies as well as individual variability. Anecdotally, African grey parrots (*Psittacus erithacus*) are more sensitive to itraconazole than other species of birds, even with doses as low as 2.5mg/kg orally once daily.[27,28] In several species of lizards and snakes, it has been used for the treatment mycotic infections such as *Nannizziopsis guarroi* and *Aspergillus*.

Clinical signs

Adverse effects in in African grey parrots such as anorexia, depression, and even death have been reported. A study comparing itraconazole versus voriconazole for the treatment of *Nannizziopsis guarroi* showed 5 of 7 bearded dragons die during the course of treatment with itraconazole at a dose of 5 mg/kg PO q24h. Though hepatic toxicosis was suspected based on biochemical analysis, toxicity could not be proven based on histologic examination of the livers of the bearded dragons.[29]

Treatment

Because itraconazole is metabolized by the liver and has known toxic effect on the liver, evaluation and monitoring of hepatic values before and during treatment with itraconazole is vital. Discontinue itraconazole if patient becomes anorectic, lethargic, or weak and/or if hepatic values begin to rise. Milk thistle may be beneficial as a hepatoprotectant though studies are lacking proving their efficacy in birds and reptiles.[30] An alternative antifungal, voriconazole has been evaluated in bearded dragons and in African grey parrots (*Psittacus erithacus timneh*) and may be a more suitable alternative for susceptible fungal diseases in these species.[31]

Anti-Inflammatory and Pain Medications

Glucocorticoids

Glucocorticoids are used for their anti-inflammatory, immunosuppressive, and anti-neoplastic effects, in both human and veterinary medicine. Birds and certain mammalian species, such as the mouse, rat, and rabbit, are considered steroid-sensitive species, while dogs, ferrets, and guinea pigs are considered steroid-resistant species.[32,33] Steroid-sensitive species exhibit an immediate and profound lymphopenia due to lymphocytolysis induced by activation of a resident endonuclease when administered exogenous glucocorticoids.[32] Steroid-resistant species can also be adversely affected by steroid administration, but usually only with high doses and/or chronic use.

Systemic administration of steroids have been used as an adjunct to chemotherapy and radiation for treatment of hematopoietic neoplasia and lymphoma in several avian species.[34-37] A short course of prednisolone has been used in conjunction with radiation therapy for the treatment of lymphocytic thymomas in rabbits.[38] Additionally, long-term prednisolone as a sole agent has been used successfully as a palliative measure, where the rabbits' clinical signs improved and a significant decrease in the size of their thymomas were noted radiographically. Adverse effects, including mild to moderate pododermatitis and mild upper respiratory infections, did develop in some, but not all rabbits.

Clinical signs

In rabbits, most adverse effects are secondary to the immunosuppressive effects of steroids, such as mucopurulent ocular and nasal discharge associated with an upper respiratory infection. Pododermatitis, progressing to hock abscesses, may be due to a combination of immunosuppression and thinning of the skin that occurs with administration of long-term glucocorticoids. Anecdotally, otitis externa and pinna abscessation has been observed following a short course of topical otic medication containing a steroid in rabbits. In birds, peer-reviewed literature documenting the adverse effects of steroid use include immunosuppression leading to secondary bacterial and fungal infections, delayed wound healing, diabetes mellitus, vaculolar hepatopathy, and gastrointestinal ulceration.[39] Involution of the cloacal bursa, thymus, and spleen with resulting suppression of both cell-mediated and humoral immunity has been reported following corticosteroid use in birds.[40,41] In humans, the function of respiratory macrophages is negatively impacted by exposure to systemic corticosteroids.[42] Because these immune cells provide the initial defense against Aspergillus infections in avian species and humans, immunocompromised patients are predisposed to fungal disease.[15] Even the use of topical corticosteroids should be used cautiously, as adverse effects have also been reported in some species.[43]

Steroid-induced immunosuppression in animal models is common in the laboratory setting; however, there are no pharmacokinetic, efficacy, or safety studies of the use of any steroid for treatment of any disease in any exotic companion species to date. In specific pathologies where the use of steroid therapy is indicated, the clinician must weigh the benefits against potential adverse effects and they must clearly communicate the pros and cons with the client. The patient's white cell count should be evaluated for leukopenia prior to and during treatment. A biochemical profile should be performed to evaluate for hyperglycemia and hepatic dysfunction.

Treatment

Antifungal medications are commonly prescribed prophylactically to help reduce the risk of systemic aspergillosis in certain avian species;[36,44] however, this is clinician preference, and to the authors' knowledge, there are no studies which have evaluated the efficacy of anti-fungal medication(s) during treatment with topical or systemic steroid administration; however, if a bacterial or fungal infection develops during the course of treatment with steroids, administration of concurrent antibiotics or antifungals should be administered. Acid reducers, such as H2-blockers and proton-pump inhibitors may reduce the risk of gastrointestinal ulceration.

Exotic animal practitioners must extrapolate what is known for other domestic animals, and assume some degree of risk when prescribing steroids to their patients, especially if the steroid sensitivity of that species is unknown. There is no right or wrong answer to the question, "Should glucocorticoids be used in avian and exotic species?"[45,46] Opinions are forged after apparent success or failure of a glucocorticoid in an exotic patient.

While formation of these opinions are inevitable for those of us who work with non-traditional species due to a paucity of per-reviewed scientific data, practitioners should critically evaluate case outcome with as much evidence-based medicine as possible. Was the positive case outcome directly related to the use of a steroid, or were there other factors involved? Conversely, was the deterioration of a patient related to the use of a steroid, or was that simply the natural progression of the patient's disease? Corticosteroids, have well documented side effects in all species, and these risks should be weighed against their potential benefits for each patient.

Ibuprofen

Ibuprofen is a widely used nonsteroidal anti-inflammatory (NSAID) in human medicine for its analgesic, anti-inflammatory, and antipyretic effects. It has a narrow safety margin in dogs, cats, and ferrets. Because of a ferret's relatively small size, even a single over-the-counter 200-mg tablet can result in a toxic dose of 100-400 mg/kg in a ferret.

Ibuprofen non-selectively inhibits cyclooxygenase enzymes by blocking the conversion of arachidonic acid into various prostaglandins. Inhibition of COX-1 enzymes target the prostaglandins that have a cytoprotective effect on the stomach and the regulation of gastric and renal blood flow, leading to gastrointestinal upset, ulceration, and hemorrhage, especially in cases of overdosage and/or chronic usage.[47] The conversion of arachidonic acid to thromboxane A2 is also blocked, which can lead to impaired platelet aggregation and clot formation.

Clinical signs

A retrospective study of cases reported to the ASPCA found that 93% of ferrets developed neurologic symptoms, ranging from depression to being comatose and displaying ataxia, tremors, weakness, and seizures.[48] Gastrointestinal signs (vomiting, anorexia, diarrhea, melena) were noted in 55% of the ferrets and renal dysfunction (polyuria, polydipsia, dysuria) were noted in 13% of the cases. Weight loss, shallow breathing, metabolic acidosis, dehydration, hypothermia, and death were other reported clinical signs. Clinical signs usually occur within 8 hours of ingestion, but may occur with 4 hours.[48]

Treatment

Consider emesis if the ferret is not showing neurologic signs or vomiting if ingestion of ibuprofen occurred within 2 hours of presentation.[48,49] Activated charcoal can be administered every 6 to 8 hours to prevent enterohepatic recirculation. Biochemical analysis should be performed to evaluate renal parameters. Diuresis with intravenous fluids is strongly recommended for a minimum of 48 hours and should be continued until renal values are back to normal. The urine specific gravity should also be assessed and the urine should be monitored for tubular casts, which can be seen within 18 hours of exposure.[48,49] Gastroprotective medications should be administered to treat and prevent gastric ulcers. Misoprostol, a synthetic prostaglandin (PGE1) has a cytoprotective effect on the gastric mucosa; an anecdotal dosage 1-5 μg/kg PO q8h has been suggested.[48] Seizures should be controlled with anticonvulsants; however, once the patient develops seizures, the prognosis is guarded. Toxicologic analyses for ibuprofen can be performed on serum, urine, and hepatic samples using gas chromatography and mass spectrophotometry.

Ketoprofen

Ketoprofen is also a nonselective NSAID used for its analgesic, anti-inflammatory, and antipyretic effects in a variety of veterinary species, including ferrets, birds, reptiles, and other small mammals.[3] One study found significant mucosal damage in the intestinal tract of rats within 24 hours after a single dose of ketoprofen

(5 mg/kg).[50] These rats had significant drops in their PCV, positive fecal occult blood tests, and varying grades of intestinal ulcers on necropsy when compared to control rats. The severity of the lesions were exacerbated when ketoprofen was administered in conjunction with anesthesia.[50]

Clinical signs

Related to ulceration of the gastrointestinal tract, including lethargy, pale mucus membranes, abdominal pain, and melena. Survivors of the acute gastrointestinal ulceration may die 1-2 weeks later, possibly due to bacterial translocation and sepsis. On necropsy, intestinal ulcers, diffuse bleeding into the intestinal lumen, adhesions, fibrosis, and ascites secondary to peritonitis were noted.

Treatment

Gastroprotectants to treat and protect the gastrointestinal mucosa should be administered. In cases of possible sepsis, antibiotic therapy and supportive measures such as parental fluids and nutritional support should be instituted.

Tramadol

Tramadol is a synthetic mu-receptor agonist, but also inhibits the reuptake of serotonin and norephinephrine for its analgesic properties. This pain medication has been used in variety of species, with pharmacokinetic and/or antinociceptive studies in rats,[51] rabbits,[52] turtles,[53] red-tailed hawks (*Buteo jamaicensis*),[54] Hispaniolan Amazon parrots (*Amazona ventralis*),[55,56] and American kestrels (*Falco sparverius*).[57] Based on these studies, there appears to be significant interspecies differences in terms of bioavailability and efficacy in providing pain relief.

Clinical signs

A recent study found that subcutaneous administration of tramadol 10-40 mg/kg in chinchillas (*Chinchilla lanigera*) not only did not provide thermal antinociception, but resulted in adverse gastrointestinal signs such as hyporexia and decreased fecal production for up to 72 hours. The study also found dose-limiting neurologic symptoms, such as whole body tremors and facial muscle fasciculation, at 60 mg/kg subcutaneously.[58] These neurologic symptoms are consistent with other species suffering from serotonin syndrome, where there are overactive central and peripheral serotonin receptors from inhibition of serotonin reuptake.

Treatment

Chinchillas exhibiting neurologic symptoms recovered within 30-45 min post-exposure. Benzodiazepines may help with tremors and seizures. Active cooling may be necessary if patients are hyperthermic from the muscle tremors.

References

1. Peterson M, Talcott P. *Small Animal Toxicology*. St. Louis, MO: Elsevier Saunders;2006.

2. Teare JA, Bush M. Toxicity and efficacy of ivermectin in chelonians. *Red.* 1983;2:1.

3. Carpenter JW. *Exotic Animal Formulary*. 4th ed. St. Louis, MO: Elsevier Saunders;2013.

4. Chen L-J, Sun B-H, ping Qu J, et al. Avermectin induced inflammation damage in king pigeon brain. *Chemosphere.* 2013;93:2528-2534.

5. Qu J, Li M, Zhao F, et al. Autophagy is upregulated in brain tissues of pigeons exposed to avermectin. *Ecotoxicol Environ Saf.* 2015;113:159-168.

6. Rosskopf WJ. Common conditions and syndromes of canaries, finches, lories and lorikeets, lovebirds, and macaws. *Semin Avian Exot Pet Med.* 2003;131-143.

7. Bennett R, Mehler S. Neurology In: Mader D, ed. *Reptile Medicine and Surgery*. 2nd ed. St. Louis, MO: Elsevier Saunders;2006:239-250.

8. Fitzgerald KT, Newquist KL. Poisonings in reptiles. *Vet Clin North Am Exot Anim Pract.* 2008;11:327-357.

9. Plumb DC. *Plumb's Veterinary Drug Handbook*. 7th ed. John Wiley & Sons, Inc;2011.

10. Stern LA. Fipronil toxicosis in rabbits. Available at: http://veterinarymedicine.dvm360.com/fipronil-toxicosis-rabbits. Accessed July 1, 2018.

11. Johnston MS. Clinical toxicoses of domestic rabbits. *Vet Clin North Am Exot Anim Pract.* 2008;11:315-326.

12. Strolin Benedetti M, Coupez R, Whomsley R, et al. Comparative pharmacokinetics and metabolism of levetiracetam, a new anti-epileptic agent, in mouse, rat, rabbit and dog. *Xenobiotica.* 2004;34:281-300.

13. Martin R. Modes of action of anthelmintic drugs. *Vet J.* 1997;154:11-34.

14. Lindemann DM, Eshar D, Nietfeld JC, et al. Suspected fenbendazole toxicity in an American white pelican (*Pelecanus erythrorhynchos*). *J Zoo Wildl Med.* 2016;47:681-685.

15. Pendl H, Tizard I. Immunology In: Speer BL, ed. *Current Therapy in Avian Medicine and Surgery*. St. Louis, MO: Elsevier Saunders;2016:400-432.

16. Graham JE, Garner MM, Reavill DR. Benzimidazole toxicosis in rabbits: 13 cases (2003 to 2011). *J Exot Pet Med.* 2014;23:188-195.

17. Bonar CJ, Lewandowski AH, Schaul J. Suspected fenbendazole toxicosis in 2 vulture species (*Gyps africanus*, *Torgos tracheliotus*) and marabou storks (*Leptoptilos crumeniferus*). *J Avian Med Surg.* 2003;17:16-19.

18. Howard LL, Papendick R, Stalis IH, et al. Fenbendazole and albendazole toxicity in pigeons and doves. *J Avian Med Surg.* 2002;16:203-210.

19. Gozalo AS, Schwiebert RS, Lawson GW. Mortality associated with fenbendazole administration in pigeons (*Columba livia*). *J Am Assoc Lab Anim Sci.* 2006;45:63-66.

20. Weber MA, Terrell SP, Neiffer DL, et al. Bone marrow hypoplasia and intestinal crypt cell necrosis associated with fenbendazole administration in five painted storks. *J Am Vet Med Assoc.* 2002;221:417-419.

21. Neiffer DL, Lydick D, Burks K, et al. Hematologic and plasma biochemical changes associated with fenbendazole administration in Hermann's tortoises (*Testudo hermanni*). *J Zoo Wildl Med.* 2005;36:661-672.

22. Weber MA, Miller MA, Neiffer DL, et al. Presumptive fenbendazole toxicosis in North American porcupines. *J Am Vet Med Assoc.* 2006;228:1240-1242.

23. Perkins SE, Fox JG, Taylor NS, et al. Detection of *Clostridium difficile* toxins from the small intestine and cecum of rabbits with naturally acquired enterotoxemia. *Lab Anim Sci.* 1995;45:379-384.

24. Lowe B, Fox J, Bartlett J. *Clostridium difficile*-associated cecitis in guinea pigs exposed to penicillin. *Am J Vet Res.* 1980;41:1277-1279.

25. Lipman N, Weischedel A, Connors M, et al. Utilization of cholestyramine resin as a preventive treatment for antibiotic (clindamycin) induced enterotoxaemia in the rabbit. *Lab Anim.* 1992;26:1-8.

26. Oglesbee B, Jenkins J. Gastrointestinal diseases. In: Quesenberry KE, Carpenter JW, eds. *Ferrets, Rabbits, and Rodents: Clinical Medicine and Surgery.* 3rd ed. St. Louis, MO: Elsevier Saunders;2012:193.

27. Orosz SE, Frazier DL. Antifungal agents: a review of their pharmacology and therapeutic indications. *J Avian Med Surg.* 1995:8-18.

28. Orosz SE, Frazier DL, Schroeder EC, et al. Pharmacokinetic properties of itraconazole in blue-fronted Amazon parrots (*Amazona aestiva aestiva*). *J Avian Med Surg.* 1996:168-173.

29. Van Waeyenberghe L, Baert K, Pasmans F, et al. Voriconazole, a safe alternative for treating infections caused by the *Chrysosporium* anamorph of *Nannizziopsis vriesii* in bearded dragons (*Pogona vitticeps*). *Med Mycol.* 2010;48:880-885.

30. Grizzle J, Hadley TL, Rotstein DS, et al. Effects of dietary milk thistle on blood parameters, liver pathology, and hepatobiliary scintigraphy in white carneaux pigeons (*Columba livia*) challenged with B1 aflatoxin. *J Avian Med Surg.* 2009;23:114-124.

31. Flammer K, Osborne JAN, Webb DJ, et al. Pharmacokinetics of voriconazole after oral administration of single and multiple doses in African grey parrots (*Psittacus erithacus timneh*). *Am J Vet Res.* 2008;69:114-121.

32. Cohn LA. The influence of corticosteroids on host defense mechanisms. *J Vet Intern Med.* 1991;5:95-104.

33. O'Malley B. *Clinical Anatomy and Physiology of Exotic Species*. St. Louis, MO: Elsevier Saunders;2005.

34. Newell SM, McMillan MC, Moore FM. Diagnosis and treatment of lymphocytic leukemia and malignant lymphoma in a Pekin duck (*Anas platyrhynchos domesticus*). *J Assoc Avian Vet.* 1991:83-86.

35. Yu PH, Chi CH. Long-term management of thymic lymphoma in a java sparrow (*Lonchura oryzivora*). *J Avian Med Surg.* 2015;29:51-54.

36. Sinclair KM, Hawkins MG, Wright L, et al. Chronic T-cell lymphocytic leukemia in a black swan (*Cygnus atratus*): Diagnosis, treatment, and pathology. *J Avian Med Surg.* 2015;29:326-335.

37. Hammond EE, Guzman DS-M, Garner MM, et al. Long-term treatment of chronic lymphocytic leukemia in a green-winged macaw (*Ara chloroptera*). *J Avian Med Surg.* 2010;24:330-338.

38. Huston S, Lee P, Quesenberry K, et al. Cardiovascular disease, lymphoproliferative disorders, and thymomas In: Quesenberry K, Carpenter JW, eds. *Ferrets, Rabbits, and Rodents: Clinical Medicine and Surgery.* St. Louis, MO: Elsevier Saunders;2012;265-266.

39. Rosenthal KL. Therapeutic contraindications in exotic pets. *Semin Avian Exot Pet Med.* 2004;13:44-48.

40. Leili S, Scanes CG. The effects of glucocorticoids (dexamethasone) on insulin-like growth factor-I, IGF-binding proteins, and growth in chickens. *Proc Soc Exp Biol Med.* 1998;218:329-333.

41. Lumeij J. Endocrinology In: Ritchie B, Harrison G, Harrison L, eds. *Avian Medicine; Principles and Application.* Lake Worth, FL: Wingers Publishing;1994:599-601.

42. Safdar A. Immunotherapy for invasive mold disease in severely immunosuppressed patients. *Clin Infect Dis.* 2013;57:94-100.

43. Hess L. Corticosteroid synthesis and metabolism in birds. *Semin Avian Exot Pet Med.* 2002;65-70.

44. Carter RT, Murphy CJ, Stuhr CM, et al. Bilateral phacoemulsification and intraocular lens implantation in a great horned owl. *J Am Vet Med Assoc.* 2007;230:559-561.

45. Flinchum G. Corticosteroids may have some use in birds. *Exot DVM.* 2001;3.5:43.

46. Orcutt CJ. Risks of corticosteriod use in birds. *Exot DVM.* 2001;3.5:42.

47. Hunter LJ, Wood DM, Dargan PI. The patterns of toxicity and management of acute nonsteroidal anti-inflammatory drug (NSAID) overdose. *Open Access Emerg Med.* 2011;3:39.

48. Richardson JA, Balabuszko RA. Ibuprofen ingestion in ferrets: 43 cases January 1995–March 2000. *J Vet Emerg Crit Care.* 2001;11:53-58.

49. Dunayer E. Toxicology of ferrets. *Vet Clin North Am Exot Anim Pract.* 2008;11:301-314.

50. Shientag LJ, Wheeler SM, Garlick DS, et al. A therapeutic dose of ketoprofen causes acute gastrointestinal bleeding, erosions, and ulcers in rats. *J Am Assoc Lab Anim Sci.* 2012;51:832-841.

51. Loram LC, Mitchell D, Skosana M, et al. Tramadol is more effective than morphine and amitriptyline against ischaemic pain but not thermal pain in rats. *Pharmacol Res.* 2007;56:80-85.

52. Souza MJ, Greenacre CB, Cox SK. Pharmacokinetics of orally administered tramadol in domestic rabbits (*Oryctolagus cuniculus*). *Am J Vet Res.* 2008;69:979-982.

53. Baker BB, Sladky KK, Johnson SM. Evaluation of the analgesic effects of oral and subcutaneous tramadol administration in red-eared slider turtles. *J Am Vet Med Assoc.* 2011;238:220-227.

54. Souza M, Martin-Jimenez T, Jones M, et al. Pharmacokinetics of oral tramadol in red-tailed hawks (*Buteo jamaicensis*). *J Vet Pharmacol Ther.* 2011;34:86-88.

55. Souza MJ, Gerhardt L, Cox S. Pharmacokinetics of repeated oral administration of tramadol hydrochloride in Hispaniolan Amazon parrots (*Amazona ventralis*). *Am J Vet Res.* 2013;74:957-962.

56. Geelen S, Guzman DS-M, Souza MJ, et al. Antinociceptive effects of tramadol hydrochloride after intravenous administration to Hispaniolan Amazon parrots (*Amazona ventralis*). *Am J Vet Res.* 2013;74:201-206.

57. Guzman DS-M, Drazenovich TL, Olsen GH, et al. Evaluation of thermal antinociceptive effects after oral administration of tramadol hydrochloride to American kestrels (*Falco sparverius*). *Am J Vet Res.* 2014;75:117-123.

58. Evenson E, Mans C. Antinociceptive efficacy and safety of subcutaneous tramadol in chinchillas (*Chinchilla lanigera*). *J Exot Pet Med.* In press.

Comparative Clinical Avian and Reptile Hematology

Vickie Joseph, DVM, DABVP (Avian Practice)

Session #2087

Affiliation: From Clinical Avian Pathology Services, 3580 Garden Bar Rd, Lincoln, CA 95648, USA, and Bird & Pet Clinic of Roseville, 3985 Foothills Blvd Roseville, CA 95747, USA.

Abstract: Interpretation of the avian and reptile hemogram can be complex and confusing. While subtle differences exist in the non-diseased patient, their response to disease is quite varied. Comparative hematologic changes in the response to bacterial, fungal, viral, parasitic, and toxins will be presented. Interpretation of the hemogram with prognostic evaluation will also be addressed.

Introduction

Evaluating the avian and reptile hemogram is challenging. Extrinsic and intrinsic factors will affect the hemogram numbers, distribution of the leukocytes and the patient's response to disease. A brief review of normal cell morphology is provided for microscopic identification and comparison between the bird and reptile. Detailed reviews of abnormal changes seen in the hemogram in response to selected disease processes are presented for microscopic evaluation and interpretation.

Hematology

Blood collection and preparation

Although it is not the scope of this paper to address sites of blood collection and sample preparation, a few important points need to be explained. To preserve cell morphology and decrease sample artifacts, hematologic testing should be performed as soon as possible following blood collection. Prolonged exposure to any anticoagulant (EDTA or lithium heparin) will distort cell morphology and increase sample artifacts such as smudge cells.

The anticoagulant EDTA is not recommended for use in certain avian species (crows, jays, ravens, magpies, ducks, cranes and hornbills) and reptiles (chelonians). Incomplete anticoagulation or partial hemolysis may occur. In general, heparin is recommended and has the advantage of being used as an anticoagulant for hematology and plasma for blood chemistry evaluation; however, heparinized samples may result in inadequate staining of the white blood cells and affecting cell morphology or clumping of the white blood cells and thrombocytes.

Regardless of the anticoagulant of choice, blood smears should be made immediately without exposure to anticoagulants. This preserves cell morphology and helps to prevent clumping of the white blood cells and sample artifact. Also important to remember is to air dry the blood smears with a rapid fanning of the slides. Smears allowed to dry to slowly may result in perinuclear ring artifacts. These artifacts may be mistaken for *Hemoproteus* sp. or *Plasmodium* sp. in the avian or reptile patient. In reptiles, lymphatic vessels often accompany blood vessels, which may result in a collection of a mixture of blood and lymph. Lymphatic dilution of the sample will affect the hemogram by decreasing the hematocrit, hemoglobin concentration, total red and white blood cell counts.

The total blood volume of reptiles and birds will vary between species. The reported reptile range is estimated at 5-8% of body weight in grams, while in birds it is estimated at 6-12% of body weight. In the healthy reptile or bird the amount of blood that can safely be drawn is 0.5-0.8% and 0.6-0.12% of body weight in grams, respectively. Reported packed cell volume (PCV) in reptiles ranges from 20-40% and birds from 35-55%. In general due to the lower PCV in reptiles, the yield of a plasma sample is often greater versus the bird.

Erythrocytes

Reptile erythrocytes are generally larger than avian erythrocytes, but smaller than amphibian erythrocytes. Depending on the species, the average life span of the reptile erythrocyte in the peripheral blood is 600-800 days and only 28-45 days in the bird. The slow metabolic rate of reptiles and age may play a role in the red blood cell life span. In reptiles, erythrocytic indices are at their highest prior to hibernation and lowest immediately following hibernation. The PCV and the total erythrocyte count increases with age in both the avian and reptile species.

The color of the adult erythrocyte is orange and contains a nucleus. Circulating erythrocytes that have a polychromatic color to the cytoplasm are termed polychromatophilic erythrocytes. They are similar in size and shape to the mature erythrocytes. These polychromatic erythrocytes reflect the final stage of maturation and are often termed reticulocytes; however, a new methylene blue stain is required to verify the aggregate form of the reticulum encircling the erythrocyte nucleus to properly classify. In the avian species, up to 5% percent of the circulating erythrocytes are polychromatic cells, which reflect the short life span of the avian erythrocyte. In contrast the slower turnover rate of reptilian erythrocytes and the long erythrocyte life span result in less than 1% percent of the circulating erythrocytes as polychromatic cells. For the avian patient, polychromasia is graded on a scale of 1+ to 4+, with 1+ considered normal in the non-anemic patient. In the reptile, polychromasia is usually not graded due to the fact that polychromasia is less than 1% of the total erythrocyte population. In either species, the presence of increased polychromatic erythrocytes and reticulocytes is evidence of active erythropoiesis. In most instances, the classification of erythrocyte maturation is not detailed as it is in mammals. The different stages of the immature erythrocytes are just referred to as immature erythrocytes. The presence of immature erythrocytes indicates a marked erythropoietic response to anemia. Immature erythrocytes are occasionally seen in the peripheral blood of very young birds and reptiles or reptiles going through ecdysis. The response to anemia is slower in the reptile compared to birds, requiring up to 30 days to be visible in the hemogram and up to 8 weeks before the maximal response is noted. This may be due to prolonged development time of the rubriblasts in the bone marrow and other hematopoietic sites.

Hypochromasia increases in either species may be a result of nutritional deficiencies leading to iron deficiencies, lead toxicity or chronic inflammatory disease. Occasional anisocytosis and poikilocytosis are considered normal in the peripheral blood of both species. Increases of anisocytosis with an increase of polychromasia suggest a regenerative anemia. Increases of poikilocytosis may suggest erythrocyte dysgenesis. In addition, anisocytosis and poikilocytosis may also be found in the post hibernation reptile or those with severe inflammatory disease. Basophilic stippling may be associated with erythrocyte regeneration. It is rarely associated with lead toxicity; however, in reptiles it may suggest iron deficiency

White blood cells

Lymphocytes

The function of the lymphocyte is similar to mammals and reflects immune stimulation. In many reptile species, the lymphocyte can be the dominant cell with ranges up to 80% of the leukocyte numbers.

Heterophils

Functionally is equivalent to the mammalian neutrophil. They are highly phagocytic with bactericidal activity. Heterophils also participate in viral and parasitic infections.

Monocytes

Monocytes have phagocytic capabilities and are active with chronic inflammatory conditions, chronic antigenic stimulation, bacterial, fungal or parasitic diseases. Unique to reptiles, circulating monocytes and macrophages that contain melanin pigment (melanomacrophages), nucleoproteinaceous debris, or lipid vacuoles and erythrophagocytic macrophages can be observed in the peripheral blood.

The reptile monocytic cell has been referred to as monocytes, monocytoid azurophils, azurophilic monocytes or azurophils.

Reptile azurophils

An azurophilic leukocyte, the azurophil causes confusion in classification and identification. The granulopoietic origin of the azurophil has not been documented or confirmed. Many researchers believe this cell should be considered a monocyte and not a separate cell type. They may represent an immature form of the monocyte. The azurophil is slightly smaller than the monocyte, has an eccentrically placed round to oval to bilobed nucleus, basophilic cytoplasm and contains small numbers of cytoplasmic azurophilic granules. Similar to the monocyte, this cell has phagocytic capabilities. Increased numbers are often seen with acute bacterial and inflammatory conditions of the snake, and chronic inflammatory conditions of the other types of reptiles.

Plasma cells

These cells may occasionally be seen in the peripheral blood smear. With immune stimulation, the cell will become distended with round clear structures (Russell bodies), and is now referred to as a Mott cell. May be seen with severe infection or inflammatory disease.

Eosinophils

Eosinophils are similar in size to heterophils. In both species, the granules are primarily round; however, in the avian species, the nucleus is lobed while in reptiles the nucleus is round to oval and eccentrically placed (lobed in some lizards). Although their function is not entirely known, these cells may play a role in parasitic infections or hypersensitivity reactions. Eosinophils may compose 7-20% of the leukocytes in the healthy reptile, with lower percentages in the lizard and higher percentages in the chelonians. This is in contrast to birds, where eosinophils are not a common occurrence in the peripheral blood.

Basophils

These are round cells often smaller versus the eosinophil. The cytoplasmic granules contain histamine and their function is similar to the mammalian basophil by processing surface immunoglobulins and releasing histamine. Basophils may participate in acute inflammation and type 1 hypersensitivity reactions in avian and reptile species. In addition, reptile basophils have been associated with parasitic and viral infections. Whereas basophils may be found in small numbers in avian peripheral blood, the normal basophil percent in reptiles may range from 0-40% (tortoises) depending on the species and up to 65% in healthy freshwater turtles.

Thrombocytes

A small nucleated cell, the thrombocyte is the second most numerous cell in the peripheral blood. They are derived from a mononuclear precursor in the bone marrow. The primary function of the thrombocyte is hemostasis. Thrombocytes are also considered to be nonspecific scavengers, capable of removing foreign material including bacteria from the blood.

A guide to avian and reptile hematology is provided in Table 1. Potential causes of increased and decreased values are outlined.

Table 1. Avian and reptile hematology guide.

Value	Increase	Decrease
Packed cell volume (PCV)	Dehydration	Anemia (nonregenerative, hemolytic, hemorrhagic)
	Abnormal myeloid production	
Total protein (TP)	Dehydration (elevated albumin)	Metabolic disorder (liver, kidney, GI)
	Chronic inflammatory disease (elevated globulins	Parasites
		Nutrition
White blood cells (WBC)	Bacterial, parasitic, fungal infections	Viral (except with inflammatory viruses)
	Neoplasia, stress	Severe chronic infection
		Overwhelming septicemia
		Toxic chemicals
Thrombocytes	Hemostasis regenerative response following anemia	Decreased production
	Chronic inflammation	Increased peripheral utilization (DIC, septicemia)
Heterophils	Chronic inflammatory disease	Viral infections
	Bacterial or parasitic infections, neoplasia	Overwhelming septicemia
	Stress	
Lymphocytes	Viral, parasitic infections	Bacterial
	Wound healing, immune stimulation	Chronic stress, chronic malnutrition
	Ecdysis	Immunosuppression
Monocytes	Chronic inflammatory or bacterial disease	Significance unknown
Azurophils		
	Granulomatous disease, parasites	
Eosinophils	Parasites, allergic reaction, immune stimulation	Significance unknown
Basophils	Acute inflammation, hypersensitivity	Significance unknown
	Viral and parasitic disease in reptiles	

Hematology Changes in Response to Disease

When interpreting reptile leukocyte parameters, it is important to remember that they are affected by many intrinsic and extrinsic factors and at times may be difficult to properly interpret, versus birds.

Leukocytosis

Stress leukograms show an absolute leukocytosis with moderate mature heterophilia (H) and lymphopenia (L). The H/L ratio is a reliable indicator of stress. One must also identify the lymphocytic species when interpreting the leukogram. Amazon parrots, canaries, iguanas, and some species of sea turtles are often considered to be lymphocytic species; however, in the reptile extrinsic factors including season, temperature, habitat, diet and stress can influence the blood parameters. The cell morphology is considered normal.

Inflammatory leukograms are variable depending on the severity of the illness or condition. Heterophils should predominate and monocytes/azurophils may be present. Localized infections often show a mild inflammatory leukogram. Reptiles are capable of a strong inflammatory response with high numbers of reactive monocytes/azurophils. This at times may be mistaken as a neoplastic process, but with treatment of the underlying disease, the leukogram can return to normal.

Birds or reptiles with a severe inflammatory disease (usually bacterial etiology) develop a marked leukocytosis with the presence of immature heterophils (bands, metaheterophils, myeloheterophils, progranulocytes) and toxicity. A regenerative left shift is when the mature heterophils outnumber the immature heterophils, whereas a degenerative left shift reflects the predominance of immature heterophils. The latter can be a grave situation.

Lymphocytosis may occur with wound healing, parasitic infections, viral infections and inflammatory disease. A lymphocytosis may also be present during the reptile ecdysis. The presence of reactive lymphocytes and plasma cells reflects stimulation of the immune system. Lymphocytic leukemia is also a possibility with very elevated white blood counts and lymphocytosis

A marked leukocytosis and heterophilia are often associated with diseases such as *Chlamydia* psittaci, *Aspergillus* spp., *Mycobacterium* spp. or some other granulomatous disease. These diseases may also show a non-regenerative anemia, monocytosis/azurophilia and thrombocytosis.

Leukopenia

Leukopenia reflects the consumption or decreased production of the peripheral leukocytes. The presence of an absolute heteropenia in conjunction with immature heterophils and toxicity represents overwhelming demand for heterophils in the periphery due to bacterial sepsis or viral disease. This is a degenerative response and carries a grave prognosis.

Leukopenia with heteropenia, anemia and thrombocytopenia suggest injury to the bone marrow.

Lymphopenia may occur secondary to diseases that cause immunosuppression, chronic malnutrition and chronic stress.

Neoplasia

When atypical large lymphocytes with scalloped edges are present, an underlying neoplasia (lymphosarcoma) process needs to be considered; however, some viral infections that are considered inflammatory (West Nile virus) may also present with atypical lymphocytes.

Blast leukemia

Blast leukemia reveals unclassified cells of the mononuclear line with immature cells.

Chronic lymphocytic leukemia

High white blood cell counts are seen with the predominant cell line being adult lymphocytes.

Myeloid leukemia

Myeloid leukemia is characterized by abnormal erythrocytes, mitotic figures, and a large number of rubriblasts.

References

1. Campbell T. *Exotic Animal Hematology and Cytology.* Ames, IA: John Wiley & Sons, Inc;2015.

2. Giordano N, Stefania L, Mattia B. Clinical hematology in reptilian species. *Vet Clin North Am Exot Anim Pract.* 2013;16:1-30.

3. Nicole S, Rick A, Katherine AS. Diagnostic hematology of reptiles. *Clin Lab Med.* 2011;31:87-108.

4. Raffaella C, Lorenzo C. Overview of psittacine blood analysis and comparative retrospective study of clinical diagnosis, hematology and blood chemistry in selected psittacine species. *Vet Clin North Am Exot Anim Pract.* 2013;16:71-120.

Alternative Vet for Alternative Pets: Integrating Holistic Modalities in Exotic Pet Practice

Robert D. Ness, DVM

Session #2380

Affiliation: From Ness Exotic Wellness Center, 1007 Maple Ave, Lisle, IL 60532, USA.

Abstract: Exotic pet practice has always had its share of unique challenges, which becomes even more complex when coupled with the diverse realm of holistic or alternative therapy. Integrative therapy involves the unifying of conventional medicine with various holistic care modalities. Fortunately, research has been done and is currently underway studying these treatment methods and it has demonstrated many of the benefits and verified claims made by holistic practitioners. Manipulative therapies include photobiomodulation, acupuncture and chiropractic care. Medicinal treatments include nutraceuticals, herbal remedies, essential oils, flower essence formulas and homeopathy. Each of these diverse modalities from ancient Chinese medicine to New Age technology will be introduced and briefly discussed; however, each is worthy of a course dedicated to its own merits.

Introduction

Exotic pet practice poses a variety of unique challenges due to the patient size, unique anatomy and physiology and extensive diversity. As a profession, we have made huge strides within diagnostic and therapeutic fields in several species, beginning with the sharing of anecdotal clinical information and captive rearing to specific research and documentation of clinical studies. The same is true for the variety of holistic modalities in human and veterinary medicine. Integration of these modalities into the exotic realm adds another level of complexity to the practice of medicine in these patients; however, these therapies also provide potential avenues for treating cases which had limited success in western medicine. This presentation explores practical integration of various holistic modalities for common exotic pet conditions.

The holistic therapies discussed in this presentation are not meant to replace conventional avian and exotic veterinary care. Instead, these modalities should be considered adjunctive therapy integrated into the overall treatment plan, thus termed integrative medicine. Many species treated in exotic pet practice are very adept at hiding their clinical signs as instinct and evolution has taught them to survive predation. The initial symptom of disease should be considered a desperate cry for help requiring current, quality veterinary care to establish a diagnosis as well as stabilization with cutting edge conventional therapy. In addition, these holistic remedies and techniques can be lifesaving when used appropriately with standard conventional care.

Photobiomodulation (Laser Therapy)

Photobiomodulation has become more common in veterinary practice. Therapeutic lasers are being used in small animal practices for the treatment of simple surgical incisions to severe osteoarthritis and fractures. In addition, companion pet and equine practitioners routinely incorporate laser therapy in any wound management protocol. This technology is being utilized to treat various conditions that exhibit pain or inflammation. These benefits can be appreciated in all species, including avian and exotic pets, zoo animals and wildlife.

Laser protocols have been reported and incorporated for bird and exotic pets. Previously, protocol settings for use on exotic species were extrapolated from established protocols for cats and dogs. Consideration for the anatomic and physiologic differences in these exotic species required further adjustments of the protocols. For example, adjustments are made to further decrease the avian settings to take into account their thin, delicate skin and feather architecture, as well as the patient's high metabolic rates. Conversely, the dosage settings for reptiles are relatively increased to accommodate for their thicker skin and lower metabolic rates. Small mammals require fewer modifications except for body size and weight when utilizing the feline settings. The settings on most currently available units for exotic pets have a pulsed frequency and generally lower power pre-sets than comparable protocols for other species as an additional precautionary measure.

Successful treatment depends on delivering an appropriate dose of laser energy into the affected tissue. Insufficient dosing results in treatment failure. Target doses are expressed in J/cm^2. One Joule (J) is the energy delivered with one Watt of power in one second. Established target doses are 3-4 J/cm^2 for superficial conditions and 8-10 J/cm^2 in deep conditions; however, much higher doses have been utilized.[1] Power and time are inversely proportional in calculating the total energy (Joules) delivered. Treatment time can be reduced by increasing power. Likewise, treatment time increases if the power (Watts) is reduced. When treating superficial conditions with a lower target dose, power usually remains relatively low since achieving a target dose of 3-4 J/cm^2 does not take long. Many superficial conditions affect sensitive tissue and a lower power setting is more comfortable for the patient. When treating deep tissue conditions, power is increased to reduce the time required to achieve a relatively higher target dose of at least 8-10 J/cm^2. Some photobiomodulation protocols, such as wound management or abdominal discomfort, require an estimation of the treatment area when calculating the target dose of a treatment. A simple technique for estimating the patient's surface treatment area is by comparing the size of the area to be treated relative to a 3×5 inch index card, which has an approximate area of 100 cm^2.

Chiropractic Care

Chiropractic care of animals dates back to when BJ Palmer, the founder of modern chiropractic medicine, referenced adjustments of vertebral subluxations in sick cows, horses, dogs and cats in 1944. The first attempt to establish a chiropractic organization to promote chiropractic in animals was in the early 1980s in New Jersey, called Options for Animals. This original group disbanded and the organizations name was transferred to Dr Sharon Willoughby, who had earned the titles as veterinarian and chiropractor. By 1986, Dr Willoughby established an animal chiropractic school under the name Options for Animals as well as founded the American Veterinary Chiropractic Association (AVCA). This course was offered to veterinarians and chiropractors as a four module course focusing on chiropractic treatment of horses and dogs. The AVCA has grown and matured as an educational and legislative organization. Several chiropractic colleges and training programs have evolved from this original group founded by Sharon Willoughby.

Many exotic pet species treated in veterinary practice are appropriate patients for chiropractic care, since anything possessing a vertebral column can be adjusted. Chiropractic care involves the use of specific force applied through an intervertebral unit in a precise angle and direction in an effort to normalize spinal function. Species as diverse as a mouse to an elephant have benefited from the chiropractic therapy. Each species varies a bit in their precise vertebral formula and facet angles between vertebrae, but the general concepts remain the same.

As with any therapeutic plan, a thorough physical examination with appropriate diagnostic tests should precede a chiropractic adjustment. Whole body radiographs are indicated for ruling out fractures or potential intervertebral disc rupture and routine blood work is often indicated to assess underlying health status of the patient. Concurrent medical conditions must be assessed and properly addressed with appropriate medical therapy in conjunction with chiropractic care. Although chiropractic adjustments may be indicated in a variety of conditions seen in veterinary medicine, it is seldom the only therapy necessary.

Rabbits are one of the most conducive species to chiropractic care in exotic pet practice. Although the rabbit's vertebral column is delicate and fragile, it is relatively easy to motion palpate and evaluate. Rabbits are responsive to chiropractic adjustments and manipulation, both in regards to acceptance of the manipulation as well as the clinical response to therapy in the authors professional experience. The vertebral formula for a domestic rabbit is C7, T12-13, L7, S4, Ca15-16. The skeletal mass accommodates for only 6% of the New Zealand rabbit's total body mass, while the muscles comprise approximately 56%.[2] This leads to a delicate spine that is susceptible to injury with improper handling or trauma. The most common region for spinal injury is at L6-7 due to the heavy and powerful hindquarters twisting at the lumbosacral junction. Hind limb paresis to paralysis is a common presenting complaint in pet rabbit medicine, which is one of the more common conditions that may benefit from chiropractic therapy in many species.

Various medical conditions commonly encountered in exotic pet medicine can benefit from chiropractic care regardless of the species. Similar to chiropractic care in people, several musculoskeletal conditions that may respond to chiropractic care include spinal spondylosis, degenerative joint disease and orthopedic trauma. Precautions must be taken to avoid adjustments in cases involving intervertebral disk disease and fractures to avoid worsening of the patient's condition. Certain systemic and metabolic conditions that have underlying benefits from chiropractic care include gastrointestinal mobility and urinary incontinence. In all of these conditions and disorders, chiropractic has the primary effect of establishing proper nerve innervation to the tissues, thus leading to proper function.

Torticollis is characteristic of several disorders in many species, including trauma, otitis media, cerebral vascular accidents, aberrant parasitic larval migrans (*Encephalitozoon cuniculi* in rabbits and *Baylisascaris procyonis* in rabbits and rodents) and brain tumors. Although each underlying condition requires specific treatment, any case presenting with torticollis may benefit from chiropractic care. Cervical subluxations and fixations are a common consequence of the prolonged tilting of the head and associated rolling and falling from the vestibular imbalance.[3] In addition to the expected cervical vertebral adjustments, hyoid and temporomandibular joint (TMJ) manipulations should also be considered in the thorough chiropractic assessment and treatment. In addition, certain physical therapy and massage techniques that the client can perform on their pet while at home have been shown to be very beneficial for a successful recovery from torticollis. One such technique involves gently placing one's index finger under the "down-side" of the head tilt to lift the patient head up into normal position. Then simply sitting with the patient's head in more normal orientation and gently rubbing or massaging the tension stretched cervical musculature on the opposite side of the neck. This is soothing to most pets with this condition and also establishes a visual plane in proper orientation which improves neuroplasticity.

Chiropractic care can be incorporated as part of standard preventive medicine protocol as well, especially in the elderly patient. Routine chiropractic examination will support proper nerve function and spinal mobility. With proper nerve function, the body's organ systems will operate more efficiently and be less susceptible to disease and metabolic imbalance. Normal spinal mobility can lead to reduction in degenerative joint diseases and arthritic changes. Relief of the day-to-day subluxations resulting from the rambunctious lifestyle of the typical house rabbit or other pets will improve the overall quality of life, with less pain and discomfort. Maintenance chiropractic therapy can be incorporated into a geriatric health plan for the older rabbit to minimize the discomfort experienced by arthritis and spondylosis often experienced in the elderly patient.

Acupuncture

Acupuncture has been used for at least 5000 years, with China being the site of origin. About 500 AD, the practice of acupuncture spread to Japan and Korea, who established their own unique forms. By the 6th century, acupuncture had spread throughout Asia. By the 17th century it was found in Europe, and finally arrived in North America during the 19th and 20th century. It was not until 1971 that acupuncture made its way into American culture. This was the result of a New York Times journalist being treated with acupuncture while on assignment in China. He had his appendix removed and was treated with acupuncture for postoperative pain. Over the past 30 years, acupuncture has slowly become more mainstream in American culture.

Veterinary acupuncture also has a long history. Elephant acupuncture dates back about 3000 years in Sri Lanka. Several veterinary applications were recorded by the Chinese Chou Dynasty dating back to 1066 to 221 BC. The father of Chinese veterinary medicine is Shun Yang (Pao Lo) who was the first full time practitioner of Chinese Veterinary Medicine in 430 BC. Veterinary acupuncture has developed in various parts of the World, especially in Asia, over the past 2000 years. In 1974, the International Veterinary Acupuncture Society (IVAS) was established, and has since become a core association for veterinary acupuncture in the United States and the World. Since 1998, two other teaching organizations in the United States (Chi-Institute of Traditional Chinese Veterinary Medicine and Colorado Veterinary Medical Association) have offered training in veterinary acupuncture.

Acupuncture is one part of the holistic health system known as Traditional Chinese Medicine (TCM). Other TCM components include proper nutrition, exercise, herbal remedies and appropriate lifestyle. The main premise of TCM is that we are all part of nature and health is achieved by establishing balance with the natural world. This balance of nature is characterized by the Chinese concept of Yin-Yang, which is the balance between such things as light and dark, wet and dry, and hot and cold.

Acupuncture involves the placement of fine needles into specific points on the body to elicit a physiologic and energetic response along special pathways known as meridians. Meridians are interconnected energetic pathways that run throughout the body. The theory behind acupuncture is that these pathways carry the body's Qi, defined as vital life force or energy. The presence of Qi is what defines the existence of life. The placement of acupuncture needles into points along these meridians enables the body to restore itself to homeostasis by affecting the Qi flow. These meridian pathways are currently thought to run within the myofascial plane between the skin and underlying muscles.

The physiologic effects of acupuncture are being studied and verified by scientific methods. Physically, there is an increase in nerve endings, small capillary beds and aggregations of mast cells at the site of acupuncture points.[4] As a result, a measurable physiologic effect in beta endorphin release, stimulation of circulation by vasodilation, which leads to a decrease in inflammation results from acupuncture stimulation. In pain control, experiments have shown a modification in neural impulse transmission from the spinal cord to the brain after acupuncture.[5,6] This effect is known as "gate control" theory, which proposes that acupuncture can block the action of pain fibers in the spinal cord.[7]

A variety of acupuncture techniques exist. The use of the different techniques depends on the species and general cooperation of the patient, type and severity of the condition being treated and personal preference of the acupuncturist. Traditional dry needling is commonly used in animals. Needle size depends on the size of the patient and the specific acupoint. The use of 20 gauge × 1.5 inch needles is appropriate for large dogs, while 36 gauge × 0.5 inch may be used in birds or small rodents. Electro acupuncture involves stimulation of acupoints with a mild electrical current to elicit a stronger stimulation than dry needling alone. Another alternative to traditional needling is aquapuncture. This technique involves the injection of cyanocobalamin (vitamin B12) or saline into the acupoint using a 27- to 29-gauge hypodermic needle and 0.5- to 1-ml syringe. The aquapuncture technique has the added advantage of providing a longer lasting effect at the site. Another technique for potential use in veterinary patients is laser therapy. Low intensity laser lights are effective in penetrating the skin to stimulate shallow acupoints. Disadvantages of laser therapy include the lack of specificity for acupoint stimulation in areas where multiple points are close together and lack of stimulation of deeper points. Gold beads or wire implants have been used for chronic cases requiring much longer periods of stimulation. Moxabustion utilizes the burning of the Chinese herb moxa over the acupoint. This can be used by itself over the area of the point or in conjunction with needles.

The clinical applications of veterinary acupuncture include everything from pain management to treatment of systemic diseases. Acupuncture is effective for many chronic disorders, such as allergies, urinary incontinence, inflammatory bowel disease, reproductive disorders and cancer.[8,9] Traditionally, acupuncture is usually combined with Chinese herbs and proper nutrition to achieve the greatest effect.

Herbal Formulations

The use of specific herbs for medicinal purposes dates back thousands of years. Several herbs are mentioned in the Bible and archeologists have documented herbal usage back to prehistoric times. Herbs are used around the world, including Western herbs from North America, Aryuveic herbs from India and traditional Chinese herbs.

Approximately 25% of our conventional drugs are derived from plants. Conventional drugs typically contain a single active constituent from the plant, whereas herbs provide a broader and balanced effect on the body through the synergistic actions of the herbal components.[10] Herbs are best prescribed to treat the entire individual and not only the clinical signs. Herbal blends and formulations combine the benefits of multiple herbs, which typically produce a synergistic action while minimizing the potential toxic effects of a single herb.[11] Herbs provide many unique qualities that are very limited in conventional medicine, such as anti-cancer, anti-viral and immunoregulation properties.[12,13]

Standardized extracts are available for certain herbs through concentrating the active ingredients, resulting in more of a plant drug than an herbal medicine. Standardization alters the physical and energetic nature of the herb. This process also eliminates the synergistic effects of the myriad of chemical components in the plant. Standardization is advantageous for some herbs, such as milk thistle, since the specific active constituent is known and concentrated in the process. Other factors that affect the potency and medicinally active components of the herb include the method and time of harvest, the part and preparation of the plant included and the handling and processing of the finished product. Only well-known and respected herbal companies should be considered when purchasing herbal products. Whenever possible, fresh herbs or glycerin-based extracts should be used.

Herbs are effective in the treatment of many conditions in pets. Herbal remedies can be more effective than conventional therapy in treating metabolic conditions, such as liver and kidney diseases. Milk thistle seed is a potent hepatoprotectant and antioxidant herb, with Silymarin being one of its active constituents that has been isolated and promoted in mainstream medicine as a liver therapy. Dandelion leaf is used as a diuretic while the root is a liver tonic, demonstrating the different benefits from specific parts of the plant. Herbs are an excellent alternative to antibiotics in the treatment of some infectious diseases, with wider antibacterial effects as well as various antifungal and antiviral actions, such as olive leaf.[14,15] Many of these herbal remedies also support the immune system to assist in the recovery of the patient. Some herbal formulations serve as cardioprotective agents, detoxification agents, antioxidants, anti-inflammatory and anti-cancer therapies.[16]

Nutraceutical Supplements

Nutraceuticals are micronutrients, macronutrients, and other nutritional supplements that are used as therapeutic agents. Examples include vitamins and minerals, probiotics, digestive enzymes, and anti-oxidants. This is the clinical application of nutrition in the treatment of disease and metabolic disorders. It is commonly stated that malnutrition is the underlying cause of many of the disease syndromes encountered in birds and exotic pets. Significant advances have been made in pet nutrition with formulated diets, but it is only the beginning. Specific nutritional requirements have not been established for many of the exotic species commonly kept as pets, therefore current recommendations and diets are based on anecdotal experience and limited nutritional studies. The specific diet and variety of foods depends on the species, age, health status, and activity level of the individual pet, among other things.

Nutraceuticals are used for various digestive disorders and other metabolic conditions in pets. Some of the commonly used supplements are aloe juice, apple cider vinegar, probiotics, and digestive enzymes. Aloe vera provides an effective boost to the immune system, soothing anti-inflammatory effect on the GI tract, and is an excellent source of vitamins, minerals, and amino acids.[17,18] Aloe vera can be given orally in the form of a gel or juice at the dose of 1 drop per 100 grams body weight (0.50 ml/kg) 3-6 times daily, or in the drinking water at the rate of 2 ml/4 oz drinking water.[19] Apple cider vinegar is an acidifier of the intestinal tract and entire body. Specific indications for apple cider vinegar include chronic bacterial or yeast infections, chronic diarrhea or foul smelling stools. It is dosed at 1-2 tbsp/8 oz of drinking water, as the only water source for 1-2 weeks.[19] Probiotics are supplements containing beneficial bacteria given to re-establish the normal bacterial flora in the digestive tract. These bacteria include lactobacillus and bifidobacterium. They are often given after antibiotic therapy or severe GI disturbances. Digestive enzymes are beneficial for certain pancreatic diseases or primary digestive disorders leading to maldigestion. The classic essential enzymes provided in most formulations include protease, lipase and amylase. These may be combined with other specific enzymes or herbs depending on the condition.

Supplements used in the treatment of inflammatory conditions and arthritis include glucosamine, Methylsulfonylmethane (MSM), and oligomeric proanthocyanidins. Proanthocyanidins are a group of strong anti-oxidants which scavenge destructive free radicals, and include grape seed extract, pine bark extract, bilberry, and citrus bioflavonoids. These substances provide excellent anti-oxidant effects that result in less inflammation, improved cellular integrity, and elimination of free radicals from the body. Glucosamine sulfate is the preferred and most effective form of glucosamine products.[20] It has reparative effects on the joints and reduces inflammatory effects of arthritis. Some formulations of glucosamine contain chondroitin or MSM for further joint support, but glucosamine is shown to be useful alone, but more effective when in combination. MSM is a sulfur based supplement that has anti-inflammatory effects on joints and generally supports healthy tissue and cells. Sulfur is suspected to be an important mineral for the body to prevent degradation of tissues at the cellular level.

Essential Oil Therapy

Essential oils are volatile aromatic compounds harvested from the leaves, bark, fruit, flowers, stems, roots or other plant parts. The plant material is steam distilled or cold pressed to produce the medicinal oils. These compounds can be used aromatically, topically, and internally in some cases with precaution. Currently, the most common use is through water diffusion in the vicinity of the patient. It is vitally important to only use quality, high grade, pure volatile oils for birds and animals. Some companies produce oils with synthetic components or lower quality ingredients that may smell good initially, but can be irritating when inhaled or topically administered.

The topic of using essential oils in birds is controversial. Due to the unique respiratory system of birds, they are very susceptible to airborne irritants or aromas. If an essential oil is used that contains synthetics or poor quality volatile oils, then it may prove fatal to the avian patient while being safe for people and other animals. Some reputable suppliers have developed formulas that claim to be completely safe for all species, including birds. Therefore, it is critical to know the source and trust the formulations administered to patients and evaluate their studies and processing. Always remain cautious when administering anything aromatic to an avian patient.

Lavender is one of the most common and safest essential oils. Lavender oil is steam distilled from the flowers of the lavender plant, *Lavandula angustifolia.* It can be water diffused, applied topically or even taken internally in some cases. This oil has many calming properties and can be safely used in all species. This oil is given to calm anxious pets or soothe those encountering stressful situations, such as loud noises, visitors at home or trips to the vet. It can also be useful in adjusting to changes in the pet's life, such as a new home or addition of another pet or child.[21,22]

Peppermint oil is steam distilled using the whole peppermint plant, *Mentha piperita.* The menthol content of peppermint is high, yielding a cool, refreshing scent. Peppermint has cooling properties and an energizing effect, which can be used to cool sore muscles and aching joints. It is also used to soothe upset stomachs and ease airway congestion. This oil can also be administered as diffusion, topical and orally.

Essential oils can be given individually, as described in the above couple examples, or also in combination as formulations with other essential oils. Many of the essential oil companies have their own concoctions with proprietary blends, which carry their own unique product names. This can be confusing until a relationship is established with comprehension of their formulary of products and associated uses.

Flower Essence Formula

Flower essence therapy proposes similar principles applied to homeopathy. Both forms of therapy are based on curing the patient by restoring the body through energy pattern and vibrational characteristics. The underlying premise is that all life forms possess an innate vibrational energy force that is disrupted by conditions and circumstances of our environment leading to disease and illness. These disruptions are further related to emotional and specific behavioral traits, which can be characterized and treated with the vital energy or essence of certain flowers. The aroma or essence of a flower naturally elicits an emotional response, similar to the way music affects a person's mood.

Dr Edward Bach is credited for the development of the first modern therapeutic system of flower essence therapy. Dr Bach was a distinguished British physician in the early 1900's with a strong influence from Hahnemann and the concepts of homeopathy. Dr Bach developed various vaccines during his tenure in immunology and then developed some of the first nosodes, oral homeopathic vaccines. In his clinical experience, Dr Bach realized the importance of the mental and emotional state of mind in the recovery from illness. In 1930, he embarked on a quest to develop a treatment method that did not depend on the destruction or alteration of one living thing to benefit another, which ultimately lead to the discovery of his first twelve healing herbs with a natural affinity to mental traits. In all, 38 healing remedies were identified which he believed would remedy all the negative states of mind that afflict mankind.

A variety of other flower essences have been described throughout the years. In the 1970s, Richard Katz and Patricia Kaminski developed the California Flower Essences. Ian White described the Australian Bush Flower Essences in the1980s, influenced by the Australian Aborigines traditional knowledge and experience with native plants. Other Flower Essence lines include the Alaskan (1980s), Bailey in Britain, and Celestial Remedies (1990s) to name a few. All of these flower essence lines are based on the same premises described by Dr Bach.

Flower essences can be effective in the treatment of clinical and behavioral issues of birds and other exotic pets. The underlying premise in using flower essences to treat conditions in pets is the presence of an emotional component to the problem. These formulations act on the energetic signature of different emotions that produce the outward behaviors. It is commonly accepted that emotional and psychological stress can lead to physical illness, therefore the flower essences can be incorporated into a holistic treatment plan.

The choice of remedies is individualized for each patient. Oversimplification by using a single remedy for a particular problem is much less effective than thoroughly evaluating the patient and formulating a remedy of various flower essences. Of course, there are certain remedies that are classic fits for a particular condition, such as Rescue Remedy for the stress and anxiety of a veterinary visit or treatment. This is a classical formulation of five flower remedies, consisting of star of Bethlehem, rock rose, impatiens, cherry plum and clematis. This remedy can be sprayed in the exam room, sprayed directly on the pet or given directly by mouth. Extensive repertories of other flower essences exist for man and animals.

Homeopathy

Homeopathy as practiced today is credited to Dr Samuel Hahnemann, a German medical doctor from the mid-1800s. The governing principle of homeopathic medicine is "similia similibus curentur" or "like cures like." This concept is based on using an extremely diluted form of a substance to treat a condition or group of symptoms, which in its full strength would cause similar symptoms in the patient. These remedies are made from plants, minerals, drugs, viruses, bacteria or animal substances. Homeopathic remedies work on the deep energetic level of the patient to undermine the constitutional cause of the disease, rather than mask symptoms.[23]

Homeopathy can be effective in treating pets.[24,25] In choosing an appropriate homeopathic remedy, the practitioner must be thoroughly acquainted with the western medical examination, conventional diagnosis, particular behavior characteristics and situational condition of the patient. The mental and emotional disturbances may be difficult to discern, since most pet owners do not fully understand the normal behavior and nature of the species.

The practice of homeopathy involves matching the patient's symptoms with an appropriate remedy. The first step involves making a list of the clinical signs from an evaluation and thorough history of the patient. This list is then used to look up rubrics, or lists of potential remedies, for each clinical sign from a homeopathic repertory. The rubrics are compared for overlapping remedies, which are selected as possible treatments. These are then compared in a homeopathic materia medica, which describes all the symptoms potentially treated with that remedy. The remedy that matches the symptoms or clinical signs most accurately is selected as the first remedy of choice. The most detailed repertories and materia medicas are based on human symptoms and proofs; however, limited veterinary references are available and continue to be developed. The potency and frequency are selected based on the severity of the condition and the characteristics of the patient.

Homeopathic remedies are made by serial dilutions of toxic substances that would cause symptoms similar to those being treated if used in full strength. Substances are diluted serially, either in 1/10 (X) potency, 1/100 (C) potency stages or 1/1000 (M) potency. Therefore, a 30C potency is a tincture of the homeopathic substance diluted 1/100, thirty times. Succession is carried out at each stage to release the curative energy of the substance to imprint on the memory of the water at the energetic level, as well as removes the toxic and harmful effects of the substance. The end result is a homeopathic substance that only contains the energetic signature of the toxic substance, but no physical amount of the substance itself.

Numerous homeopathic remedies have been prescribed for a huge array of conditions commonly encountered in exotic pet medicine. Traumatic conditions can be treated with homeopathic dosing of *Arnica* to minimize bruising, *Hamamelis* for bleeding, symphytum for fracture healing and *Hypericum* for nerve injuries. Emergency therapy using homeopathics include *Aconitum* for shock, belladonna for fevers and carbo vegetabilis for collapse. Nux vomica is appropriate for many digestive disorders as *Euphrasia* helps many eye conditions. Arthritic patients may require *Bryonia*, rhus tox or ruta grav depending on their disposition. This is by far not an exhaustive list of remedies, but a sampling of the breadth of potential.

Conclusions

This introduction to holistic therapy in veterinary medicine provides a brief overview of several alternative therapy modalities available to exotic pet practice. Research is being conducted in many fields of alternative medicine, which is further validating their effectiveness and verifying many of the anecdotal claims. As a starting point, some of these treatments can be integrated into practice as alternatives or adjuncts to conventional medicines. Advanced professional training is available to further delve into each of these modalities as interest and comfort levels build once the concepts are incorporated into practice. The American Holistic Veterinary Medicine Association (AHVMA) is a professional resource for the spectrum of modalities and applications in practice. The American Veterinary Chiropractic Association (AVCA) provides educational and legislative support to chiropractic care of animals. International Veterinary Acupuncture Society (IVAS) and American Academy of Veterinary Acupuncture (AAVA) provide similar support to veterinary acupuncture. The Chi-Institute is an example of an organization that provides educational programs for the field of Traditional Chinese Veterinary Medicine (TCVM). This list of reputable organizations goes on, which all serve to promote their specialized fields of alternative therapies in veterinary medicine.

References

1. Bjordal JM. Low level laser therapy (LLLT) and World Association for Laser Therapy (WALT) dosage recommendations. *Photomed Laser* Surg. 2012;30(2):61-62.

2. Jelenko C, Anderson AP, Scott TH Jr, Wheeler ML. Organ weights and water composition of the New Zealand albino rabbit (*Oryctalagus cuniculus*). *Am J Vet Res.* 1971;32(10):1637-1639.

3. Williams JE, Gloar CD. Chiropractic care of a six-year-old child with congenital torticollis. *J Chiropr Med.* 2006;5(2):65-68.

4. Wu ML, Xu DS, Bai WZ, et al. Local cutaneous nerve terminal and mast cell responses to manual acupuncture in acupoint LI4 area of the rat. *J Chem Neuroanat.* 2015;68:14-21.

5. Lee BC, Kim KW, Soh KS. Characteristic features of a nerve primo vessel suspended in rabbit brain ventricle and central canal. *J Acup Meridian Stud.* 2010;3(2):75-80.

6. Lee HS, Lee BC. Visualization of the network of primo vessels and primo nodes above the pia mater of the brain and spine of rats by using Alcian blue. *J Acup Meridian Stud.* 2012;5(5):218-225.

7. Man SC. A combination theory on the mechanism of acupuncture analgesia. *Can Fam Physician.* 1979;: 482-484.

8. Singh P, Chaturvedi A. Complementary and alternative medicine in cancer pain management: A systematic review. *Indian J Palliat Care.* 2015;21(1):105-115.

9. Hu C, Zhang H, Wu W, et al. Acupuncture for pain management in cancer: A systematic review and meta-analysis. *Evid Based Complement Alternat Med.* 2016;1720239.

10. Chen LL, Verpoorte R, Yen HR, et al. Effects of processing adjuvants on traditional Chinese herbs. *J Food Drug Anal.* 2018;26(2S):S96-S114.

11. Pruitt R, Lemanski A, Carroll A. Herbal supplements: Research findings and safety. *Nurse Pract.* 2018;43(5):32-37.

12. Wang X, Liu F, Wei K. Myocardial protection and immunoregulation of minor bupleurum decoction and its decomposed preparations on coxsackie B3m viral myocarditis in mice. *Zhongguo Zhong Xi Yi Jie He ZaZhi.* 2000;20(8):599-602.

13. Chen X, Murakami T, Oppenheim JJ, et al. Tripolide, a constituent of immunosuppressive Chinese herbal medicine, is a potent suppressor of dendritic-cell maturation and trafficking. *Blood.* 2005;106(7):2409.

14. Lee OH, Lee BY. Antioxidant and antimicrobial activities of individual and combined phenolics in *Olea europaea* leaf extract. *Bioresour Technol.* 2010;101(10):3751-3754.

15. Liu Y, McKeever LC, Malik NSA. Assessment of the antimicrobial activity of olive leaf extract against foodborne bacterial pathogens. *Front Microbiol.* 2017;8(113):1-8.

16. Vogel P, Machado IK, Garavaglia J, et al. Polyphenols benefits of olive leaf (*Olea europaea L*) to human health. *Nutr Hosp.* 2015;31(3):1427-1433.

17. Surjushe A, Vasani R, Saple DG. Aloe vera: A short review. *Indian J Dermatol.* 2008;53(4):163-166.

18. Cohen SM, Rousseau ME, Robinson EH. Therapeutic use of selected herbs. *Holist Nurs Pract.* 2000;14(3): 59-68.

19. McCluggage D, Higdon PL. *Holistic Care for Birds: Manual of Wellness and Healing.* New York, NY: Howell Book House;1999.

20. Saengnipanthkul S, Waikakul S, Rojanasthien S, et al. Differentiation of patented crystalline glucosamine sulfate from other glucosamine preparations will optimize osteoarthritis treatment. *Int J Rheum Dis.* 2017; doi:10.1111/1756-185X.13068.

21. Diego MA, Jones NA, Field T, et al. Aromatherapy positively affects mood, EEG patterns of alertness and math computations. *Int J Neurosci.* 1998;96(3-4):217-224.

22. Field T, Cullen C, Largie S, et al. Lavendar bath oil reduces stress and crying and enhances sleep in very young infants. *Early Hum Dev.* 2008;84(6):399-401.

23. Hahnemann S. *Organon of Medicine.* 6th ed. W Boericke translation. New Dehli, India: B Jain Publishers;2002.

24. Mathie RT, Hacke D, Clausen J. Randomised controlled trials of veterinary homeopathy: Characterising the peer-reviewed research literature for systematic review. *Homeopathy.* 2012;101(4):196-203.

25. Hektoen L. Review of the current involvement of homeopathy in veterinary practice and research. *Vet Rec.* 2005;157(8):224-229.

Additional Reading

Photobiomodulation

1. Mayer J, Ness RD. Laser therapy for exotic small mammals. In: Riegel RJ, Godbold JC, eds. *Laser Therapy in Veterinary Medicine: Photobiomodulation.* West Sussex, UK: Wiley-Blackwell;2017.

2. Mayer J, Ness RD. Laser therapy for reptiles. In: Riegel RJ, Godbold JC, eds. *Laser Therapy in Veterinary Medicine: Photobiomodulation.* West Sussex, UK: Wiley-Blackwell; 2017.

3. Ness RD, Mayer J. Laser therapy for birds. In: Riegel RJ, Godbold JC, eds. *Laser Therapy in Veterinary Medicine: Photobiomodulation.* West Sussex, UK: Wiley-Blackwell; 2017.

Chiropractic Care

1. Ness RD. Chiropractic care for exotic pets. *Exot DVM.* 2000;2(1):15-18.

2. Ness RD. Chiropractic care of pet rabbits. *Proc Annu Conf Am Vet Chiropr Assoc.* 2016.

Acupuncture

1. Xie H, Eckermann-Ross C. Introduction to traditional Chinese veterinary medicine in pediatric exotic animal practice. *Vet Clin North Am Exotic Animal Pract.* 2012;15(2):311-329.

2. Xie H, Trevisanello L. Application of traditional Chinese veterinary medicine in exotic animals. *Proc Annu Conf Int Trad Chinese Vet Med.* 2011.

Herbal Formulations

1. Wulff ML, Tilford GL. *Herbs for Pets: The Natural Way to Enhance Your Pet's Life.* Irving, CA: Bow Tie Press;2009.

2. Ness RD. Integrative medicine. In: Harrison GJ, Lightfoot T, eds. *Clinical Avian Medicine.* Vol 1. Palm Beach, FL: Spix Publishing;2006.

Nutraceutical Supplements

1. Mejia-Fava J, Colitz CM. Supplements for exotic pets. *Vet Clin North Am Exotic Animal Pract.* 2014;17(3):503-525.

Essential Oil

1. Shelton M. *The Animal Desk Reference II: Essential Oils for Animals.* 2nd ed. Howard Lake, MN: CreateSpace Independent Publishing;2018.

Flower Essence

1. Graham H, Vlamis G. *Bach Flower Remedies for Animals.* Forres, Scotland: Findhorn Press;1999.

Homeopathy

1. Day, C. *The Homeopathic Treatment of Small Animals, Principles and Practice.* Ashingdon, England: CW Daniel Company;1998.

2. MacLeod G. *A Veterinary Materia Medica and Clinical Repertory.* Ashingdon, England: CW Daniel Company;2004.

Mycobacteriosis in Exotic Pet Practice

Minh Huynh, DVM, MRCVS, DipECZM (Avian), Dipl ACZM

Session #2321

Affiliation: From Centre Hospitalier Vétérinaire Frégis, 43 Avenue Aristide Briand, 94110 Arcueil, France.

Abstract: Mycobacterial infections are uncommon in companion pet animals; however, due to public health implications, the exotic practitioner should be aware of this entity, especially regarding the potential emergence of antibiotic resistance associated with empiric treatment. This review will focus on epidemiologic and clinical aspects of mycobacteriosis in small mammals (ferrets, lagomorph, rodents), birds, reptiles, amphibians and fish.

Clinical Signs

Mycobacteriosis is typically polymorphic. Chronicity is usually a landmark of the disease with progressive weight loss. Infection via ingestion disseminates from the intestinal tract and diffuses to the liver and to abdominal lymph nodes and spleen. Inhalation usually produces lesions in the lungs. Skin infection is common especially in reptiles, amphibians and fish. In ferrets, typical clinical signs seem to be ophthalmic presentation with eyelid edema.[1,2] In a case series, 8 of 17 affected ferrets (47%) presented with ophthalmic signs.[1] In birds, bony lesion can be seen, which does not seem to occur in exotic pet mammals.[3,4] One report of neural tropism has been described in birds.[5] Reports of spontaneous mycobacterial infection are scarce in lagomorphs, pet rodents, and reptiles.[6-10] Mycobacterial infection in aquatic organisms such as amphibians and fish is reported to be very common.[11] Bloat, buoyancy problems, and spinal deformities are clinical features that are relatively specific.[11]

Diagnosis

Diagnosis is difficult to achieve since cultures or a polymerase chain reaction (PCR) assay needs to be performed on tissue samples. Comparison of sampling methods and diagnosis have been studied in birds and fish. A study comparing sampling methods and diagnostic methods in naturally infected doves was performed.[12] The birds were part of a flock with confirmed mycobacteriosis and were lethargic and emaciated. Sensitivity was best for culture (100%) compared to Ziehl-Neelsen stain (94,4%) and PCR assay (84,6%).[12] The best sample was the spleen (100%) compared to liver (83,3%), bone marrow (72,2%), or duodenal aspirate (35,5%).[12] One of the main drawbacks of culture is the long delay between sampling and result. Ziehl-Neelsen is fast and cost-effective to perform; however, species identification cannot be achieved. In clinical situations where minimally invasive procedures such as endoscopy cannot be performed, bone marrow aspirate is accessible and PCR assay may achieve a sensitivity of 22.2%. [12] Otherwise, coelomic endoscopy and spleen biopsy combined with Ziehl-Neelsen stain would be the most likely to achieve a correct diagnosis based on this study.[12] In fish, Ziehl-Neelsen staining allowed detection in 14.3% of tissue sample, whereas bacterial culture allowed isolation of mycobacteria in 42.9% of tissue samples in one study examining healthy fish in aquaria. Regarding isolation in the water, 24.6% and 75.4% of the samples were positive with Ziehl-Neelsen and culture respectively.

In practice, evidence of "negative images" in cytologic examination stained with Diff quick or Wright-Giemsa stain of fine needle aspirate from lesions allows a rapid diagnosis. Hematology and biochemistry findings are unspecific and reflect the chronicity of an inflammatory process. Very severe leucocytosis with monocytosis may orient the clinician toward a diagnosis of mycobacterial infection.[13,14]

Risk Assessment

In cases of suspected mycobacteriosis, the clinician may have to evaluate the zoonotic risk. In the first place, variation in species susceptibility has to be considered (Table 1). Among companion mammals, ferrets seem to be specifically susceptible compared to rabbits and rodents. Surprisingly, although guinea pigs are considered very susceptible in experimental conditions, there has been no report of spontaneous mycobacteriosis in pet guinea pigs.[15] Birds from all species are susceptible, whereas reptile infection is a rare occurrence. Fish infection is very common with infection ranging from 46.8-79.4% of all sampled fish.[16,17]

Table 1. Documented cases of mycobacteriosis in exotic pet species.

Animal	Mycobacterium
Ferret	*M. celatum,*[18,19] *M. genavense,*[2] *M. avium,*[20-22] *M. xenopi,*[23,24] *M. abscessus,*[25] *M. chelonae,*[1] *M. gordonae, M. bovis*
Rabbit	*M. avium,*[6] *M. genavense*[7]
Rodent	
Chinchilla	*M. avium,*[8] *M.genavense*[9]
Squirrel	*M. avium*[26]
Hamster	*M. chelonae*[27]
Birds	*M. tuberculosis, M. avium, M. genavense*
Reptiles	*M. marinum,*[28-30] *M. kansasii*[11] *M. chelonae,*[11] *M. haemophilum,*[11] *M. kumamotonerse*[11]
Amphibian	*M. marinum, M. chelonae, M. fortuitum, M. xenopi, M. abscessus, M. avium, M. szulgai*
Fish	*M. marinum, M. fortuitum, M. chelonae, M. peregrinum, M. gordonae*

Transmission from animal to human is rarely documented (Table 2). *M. tuberculosis* infection in birds is a notable exception since all cases described are reverse zoonosis (eg, the humans were suspected or determined to be the source of contamination). Two case reports of *M. chelonae* and *M. bovis* infection in humans was related to a ferret bite. The most concerning threat to owners remains fish handling with numerous reports from *M. marinum*. Notably, aquarium and fish have been suspected as a primary source of mycobacteriosis in other species contamination such as ferret,[24] bearded dragon,[30] spiny tail lizard,[29] and a blue-fronted Amazon parrot.[31]

Table 2. Documented cases of zoonosis in exotic pet species.

Animal	Zoonosis
Ferret	*M. celatum* (possible),[32] *M. chelonae,*[33] *M. bovis*[34]
Birds	*M. tuberculosis* (likely reverse zoonosis)[35-38]
Fish	*M. marinum* (32 case reports + 16 case series)[39]

Conclusions

Mycobacteriosis remains a diagnostic challenge both from the animal and the owner perspective. Although prompt diagnosis may not always be feasible, clinicians and owners should be encouraged to investigate as much as possible the disease to rule out a potential mycobacteriosis. Furthermore, empirical use of antibiotics may raise concern about useless treatment and emergence of resistant mycobacterial strains.

References

1. Mentré V, Bulliot C. A Retrospective study of 17 cases of mycobacteriosis in domestic ferrets (*Mustela putorius furo*) between 2005 and 2013. *J Exot Pet Med.* 2015;24:340-349.

2. Lucas J, Lucas A, Furber H, et al. Mycobacterium genavense infection in two aged ferrets with conjunctival lesions. *Aust Vet J.* 2000;78:685-689.

3. Sadar MJ, McRuer D, Hawkins MG, et al. Multifocal respiratory and vertebral mycobacteriosis in a red-tailed hawk (*Buteo jamaicensis*). *J Zoo Wildl Med.* 2015;46:150-154.

4. Palmieri C, Roy P, Dhillon AS, et al. Avian mycobacteriosis in psittacines: A retrospective study of 123 cases. *J Comp Pathol.* 2013;148:126-138.

5. Gomez G, Saggese MD, Weeks BR, et al. Granulomatous encephalomyelitis and intestinal ganglionitis in a spectacled Amazon parrot (*Amazona albifrons*) infected with *Mycobacterium genavense*. *J Comp Pathol.* 2011;144:219-222.

6. Klotz D, Barth SA, Baumgartner W, et al. *Mycobacterium avium* subsp. *hominissuis* infection in a domestic rabbit, Germany. *Emerg Infect Dis.* 2018;24:596-598.

7. Ludwig E, Reischl U, Janik D, et al. Granulomatous pneumonia caused by *Mycobacterium genavense* in a dwarf rabbit (*Oryctolagus cuniculus*). *Vet Pathol* 2009;46:1000-1002.

8. Barthel Y, Drews S, Fehr M, et al. Concurrent infection with *Mycobacterium avium* subsp. *hominissuis* and *Giardia duodenalis* in a chinchilla (*Chinchilla lanigera* f. dom.). *Berl Munch Tierarztl Wochenschr.* 2016;129:242-246.

9. Huynh M, Pingret JL, Nicolier A. Disseminated *Mycobacterium genavense* infection in a chinchilla (*Chinchilla lanigera*). *J Comp Pathol.* 2014;151:122-125.

10. Mitchell MA. Mycobacterial infections in reptiles. *Vet Clin North Am Exot Anim Pract.* 2012;15:101-111.

11. Reavill DR, Schmidt RE. Mycobacterial lesions in fish, amphibians, reptiles, rodents, lagomorphs, and ferrets with reference to animal models. *Vet Clin North Am Exot Anim Pract.* 2012;15:25-40.

12. Saggese MD, Tizard I, Phalen DN. Comparison of sampling methods, culture, acid-fast stain, and polymerase chain reaction assay for the diagnosis of mycobacteriosis in ring-neck doves (*Streptopelia risoria*). *J Avian Med Surg.* 2010;24:263-271.

13. Dahlhausen B, Tovar DS, Saggese MD. Diagnosis of mycobacterial infections in the exotic pet patient with emphasis on birds. *Vet Clin North Am Exot Anim Pract.* 2012;15:71-83.

14. Sutherland M, Courtman N, Sacks P, et al. Severe leukemoid response associated with *Mycobacterium genavense* infection in a pet budgerigar (*Melopsittacus undulatus*). *J Exot Pet Med.* 2018;27:11-16.

15. LoBue PA, Enarson DA, Thoen CO. Tuberculosis in humans and animals: an overview. *Int J Tuberc Lung Dis.* 2010;14:1075-1078.

16. Zanoni RG, Florio D, Fioravanti ML, et al. Occurrence of *Mycobacterium* spp. in ornamental fish in Italy. *J Fish Dis.* 2008;31:433-441.

17. Kusar D, Zajc U, Jencic V, et al. Mycobacteria in aquarium fish: Results of a 3-year survey indicate caution required in handling pet-shop fish. *J Fish Dis.* 2017;40:773-784.

18. Valheim M, Djonne B, Heiene R, et al. Disseminated *Mycobacterium celatum* (type 3) infection in a domestic ferret (*Mustela putorius furo*). *Vet Pathol.* 2001;38:460-463.

19. Piseddu E, Trotta M, Tortoli E, et al. Detection and molecular characterization of *Mycobacterium celatum* as a cause of splenitis in a domestic ferret (*Mustela putorius furo*). *J Comp Pathol.* 2011;144:214-218.

20. Schultheiss PC, Dolginow SZ. Granulomatous enteritis caused by *Mycobacterium avium* in a ferret. *J Am Vet Med Assoc.* 1994;204:1217-1218.

21. Saunders GK, Thomsen BV. Lymphoma and *Mycobacterium avium* infection in a ferret (*Mustela putorius furo*). *J Vet Diagn Invest.* 2006;18:513-515.

22. Bezos J, Alvarez-Carrion B, Rodriguez-Bertos A, et al. Evidence of disseminated infection by *Mycobacterium avium* subspecies *hominissuis* in a pet ferret (*Mustela putorius furo*). *Res Vet Sci.* 2016;109:52-55.

23. Gupta A, McBride AM, Holder KA, et al. Pathology in practice. Severe pyogranulomatous pneumonia, enteritis, and lymphadenitis with numerous acid-fast bacteria (*M. xenopi*). *J Am Vet Med Assoc.* 2012; 240:1427-1429.

24. Davendralingam N, Davagnanam I, Stidworthy MF, et al. Transmission of *Mycobacterium xenopi* to a pet albino ferret (*Mustela putorius furo*) from a domestic aquarium. *Vet Rec.* 2017.

25. Lunn JA, Martin P, Zaki S, et al. Pneumonia due to *Mycobacterium abscessus* in two domestic ferrets (*Mustelo putorius furo*). *Aust Vet J.* 2005;83:542-546.

26. Moreno B, Aduriz G, Garrido JM, et al. Disseminated *Mycobacterium avium* subsp. *avium* infection in a pet Korean squirrel (*Sciuris vulgaris coreae*). *Vet Pathol.* 2007;44:123-125.

27. Karbe E. Disseminated mycobacteriosis in the golden hamster. *J Vet Med.* 1987;34:391-394.

28. Bouricha M, Castan B, Duchene-Parisi E, et al. *Mycobacterium marinum* infection following contact with reptiles: Vivarium granuloma. *Int J Infect Dis.* 2014;21:17-18.

29. Morales P, Dunker F. Fish tuberculosis, *Mycobacterium marinum*, in a group of Egyptian spiny-tailed lizards, *Uromastyx aegyptius*. *J Herpetol Med Surg.* 2001;11:27-30.

30. Girling SJ, Fraser MA. Systemic mycobacteriosis in an inland bearded dragon (*Pogona vitticeps*). *Vet Rec.* 2007;160:526-528.

31. Hannon DE, Bemis DA, Garner MM. Mycobacterium marinum infection in a blue-fronted Amazon parrot (*Amazona aestiva*). *J Avian Med Surg.* 2012;26:239-247.

32. Ludwig E, Reischl U, Holzmann T, et al. Risk for *Mycobacterium celatum* infection from ferret. *Emerg Infect Dis.* 2011;17:553-555.

33. Iyengar KP, Nadkarni JB, Gupta R, et al. *Mycobacterium chelonae* hand infection following ferret bite. *Infection.* 2013;41:237-241.

34. Jones JW, Pether JV, Rainey HA, et al. Recurrent *Mycobacterium bovis* infection following a ferret bite. *J Infect.* 1993;26:225-226.

35. Peters M, Prodinger WM, Gummer H, et al. *Mycobacterium tuberculosis* infection in a blue-fronted amazon parrot (*Amazona aestiva aestiva*). *Vet Microbiol.* 2007;122:381-383.

36. Steinmetz HW, Rutz C, Hoop RK, et al. Possible human-avian transmission of *Mycobacterium tuberculosis* in a green-winged macaw (*Ara chloroptera*). *Avian Dis.* 2006;50:641-645.

37. Washko RM, Hoefer H, Kiehn TE, et al. *Mycobacterium tuberculosis* infection in a green-winged macaw (*Ara chloroptera*): Report with public health implications. *J Clin Microbiol.* 1998;36:1101-1102.

38. Schmidt V, Schneider S, Schlomer J, et al. Transmission of tuberculosis between men and pet birds: a case report. *Avian Pathol.* 2008;37:589-592.

39. Weir M, RajiĆ A, Dutil L, et al. Zoonotic bacteria, antimicrobial use and antimicrobial resistance in ornamental fish: A systematic review of the existing research and survey of aquaculture-allied professionals. *Epidemiol Infect.* 2011;140:192-206.

Basic Wildlife Rehabilitation Triage

Renée Schott, DVM, CWR

Session #2347

Affiliation: From The Wildlife Rehabilitation Center of Minnesota, 2530 Dale St N, Roseville, MN 55113, USA.

Abstract: It can be difficult for a practitioner to know which wild animals will survive the rehabilitation process to release and which will not. This Masterclass will use data from The Wildlife Rehabilitation Center of Minnesota, one of the largest wildlife rehabilitation centers in the country, to help predict which patients will make it to release and which won't. The goal of this class is to help those in private practice know when to euthanize on admission and when to treat and transfer to a wildlife rehabilitator.

Introduction

Private practitioners are often regarded as experts across all taxa, regardless of the focus of their practice. Clients and the general public will occasionally call with questions about injured or orphaned wildlife, or may simply show up in the lobby with the animal. It behooves the clinic to be prepared to deal with these scenarios so the finders have resources and options. It is important that clinics are prepared to handle and refer these calls appropriately for the sake of efficiency, the welfare of the wild animal, and the safety of the patients in the hospital. In addition, it is important for private practitioners to be armed with basic triage and euthanasia skills to humanely end the suffering of wildlife that cannot recover.

The Wildlife Rehabilitation Center of Minnesota (WRCMN) admits ~13,000 sick and injured wild animals each year. Asterisks below are used to indicate WRCMN data. It is difficult to compare data from one rehabilitation center to another due to differences in location, resources, capacity, species, diseases, among other things. For more details about how the data were compiled, please contact renee@wrcmn.org.

Preparation

Before wildlife is in the lobby or calls start coming in, clinics should prepare by following the steps below.

Find a wildlife rehabilitator or facility to work with

It is illegal to possess native wildlife without the proper permits. Veterinarians must be working with a permitted rehabilitator in order to possess wildlife they are treating. In addition, most states require rehabilitators to work with a licensed veterinarian to keep their permits active. Rehabilitators are familiar with natural history and veterinarians know veterinary medicine; working together ensures the best outcome for the animal. Additionally, rehabilitators can determine when an animal needs to be rescued and when it should be left alone. Calls are best referred to rehabilitators or facilities set up to field these questions, as questions are complicated, and answers aren't always straight forward.

Compile resources

Where can you refer phone calls or injured patients? Do you know the addresses or hours of facilities in the area? Who is a contact at your state natural resources or fish and game department and at the US Fish and Wildlife Service? Making a resources sheet and SOPs for your reception will limit miscommunications.

Know the laws

Typically, each state regulates who can rehabilitate wildlife and how. Federal regulations exist for migratory birds as well. While it is usually legal for veterinarians to treat and euthanize wildlife, they must transfer all migratory birds to a licensed wildlife rehabilitator within 24 hours of stabilization,[1] and they cannot possess other species for more than 24-48 hours unless they are already working with a permitted rehabilitator. There are often additional guidelines and regulations surrounding threated or endangered species. Keep in mind, species that have a hunting season in your area must be treated as food animals; withdrawal times and the list of prohibited medications apply.[2] More details about legal responsibilities and restrictions can be found here in Duerr and Whittington, 2017.[3]

Euthanasia

For a wild animal to allow capture, it is usually very debilitated. Release proportions vary according to species, age and injuries, but range from 24%-85%.[4-7] Euthanasia is common, necessary and humane. In most areas, the number of animals that need rehabilitation are too numerous for the available resources. When considering the below cases, it is important to consider these resources, the anticipated number/species/age of animals needing rehabilitation, rehabilitator/facility limits, and welfare. The ability of the rehabilitator, facility, or clinic to accept animals for rehabilitation will be dependent on the above factors. If resources (time, space, money, people) are low, the overall ability of the caretakers will diminish which will lower welfare for everyone. Euthanizing the least likely to survive and basing the number of animals in care on resources available, will allow those with the best prognosis to receive the best care and ultimately the best chance at release. Because stretching resources results in a lower standard of care for all, euthanizing more animals on admission in a strategic manner often results in an increase in the overall number of animals being released. The data provided here are from WRCMN.

What types of orthopedic conditions should be euthanized?

Prognoses and treatment options are taxa specific.[8] Consider your ability, the ability of the wildlife rehabilitator, and the species. Below are generalizations that apply to most taxa unless specified otherwise.

Necrotic or old, compound fractures

Compound fractures are inherently more prone to healing complications owing to the increased likelihood of contamination and/or infection. Necrotic bone must be removed and in general, most species need to have almost perfect fracture alignment with < 10% loss of long bone length. Necrotic bone is almost always accompanied by necrotic soft tissue. Many wild animals (especially birds) have very little soft tissue over certain bones and the loss of that soft tissue means the bones underneath will not heal or will have complications.[9] Older fractures are also inherently prone to more complications; if the bone has stopped healing, then all bone ends must be freshened to be stimulated to heal again. In addition, muscles, tendons and ligaments often have contracted, making reduction difficult. Not only do the bones need to heal perfectly, but all joints need near perfect range of motion.

The definition of old fracture varies with species, age and type of fracture. For instance, a 2-day-old compound, comminuted fracture in a nestling songbird may be considered old, whereas a 7-day-old closed, simple, midshaft fracture in a fox may be fresh enough to fix.

Traumatic joint injuries

In the author's experience, wildlife that experience traumatic joint injuries, begin trying to reduce the abnormal movement in the joint by producing fibrous tissue around the joint. This reduces range of motion chronically which leaves the animal non-releasable. In the author's experience, luxations and sub-luxations (excluding coracoid luxations off the keel, which usually heal well) will almost always heal with reduced range of motion and should be euthanized. In addition, traumatic joint luxations frequently will cause osteoarthritis.[9,10]

Reports of successful surgical management of elbow luxations in raptors[11] and a woodchuck[12] have been reported, however, in both papers the possibility of osteoarthritis was not discussed. Unsuccessful treatment of luxations in wildlife are also reported.[11,13] Articular fractures need perfect alignment and rigid fixation to prevent secondary degenerative joint disease.[14] Wildlife develop reduced range of motion and signs of osteoarthritis within weeks of sustaining an articular fracture, no matter how rigid the fixation and perfect the anatomic alignment. WRCMN euthanizes all traumatic joint injuries that result in intrarticular fractures or luxations/subluxations of joints (excluding coracoid luxations off the keel) because of past experiences. The data provided here are from WRCMN.

Shoulder fractures in aerial insectivores

In North America, swallow species (including but not limited to barn swallows *Hirundo rustica* and cliff swallows *Petrochelidon pyrrhonota*), swift species (including but not limited to chimney swift *Chaetura pelagica* and Vaux's swift *Chaetura vauxi*), nightjar species (including but not limited to common nighthawk *Chordeiles minor* and Eastern whip-poor-will *Caprimulgus vociferus*) have very high metabolic demands-Vaux's swifts may need to eat > 5000 arthropods each day![15] Flight must be absolutely perfect in order to accomplish this. A bird with even the slightest flight deficit may die of starvation after release. Most of these species migrate to Central or South America for the winter which also requires sustained flight. To adequately assess pre-release flight ability, these species need a 30-foot (swallows/swifts) and 50-foot (nightjars) flight cage. Any deficit should be heavily scrutinized. In the author's and other's experiences (Veronica Bowers, Native Avian Care, personal electronic communication, 2017) < 1% of shoulder fractures in these species will recover to a releasable state. Several species do not eat in captivity and must either be trained to accept handfeeding, gavage or force fed. All are high stress species. Feather condition must be pristine. Husbandry for these species is complicated and best not undertaken by a novice.

All aerial insectivores with any shoulder girdle (coracoid, scapula, clavicle) fractures are now euthanized at WRCMN due to past experiences, other's experiences and the intense husbandry care these species require. The data provided here are from WRCMN.

What type of traumatic eye injuries should be euthanized?

In general, any condition that will result in permanent vision loss should be considered for euthanasia, even if only in one eye. Prey animals will be left unable to watch for predators, predators may have a difficult time catching enough prey (no live prey testing in captivity is equitable to catching prey in the wild). This is a controversial topic and some wildlife rehabilitation veterinarians and rehabilitators have differing opinions. This author does not release one-eyed animals, with an exception being some aquatic turtle cases. Traumatic eye injuries that result in vision loss include luxated lens (posteriorly and anteriorly luxated), ruptured globes, complete retinal detachment with incomplete reattachment, and any other condition that results in permanent vision loss.

What types of other issues should be euthanized?

Chronic pain

Wild animals have evolved to hide signs of pain and to self-repair injuries as much as possible. If they did the opposite, they would likely become prey. Therefore, any condition which has the potential to result in chronic pain in a human, should be assumed to have the potential to result in a chronic pain for the animal, should be grounds for euthanasia of the wild animal for humane reasons. Wild animals cannot receive chronic pain medications or other palliative care in the wild.

Spinal trauma with no deep pain

Most rehabilitators and veterinarians recognize spinal trauma via pelvic limb paresis or paralysis. Because of the anatomy of the spinal cord, loss of deep pain perception indicates a grave prognosis for return to function. Additionally, neuropathic pain has been reported in paralyzed companion animals.[16] Therefore, animals with chronic pain should be humanely euthanized. To test for deep pain perception, use a toe on the limb that is paralyzed/paretic. Pinch the toe; start with a superficial pinch and then slowly pinch harder. Watch for head movement, vocalization or other indication that the pain sensation made it up the spinal cord and to the brain. Ignore all movement in the limb that is being pinched as the withdrawal reflex may be intact or hyperreflexive, depending on where the lesion is. One cannot discern between the withdrawal reflex and voluntary movement when pinching a toe, and wild animals can have very strong reflexes, so it is imperative to ignore movement in the leg and watch the head of the animal. In addition to spinal trauma, brachial plexus avulsion occasionally occurs. If deep pain is not present in one forelimb of a unilaterally forelimb paralyzed animal, brachial plexus avulsion should be considered. Brachial plexus avulsion will not heal to a releasable state and these animals should be euthanized. Animals with no deep pain in alimb should be euthanized as they will not recover to a releasable state.

Seizing rabies vector species

Rabies vector species (eg, raccoon, coyote, fox, bat, skunk) with abnormal neurologic signs should be considered for euthanasia due to public health concerns. It is important to know your rabies vector species and rabies virus variants in your area and what your department of health requires you to do; they may want the animal tested for rabies or to have the contact of the finder to talk about exposure. Generally, a seizing animal bears a poor prognosis, regardless of species.

Distemper virus infection suspects

Distemper suspect species (eg, raccoon, coyote, fox) with abnormal neurologic or severe respiratory clinical signs, with or without leukopenia should be considered for euthanasia. Consider the infectious nature of this pathogen and the domestic animals in your clinic which are susceptible. Additionally, wild animals have a high probability of permanent neurologic issues even if they survive, which is rare.

Only 6.7% (25) of the 373 adult raccoons admitted to WRC from 2011-2017 were able to be released. Treatment was attempted with 68 and 25 (36.8%) of treated were released. While not all of these cases were distemper suspects, adult raccoons are difficult for the public to capture and transport; consider how debilitated this species must be to present for rehabilitation and this bears out in prognosis.

Infant babies

The younger the animal is, the lower its chance of survival when being raised in captivity by humans. Reviewing data from previous years can help facilities find which variables are best correlated with survival to release.

WRCMN data:

- Eastern cottontail rabbits (*Sylvilagus floridanus*): WRCMN received 3133 young cottontails in 2017 alone. Historically, under 65 grams, healthy Eastern cottontail rabbits have < 30% chance of release. Under 75 grams, unhealthy Eastern cottontail rabbits have a < 30% chance of release. Definition of healthy: mild-moderate dehydration, thin (not emaciated), BAR, no injuries
- Eastern grey squirrels (*Sciurus carolinensis*): WRCMN received 1285 young grey squirrels in 2017 alone. Historically, healthy grey squirrels under 40 grams have < 30% chance of release.
- Virginia opossums (*Didelphis virginiana*): WRCMN received 119 young Virginia opossums in 2017. Opossums 20-40g have a 50% chance of survival.
- Hatchling altricial birds (just out of shell) are euthanized as they have 0% chance of release.

Adult Eastern cottontails and Eastern grey squirrels with abnormal brain neurologic clinical signs

Adult Eastern cottontails and Eastern grey squirrels commonly present to WRCMN with abnormal brain neurologic clinical signs (eg, head tilt, circling, head turn, ataxia). Necropsy results most commonly conclude the etiology to be *Baylisascaris procyonis* (or a nematode tract through the brain) or head trauma.

WRCMN received 233 adult cottontails with abnormal central nervous system signs in 2017. From 2011-2017, 1727 total cottontail adults were admitted. Of these 641 had abnormal brain neurologic clinical signs. Of the 641, 3.3% were eventually released. Of the ones that were treated, 8% were released. This is compared to all other adult cottontails which had an overall release proportion of 5.5% and a treated released proportion of 19%. This author recommends euthanasia of all adult cottontails with abnormal brain neurologic clinical signs due to the poor release proportion.

WRCMN received 94 adult grey squirrels with abnormal central nervous system signs in 2017. From 2011-10.22.2017, 1295 total grey squirrel adults were admitted. Of these 414 had abnormal brain neurologic clinical signs (eg, head tilt, circling, head turn, ataxia). Of the 414, 8.7% were eventually released. Of the animals that were treated, 16.1% were released. This is compared to all other adult grey squirrels which had an overall release proportion of 10% and a treated released proportion of 29.1%. This author recommends heavily scrutinizing any adult Eastern grey squirrel with abnormal brain neurologic signs before choosing to treat.

Conditions to Not Euthanize (if you have the resources)

Mycoplasma gallisepticum conjunctivitis in house finches

Mycoplasma gallisepticum (MG) originally spread from the production poultry industry to house finches (*Haemorhous mexicanus*) in the 1990s and is now endemic in the house finch population.[17,18] When it first appeared, the poultry industry and biologist were afraid of MG recovered house finches becoming latent carriers and some antibiotics being ineffective;[19,20] however, it is hard for wild house finches to actually transmit *M. gallisepticum* to domestic poultry[21,22] and there is a treatment, that when given appropriately can eliminate the carrier state.[23] Euthanizing one or even several hundred house finches because they have this disease will not affect the population or disease status. WRCMN treats and releases house finches with presumed MG conjunctivitis according to Mashima et al 1997.

Severe mange in red fox

Juvenile red fox (*Vulpes vulpes)* occasionally are admitted in the fall with severe mange, presumably after having been forced out of maternal territory. While some of these fox are too sick or emaciated with secondary conditions to survive, many can be treated and released.[24]

Animal missing digits

Certain species can survive well in the wild with missing digits; however, the location and number of missing digits that allow successful release are species dependent.[8]

How to Euthanize Various Wildlife Species

The techniques described in Table 1 are used most frequently at WRCMN. They take into consideration the wild animal species, age, and physiology of the animal being euthanized as well as the safety of the handler; priority is given to decreasing apparent animal stress. There are other technique options. In the author's experience, the methods below have proven reliable and cause the least amount of stress to the animal.

Table 1. Euthanasia techniques.

Species or group	Age	First choice	Second choice	Notes
Eastern cottontail rabbit	Adult (250 g-1.5 kg)	0.2 ml (100 mg/ml) ketamine; 0. 4ml (5 mg/ml) midazolam; 0.2 ml (10 mg/ml) butorphanol, mix, give IM, wait 15 min, if still alive, put into isoflurane box[a] until deceased or mask at 5% isoflurane until anesthetized then IC sodium pentobarbital (390 mg/ml) 0.5 ml	2 ml (100 mg/ml) alfaxalone, give IM, wait 15 min If still alive, put into isoflurane box until deceased or mask at 5% isoflurane until anesthetized then IC sodium pentobarbital (390 mg/ml) 0.5 ml	Prolonged struggling and breath holding when placed in isoflurane box without pre-medications and even more prolonged in the isoflurane face mask/induction box
Other mammal species	Neonate, just weaned	Isoflurane box until deceased; they dislike the strong smell and will wipe their faces but are anesthetized within 10 seconds		Young mammals' brains are very tolerant to hypoxia and they hold their breath and struggle in an isoflurane mask at 5% for several minutes
Breath-holding avian species (ie, diving species, waterfowl, waterbirds)	Adults	3 mg/kg midazolam and 5 mg/kg butorphanol IM, wait 15 min then isoflurane box until deceased	Debilitated individuals are often too weak to hold their breath and do well in the isoflurane box	
Avian species	Any (except breath-holding species)	Isoflurane box until deceased		Birds have such efficient gas exchange in their lungs, they become anesthetized in ~ 2 breaths
Raccoons/similar sized canids	Adults	0.3 ml (0.5 mg/ml) dexdormitor; 0.3ml (10 mg/ml) butorphanol, mix, give IM; if seizing, start with 1 mg/kg midazolam, wait 15 min, if still alive, put into isoflurane box until deceased or mask at 5% isoflurane until anesthetized then IC sodium pentobarbital (390 mg/ml) 3 ml	Debilitated individuals are often too weak to hold their breath and do well in the isoflurane box	
Aquatic turtles	Any	15-25 mg/kg alfaxalone IM, wait 15 min, 10 mg/kg propofol IV wait 15 min, same volume (as propofol) pentobarbital IV wait 15 min, pith	Other anesthesia medications can be used but all must induce anesthesia and end in pithing	Chelonian brains are so tolerant to hypoxia (Larson et al. 2014, Jackson and Ultsch 2010, Ultsch 2006) that reanimation can happen if using anesthetics or pentobarbital alone

[a]See isoflurane box discussion in Euthanasia Techniques section.

Euthanasia Techniques

Isoflurane (or sevoflurane) via face mask

Restrain the patient and place head in mask through diaphragm or hook up vaporizer to an induction box with the animal inside. Start at maximal percentage until patient is fully anesthetized. Then administer appropriate amount of pentobarbital IV or IC.

Isoflurane box

- Liquid isoflurane is poured into a clear Rubbermaid-type container with a lid. Either do this in a vent hood or ensure container is on the ground/floor and you remain above to prevent accidental inhalation of isoflurane, which is heavier than room air. Ideally, a respirator would be worn to avoid inhalation of all fumes when not in the hood.
- If the animal is fractious, aggressive or high stress, ideally the container would be big enough for a live trap or kennel. Having several sizes of containers would be ideal for different sized animals.
- Use 1 tbsp liquid isoflurane for every 4 gallons container volume
- The box should be mostly sealed (but not completely sealed as the gas expands)
- After isoflurane is poured into the container, the animal (or animal in a live trap) can be transferred to the container and the lid replaced.
- Monitor the patient to ensure it quickly lose consciousness
- Leave patient in the tub for at least 10 minutes to ensure over dose of isoflurane has caused death Overdose of isoflurane is an approved humane euthanasia method according to the AVMA.[28]
- Rock tub before opening to ensure no movement
- Open tub outside (tub on ground, ensure you are above tub or are wearing a respirator) or in a vent hood
- Use a stethoscope to ensure heartbeat of the patient has stopped.
- Leave tub outside or in vent hood for at least 10 minutes.

Basic Wildlife Stabilization

During all of these steps, decreasing stress is the most important thing we can do for wildlife. Stress will kill wildlife and often a hands-off protocol is preferable over an invasive one.

1. Visual triage. An obvious injury that would render the animal non-releasable? Euthanize immediately. Immediately treat any breathing difficulties (place in oxygen chamber) or active hemorrhage (apply bandage to stop bleeding).
2. Treat thermoregulatory deficits. If animal is hypothermic, warm it up! If animal is hyperthermic, cool it down.

3. Leave it alone! Leave the animal in a quiet, dark room so it can de-stress and relax. This also gives the animal time to adjust internal body temperature.

4. Give fluid therapy. Assume at least 5% dehydration in all animals admitted. Sub-cutaneous fluids are fast to administer and a great choice in almost all cases.

5. Give pain medications. If there is any injury, appropriate pain management is paramount.

References

1. Code of Federal Regulations. 2018. *Electronic Code of Federal Regulations*. Available at: https://www.ecfr.gov/cgi-bin/text-idx?SID=0e9fd773e2a078e4c54d159ce4a9be80&node=50:9.0.1.1.4.2.1.2&rgn=div8. Accessed July 1, 2018.

2. Schott R. Extra-label drug use in wildlife rehabilitation medicine. *Wildl Rehabil Bull.* 2017;35:2.

3. Duerr R, Whittington JK. Legal responsibilities and restrictions on veterinarians working with wildlife. *Wildl Rehabil Bull.* 2017;35(1):25-37.

4. Kelly A, Bland M. Admissions, diagnoses, and outcomes for Eurasian sparrowhawks (*Accipiter nisus*) brought to a wildlife rehabilitation center in England. *J Raptor Res.* 2016;40(3):231-235.

5. Molina-López RA, Casal J, Darwich L. Final disposition and quality auditing of the rehabilitation process in wild raptors admitted to a wildlife rehabilitation centre in Catalonia, Spain, during a twelve year period (1995-2007). *PLoS One.* 8 2013;(4):1-8.

6. Barrows Z, Fix S. Raptors rehabilitated study in Iowa. *Wildl Res.* 1990;26(1):18-21.

7. Orós J, Montesdeoca N, Camacho M, et al. Causes of stranding and mortality, and final disposition of loggerhead sea turtles (*Caretta caretta*) admitted to a wildlife rehabilitation center in Gran Canaria Island, Spain (1998-2014): A long-term retrospective study. *PLoS One* 2016;11(2):1-14.

8. Duerr R, Schott R, Freshman J, et al. *Topics in Wildlife Medicine: Orthopedics*. Duerr R, Purdin G, eds. St Cloud, MN: National Wildlife Rehabilitators Association;2017.

9. Johnson AL. Fundamentals of orthopedic surgery and fracture management. In: Fossum TW, Hedlund CS, Johnson AL, et al, eds. *Small Animal Surgery.* St. Louis, MO: Elsevier Mosby;2007:930-982.

10. Denny HR, Butterworth S, eds. *A Guide to Canine and Feline Orthopaedic Surgery.* 4th ed. Ames, IA: Wiley-Blackwell;2000.

11. Ackermann J, Redig PT. Surgical repair of elbow luxation in raptors. *J Avian Med Surg.* 1997;11(4): 247-254.

12. Chambers JN, Hanley CS, Hernandez-Divers S. Hip luxation in a woodchuck (*Marmota monax*): Successful treatment by closed reduction and a modified Ehmer sling. *J Zoo Wildl Med.* 2004;35(4):569-571. doi:10.1638/04-012.

13. Bennett KR, Desmarchelier MR, Bailey TR. Surgical correction of bilateral patellar luxation in an american black bear cub (*Ursus americanus*). *J Zoo Wildl Med.* 2015;46(2):359-362. doi:10.1638/2013-0288R1.1.

14. Newton CD. Etiology, classification, and diagnosis of fractures. Available at: http://cal.vet.upenn.edu/projects/saortho/chapter_11/11mast.htm#b. Accessed November 4, 2017.

15. Bull EL, Beckwith RC. Diet and foraging behavior of Vaux's swifts in Northeastern Oregon. *Condor* 1993;95(4):1016-1023. doi:10.2307/1369437.

16. Granger N, Carwardine D. Acute spinal cord injury. *Vet Clin North Am Small Anim Pract.* 2014;44 (6):1131-1156. doi:10.1016/j.cvsm.2014.07.013.

17. Citti C, Blanchard A. Mycoplasmas and their host: Emerging and re-emerging minimal pathogens. *Trends Microbiol.* 2013;21(4):196-203. doi:10.1016/j.tim.2013.01.003.

18. Dhondt AA, DeCoste JC, Ley DH, et al. Diverse wild bird host range of *Mycoplasma gallisepticum* in Eastern North America. *PLoS One* 2014;9(7):doi:10.1371/journal.pone.0103553.

19. Williams ES, Yuill T, Artois M, et al. Emerging infectious diseases in wildlife. *Rev Sci Tech.* 2002;21(1):139-157. doi:10.20506/rst.21.1.1327.

20. Wellehan JF, Zens MS, Calsamiglia M, et al. Diagnosis and treatment of conjunctivitis in house finches associated with mycoplasmosis in Minnesota. *J Wildl Dis.* 2001;37(2):245-251. doi:10.7589/0090-3558-37.2.245.

21. Stallknecht DE, Luttrell MP, Fischer JR, Kleven SH. Potential for transmission of the finch strain of *Mycoplasma gallisepticum* between house finches and chickens. *Avian Dis.* 1988;45:352-358.

22. Luttrell MP, Stallknecht DE, Kleven SH, et al. *Mycoplasma gallisepticum* in house finches (*Carpodacus mexicanus*) and other wild birds associated with poultry production facilities. *Avian Dis.* 2001;45(2):321. doi:10.2307/1592971.

23. Mashima TY, Ley DH, Stoskopf MK et al. Evaluation of treatment of conjunctivitis associated with *Mycoplasma gallisepticum* in house finches (*Carpodacus mexicanus*). *J Avian Med Surg.* 1997;11:20-24.

24. Newman TJ, Baker PJ, Harris S. Nutritional condition and survival of red foxes with sarcoptic mange. *Can J Zool.* 2002;80(1):154-161. doi:10.1139/z01-216.

25. Jackson DC, Ultsch GR. Physiology of hibernation under the ice by turtles and frogs. *J Exp Zool.* 2010;313(6):311-327. doi:10.1002/jez.603.

26. Larson J, Drew KL, Folkow LP, et al. No oxygen? No problem! Intrinsic brain tolerance to hypoxia in vertebrates. *J Exp Biol.* 2014;217:1024-1039. doi:10.1242/jeb.085381.

27. Ultsch GR. 2006. The ecology of overwintering among turtles: Where turtles overwinter and its consequences. *Biol Rev Camb Philos Soc.* 2006;81:339-367. doi:10.1017/S1464793106007032.

28. Leary S, Underwood W, Lilly E, et al. 2013. *AVMA Guidelines for the Euthanasia of Animals: 2013 Edition*. Available at: https://www.avma.org/KB/Policies/Documents/euthanasia.pdf. Accessed July 1, 2018.

Case Files of the Walking Wounded: Trauma Management in Wildlife

Heather Wilson Barron, DVM, DABVP (Avian Practice)

Session #2299

Affiliation: From the Clinic for the Rehabilitation of Wildlife, 3883 Sanibel-Captiva Rd, Sanibel, FL 33957, USA.

Abstract: This masterclass will review principles of trauma and wound management for avian and chelonian patients and will use wildlife case examples to help clinicians understand the practical applications of things like bandaging, wound vacuum-assisted closure, topical medications, and emergency stabilization procedures. Multiple videos will be used to ensure understanding of more complex procedures. Questions posed using Poll Everywhere software will be used to insure audience engagement and understanding.

Introduction

The primary goal of wildlife rehabilitation is always return to normal function and release back into the wild. Chelonian and avian patients frequently present to wildlife centers for trauma and may have some unique characteristics of which clinicians need to be aware. Keeping in mind the goal of eventual release may help in assessing prognostic factors for wounded wildlife.

Handling Wildlife

Fear response is an evolutionarily advantageous adaptation in species that may fall prey to others; however, it makes the good intentions of the wildlife veterinarian and rehabilitator more complicated. In handling wildlife patients, one should always keep in mind the physiologic responses of fear and their impact on health, including endogenous release of corticosteroids (corticosterone in birds and reptiles), hyperglycemia, weight loss, immunosuppression, tachycardia, delayed wound healing, and other potentially deleterious responses. Therefore, ameliorating stress is a sound medical and surgical, as well as ethical, decision.

When handling wild animals, especially those that are wounded, a plan of action should always be in place prior to attempting restraint. All medications, food, and other supplies should be prepared ahead of time as much as possible. Physiologic response should be monitored and if heart rate, respiratory rate, or body temperature are exceeding safe parameters, restraint may need to be terminated and reattempted at a later juncture. Hyperthermia is a common complication during restraint for many species and should be avoided through minimal handling and pharmacologic intervention or otherwise addressed as required (eg, cooling fans, cool water spray, working in shade). Understanding the innate defense mechanisms of the species under care allows the handler to be well prepared and able to employ counter measures more rapidly to decrease the stress of both patient and handlers.

When handling some sea birds, such as pelicans, be aware that while the bill offers a nice handle, and pelicans do not have a lot of strength in their bill, eyewear and gloves are still recommended as personal protection equipment for the handler since the tomia, or cutting edge of the bill, is sharp and can cause something like a bad paper cut.

Also, be aware that, because of their diving lifestyle, some seabirds, like pelicans basically lack external nares, so the bill should always be left open to allow the animal to breath. In addition, covering the head and eyes of the bird with a towel will usually result in it being less stressed and more cooperative.

Housing, Husbandry, and Rehabilitation

Conventional stainless-steel cages routinely found in small animal hospitals can easily be modified to be both safe and low stress for wildlife. Towels over the front of the cage can decrease mess, increase safety, and provide barriers that will ameliorate stress; however, care should be taken to still provide a normal light cycle. For raptors and easily agitated birds prone to self-damage, carpal bumpers and tail guards should be applied to minimize trauma to the carpus and rectrices. Standard cages may be modified by using yoga mats or roof membrane hung like a curtain 1-2 inches from the wall to pad hard surfaces and minimize trauma.

Additional husbandry requirements should also be considered. For example, salt glands are present in all the Procellariformes, including pelicans. The size of the gland and the need for access to salt will vary between species. Lack of access to salt does not affect pelicans, cormorants and penguins as quickly as pelagic species. When pelagic birds are housed in fresh water and do not have access to salt for more than ten days the gland will atrophy. When a bird is released into the ocean with non-functioning salt glands the bird is unable to cope with the salt resulting in dehydration and salt toxicosis. Pelicans are more adaptable to water-salt content and may be able to go for a few weeks without gland atrophy but should still be gradually introduced to salt water during the rehabilitative phase of hospitalization.

It is important that seabirds, such as pelicans, have access to water routinely. Although large wounds may complicate this need. But generally, water birds get twice daily tub time in our hospital if they have open wounds or are not well enough to be outside in the big pool. Seabirds will often lose the waterproofing quality of their feathers if they are not allowed to swim and groom, if their feathers become soiled with their own fecal material, or due to contamination by oil from fish left in their swimming pool or from being force fed. Tub time is also important for oral hydration. Pelicans only have vestigial tongues and with their large bills obviously cannot drink from bowls. Pelican excrement is normally pretty liquid. If it looks dry and formed, the pelican is probably clinically dehydrated. This puts hospitalized pelicans at risk for a lot of secondary issues, like renal impairment, particularly with certain drugs, such as non-steroidal anti-inflammatory drugs (NSAIDs). In our hospital, seabirds will get NSAIDs for no more than 3 days before being reassessed, they never get them if they have any gastrointestinal (GI) problems, of if they have had GI surgery or have a GI foreign body, and they always get parenteral or oral fluids if they can't get tub time while on NSAIDs. They are also often on GI protectant medications.

Feeding

Many wild animals will refuse to eat in captivity and so assisted alimentation may be required. This is particularly true of many chelonians with shell trauma. Every attempt should be made to encourage the patient to eat on its own and there are a variety of tips in displaying food items to the best advantage. If gavage feeding must be done, it should always be the last treatment administered and then the patient is returned immediately to the cage to decrease stress-associated regurgitation, especially in sea birds. When force feeding whole fish to sea birds, such as pelicans use two people, one to handle fish and one to massage the fish down the esophagus to avoid getting fish oil on the plumage. If surgical stabilization and thus general anesthesia is required, this is often a good time to go ahead and place an esophagostomy tube.

Whenever possible, oral medications may be hidden in a food item to decrease the stress of medicating. If a patient is being tube fed, then medications should be front loaded into the tube to increase the likelihood that the patient will get them even if not all the food is able to be given.

Diagnostics

The minimum database for avian and reptilian wildlife patients generally includes radiographs, complete blood count (CBC), biochemical profile, and fecal exams as indicated; however, any diagnostic testing should not compound the stress or trauma to the patient. Therefore, some tests may have to wait until the patient is more stable. Assess wound severity and prognosis for return to the wild. Packed cell volume should be repeated at admission and again in 24 hours to most accurately assess acute anemia from traumatic blood loss.

Venipuncture can be problematic in some chelonian species. As well as in some avian species, such as pelicans, because the physiologic subcutaneous emphysema they normally have to cushion them from the shock of hitting the water at high speeds; however, the medial metatarsal vein is still usually a good place to be able to reliably get a blood sample and is also a good place for intravenous (IV) catheterization (as are the pouch veins).

Critical Care

Fluid support and analgesia are a critical part of early intervention in trauma cases. If the patient is dehydrated or hypothermic, it is important to take a day or so to rehydrate and warm the patient prior to giving any food or medications or metabolism of these substances may be impaired and make the animal worse. Crystalloids, such as Plasmalyte, Normasol-R, LRS or half-strength LRS + 2.5% dextrose are all acceptable empiric choices prior to obtaining bloodwork. Colloids may also be indicated, especially in cases of hypovolemic shock, common in trauma cases. Hypertonic saline may also be used to address hypovolemia and there has been increasing interest in human and small animal medicine in the use of hypertonic saline in traumatic brain injury (TBI) for its hyperosmotic potential.[1] The author uses a bolus dose of 4 ml/kg IV of 7.2 % NaCl over 10 minutes in avian and chelonian species. The increase in intravascular volume after hypertonic saline administration can be transient lasting approximately 15-30 minutes, but these effects can be prolonged by the concurrent administration of colloids.[2] Research suggests that with or without colloids, hypertonic saline-induced reductions in intracranial pressure (ICP) persist for much longer than the vascular effects and significantly longer than mannitol.[3] Hypertonic saline should be avoided in patients with significant sodium derangements or advanced dehydration.

Clinical and research data are gathering which suggests hypertonic saline may reduce ICP faster, more significantly and for longer than mannitol with less undesirable effects.[1,3,4] The use of corticosteroids in birds and chelonians is generally contraindicated, but particularly in cases of TBI. In human medicine the use of glucocorticoids in head-trauma patients has been shown to increase mortality.[5] Corticosteroids will also induce hyperglycemia which has been linked with increased free radical production, excitatory amino acid release, cerebral edema, and altered cerebral vasculature, which could potentiate further neurologic injury.[2]

Wound Care and Vacuum-Assisted Closure

Anesthesia or sedation may be necessary for adequate wound care. Analgesics are an important part of treatment and general anesthesia may be indicated. Clinical presentation may help dictate anesthetic protocols. For example, as a general guideline in chelonians, consider delaying anesthesia if the heart rate less is 15 bpm when the patient is maintained at its preferred body temperature, if blood work reveals a PCV $\leq$ 10% or a plasma TP $\leq$ 2.0 g/dl, or if there is evidence of sepsis or severe respiratory disease. Again, the patient should be warm and stable prior to anesthesia and aggressive wound management. Until then, simple bandaging may provide temporary fracture stabilization and wound coverage.

Basic care for most wounds will begin with lavage by large amounts of warm sterile 0.9% NaCl or other sterile physiologic fluid with or without dilute chlorhexidine, betadine, colloidal silver solutions, or tris EDTA as appropriate. Irrigation pressure is ideally around 8-10 PSI (example: 35 ml syringe with 19 gauge needle) to adequately clean and debride the wound. Initially this will be done once daily and then gradually less often. Debridement may also be done as needed surgically or with wet to dry bandages over the first few days. Petrolatum impregnated dressing works well to keep the tissues moist as they heal. Antimicrobials commonly used by the author with the vacuum-assisted closure (VAC) include water soluble products, such as honey or colloidal silver hydrogels. Do not use topical antimicrobials that will impair the VAC, such as creams or ointments.

Once hemorrhage is controlled and a clean wound bed established, vacuum assisted closure may reduce wound healing time. In one study in humans, VAC-treated wounds decreased in size by 78% compared to a 30% size reduction in wounds treated with wet-to-dry bandages over the same time frame. There are several publications describing the adaptations for this technique in chelonians and a couple for birds.[6-11] The VAC is usually left in place until a healthy bed of granulation tissue is noted, which may take anywhere from a few days up to several weeks. Bandage changes generally start out being done more frequently and then may stretch out to once every 5-7 days once the wound is progressing well. Aquatic species may need to be alternated in water and dry docking with the VAC if water-proofing isn't possible. Additionally, water birds that are not getting daily "tub time" will need to have hydration implemented by alternative means, such as subcutaneous or oral fluid therapy.

Shell repair

There are a myriad of ways to repair shell fractures, depending on the species and fracture site and severity. Shell fracture segments may be bridged with cable ties, hooks and suture material, screws and wire, plates, or for more minor fractures with minimal displacement, tape (or steri-strips). Older fixation methods, such as fiberglass and epoxy, inhibit healing and provide a nidus for infection, particularly in aquatic chelonians. Fractures and defects have been known to take anywhere from 1 month to many months to heal well enough to allow for release. Palpation is generally more useful than radiographs in deciding when fractures are stable enough for fixation devices to be removed. For more complicated fractures and fixators, a staged reduction can be done to gradually increase load bearing. Screw holes can be managed as any other wound and will granulate over time. One option is to flush after screw removal and then flush and pack or seal with a variety of products.

Topical medications

The use of topical antibiotics is indicated when treating infected wounds. Ointments may help to keep tissues moist, but care must be taken to be sure the pelage is not contaminated in avian patients. They generally present no danger of toxicosis, but massive topical application of antibiotic ointments or creams to large wounds may result in systemic absorption and adverse side effects. These agents most often indicated when applied to small, recent wounds. Silver sulfadiazine cream is effective against most gram-positive and gram-negative bacteria as well as most fungi. It is often used for thermal burns, as in electrocution cases, because of its ability to penetrate eschar and necrotic tissue. But it may have a negative effect on fibroblasts and thus may impede wound contraction and thus is used infrequently by the author.[12,13] Medical grade honey and silver impregnated hydrogels have been documented to be effective in a variety of chelonian and avian wounds and are generally the author's empiric preference.[14]

External coaptation and physical therapy

Keep a bandage box with appropriately sized bandage materials for small patients available to help facilitate rapid application in emergent situations. A figure-of-eight bandage may be indicated in birds for temporary stabilization of fractures of the radius, ulna or both; fractures or luxations of the elbow or carpus; or soft tissue injuries to the brachium or antebrachium. Generally, cast padding and an outer layer of vet wrap will suffice for most wild birds. Usually no more than 2-3 circuits around are needed; otherwise the bandage is needlessly bulky. A body wrap is indicated for shoulder or humeral injuries. A correctly placed bandage will have the primary feathers sitting **under** the secondary feathers, as seen in the normal perching pose.

Generally physical therapy is indicated in wing wraps to prevent patagial membrane contracture and keep a normal range of motion (ROM) in the joints. Physical therapy (PT) may begin as soon as 3 days post-admission or fixation of the limb. Initially, PT should be done daily or EOD. If ROM is good, this may be extended to every third day. Once contracture has occurred, it is more difficult to reverse it. So, it is better to do PT more often. Initially, it may be necessary to do PT under general multi-modal anesthesia (ie, injectable pain meds + inhalant anesthesia). After the first week, it may be possible to do PT awake in sedated animals given injectable analgesic drugs. After the second week, it may be possible to do PT with just oral pain medication +/- sedation. The author tries to minimize the use of intramuscular injectable medications where possible and put medications into food items. Generally, external coaptation devices are removed between 10-21 days if they are the only method of fixation. Joints should never be bandaged more than necessary.

Finally, the function of tendons and joints is assessed by manipulation of the reciprocal apparatus of the wings and legs. For example, in the wings, the carpus is depressed caudally while the elbow is depressed cranially. These combined forces should cause the elbow and carpus to pivot open and spread the wing utilizing the tendons primarily.

Heat packing prior to PT exercises may facilitate better ROM. Goniometer measurements should be used to assess ROM. In one study in green iguanas, laser therapy did not result in substantial improvement in histologic evidence of wound healing, compared with results for traditional wound management.[13]

References

1. Thongrong C, Kong N, Govindarajan B, Allen D, Mendel E, Bergese SD. Current purpose and practice of hypertonic saline in neurosurgery: A review of the literature. *World Neurosurg*. 2014;82:1307-1318.

2. Sande A, West C. Traumatic brain injury: A review of pathophysiology and management. *J Vet Emerg Crit Care*. 2010;20:177-190.

3. Sakellaridis N, Pavlou E, Karatzas S, Chroni D, et al. Comparison of mannitol and hypertonic saline in the treatment of severe brain injuries. *J Neurosurg*. 2011;114:545-548.

4. Mortazavi MM, Romeo AK, Deep A, et al. Hypertonic saline for treating raised intracranial pressure: Literature review with meta-analysis. *J Neurosurg*. 2012;116:210-221.

5. Edwards P, Arango M, Balica L, et al. Final results of MRC CRASH, a randomised placebo-controlled trial of intravenous corticosteroid in adults with head injury-outcomes at 6 months. *Lancet*. 2005;365:1957-1959.

6. Adkesson MJ, Travis EK, Weber MA, et al. Vacuum-assisted closure for treatment of a deep shell abscess and osteomyelitis in a tortoise. *J Am Vet Med Assoc.* 2007;231(8):1249-1254.

7. Bezjian M, Wellehan JF Jr, Walsh MT, et al. Management of wounds in a loggerhead sea turtle (*Caretta caretta*) caused by traumatic bycatch injury from the spines of a spotted eagle ray (*Aetobatus narinari*). *J Zoo Wildl Med.* 2014;45(2):428-432.

8. Hedley J, Woods S, Eatwell K. The use of negative pressure wound therapy following subcarapacial abscess excision in a tortoise. *J Small Anim Pract.* 2013;54(11):610-613. doi: 10.1111/jsap.12118.

9. Lafortune M, Wellehan JF, Heard D, et al. Vacuum-assisted closure (turtle VAC) in the management of traumatic shell defects in chelonians. *J Herpetol Med Surg.* 2005;15(4):4-8.

10. Grukenmeyer V. Management of avian skin wounds. *Proc Annu Conf Assoc Avian Vet.* 2009;57-62.

11. Knapp-Hoch H, de Matos R. Clinical technique: Negative pressure wound therapy-general principles and use in avian species. *J Exot Pet Med.* 2014;23(1):56-66.

12. Burke HF, Swaim SF, Amalsadvala T. Review of wound management in raptors. *J Avian Med Surg.* 2002;16:180-191.

13. Cusack LM, Mayer J, Cutler DC, et al. Gross and histologic evaluation of effects of photobiomodulation, silver sulfadiazine, and a topical antimicrobial product on experimentally induced full-thickness skin wounds in green iguanas (*Iguana iguana*). *Am J Vet Res.* 2018;79(4):465-473.

14. Mans C, Sunohara-Neilson J, Higginson G, et al. Wound management in a trumpeter swan using honey and a sustained release ionic silver hydrogel. *Exot DVM.* 2006;8(5):21-23.

Gap in pagination due to unavailable papers.

Pages 183-816

Section 32

ARAV Reproductive Session

Michelle Davis, DVM, Dipl ACZM, DipECZM (Zoo Health Management);
Alexa Delaune, DVM
Session Chairs

Developing Methods for Semen Extension in Reptiles

Sarah R. Camlic, BA,
Sean M. Perry, DVM,
Mark A. Mitchell, DVM, MS, PhD, DipECZM (Herpetology)

Session #2334

Affiliation: From the Department of Veterinary Clinical Sciences, Louisiana State University, School of Veterinary Medicine, Skip Bertman Drive, Baton Rouge, LA 70803, USA.

Semen extension is needed to advance the development of assisted reproductive technology for reptile species. Critical evaluation of semen extenders needs to occur across all reptile species, especially squamates and chelonians due to their high risk for extinction. This will allow for short term storage of spermatozoa and further development of cryopreservation protocols. No studies exist evaluating semen extenders across taxa. The purpose of this study is to determine effective semen extenders for squamate and chelonian semen stored at two temperatures. The testicles and reproductive tracts from the following species, red-eared slider (*Trachemys scripta elegans*), veiled chameleon (*Chamaeleo calyptratus*), green anoles (*Anolis carolinensis*), and banded water snakes (*Nerodia fasciata*), were harvested post-mortem to evaluate semen. The reproductive tract was separated into testicle and epididymis/ductus deferens and these parts were macerated in their respective extender treatments. Extenders evaluated include Test Yolk Buffer (Irvine scientific, Irvine, CA), Lactated Ringer's solution, Ham's Nutrient Mixture F10 (Gibco Laboratories, Gaithersburg, MD) and INRA 96 (IMV technologies, Maple Grove, MN). Semen samples were initially analyzed for concentration, morphology, motility, rate of movement, plasma membrane integrity, and acrosome integrity. A computer-assisted sperm analysis (CASA) system was used to analyze the samples. Extended semen aliquots were stored in 1.5-ml microcentrifuge tubes at room temperature (20 °C) and refrigeration (4 °C). Semen samples with extenders were evaluated using microscopy and the CASA system. Semen samples were tested at 2, 4, 6, 8, 12, 24, 48, and 72 hours from the time of collection.

Effects of Exogenous Human Chorionic Gonadotropin on Testosterone and Spermatozoa Concentrations in Veiled Chameleons (*Chamaeleo calyptratus*)

Sean M. Perry, DVM, and
Mark A. Mitchell, DVM, MS, PhD, DipECZM (Herpetology)

Session #2180

Affiliation: From the Department of Veterinary Clinical Sciences, Louisiana State University, School of Veterinary Medicine, Skip Bertman Drive, Baton Rouge, LA 70803, USA.

Exogenous hormones commonly are used to regulate the reproductive cycles of mammals, but are not routinely used in reptiles. The objectives of this study were to determine an optimal human chorionic gonadotropin (hCG) dose for male veiled chameleons (*Chamaeleo calyptratus*) and determine if exogenous hCG administration increases spermatozoa production. A prospective, longitudinal complete crossover study was performed using 11 veiled chameleons. To determine the optimal hCG dose, each animal was treated with doses of hCG at 100 IU, 200 IU and 300 IU. Blood samples were collected serially over a 24-hour period to determine peak plasma testosterone concentrations using an Arbor Assay Enzyme Immunoassay testosterone kit (Arbor Assays, Ann Arbor, MI). Following dose determination, 11 veiled chameleons were evaluated to see if the pre-determined hCG dose increased spermatozoa concentrations. A randomized complete crossover design was employed to determine if hCG has an influence on spermatozoa production. Animals were assigned to treatment and saline control injections. Semen collection with electroejaculation occurred 2 hours after the peak testosterone concentration was achieved. After a 2-week washout period, the treatment groups were crossed over and collection methods repeated. Chameleons were anesthetized with alfaxalone (10 mg/kg) for the electroejaculation period. Semen samples were evaluated grossly for volume and color prior to full evaluation. Spermatozoa concentration was evaluated using a phase contrast microscopy (magnification, 400X). Spermatozoa motility was estimated at ambient temperature and the percentage of progressively motile spermatozoa determined to the nearest 5%. None of the chameleons experienced any negative effects of the exogenous hCG treatment.

Ovarian Ultrasonographic Monitoring in a Group of Jewelled Lizards (*Timon lepidus*): Pre and Post Administration of Alpha Follitropin

Alessandro Vetere, DVM, GPCert(ExAP),
Emanuele Lubian, DVM, GPCert(ExAP),
Erika Treves, DVM

Session #2211

Affiliation: From Ambulatorio Veterinario Gaggiano, Via Galvani 42, Gaggiano, 20083, Italy.

Fifteen female jewelled lizards (*Timon lepidus*) underwent ultrasound examinations in February, March, and April of 2016 and 2017, with the purpose of monitoring the maturation of ovarian follicles in lizards that were or were not treated with exogenous alfa follitropin. The subjects were divided in 3 groups of 5 specimens each. The first group (Control 1) was subjected to a forced hibernation from October 2016 to February 2017 at a temperature of maximum 10 °C during the day and of minimum 7 °C in the night. The second (Control 2) and third (Cases) groups were maintained at the housed temperature without any temperature variation during the winter months. Every subject received alfa follitropin (1 IU total dose IM, q12h × 3 days). The dosage was subsequently increased by 1 IU every 2 days until the amount of 5 IU was reached on the tenth and last day of treatment. Ultrasonography showed a remarkable increase in size of the ovarian follicles in the alfa follitropin-treated groups compared to the 2 control groups.

References

1. Crews D. Interrelationship among ecological, behavioural and neuroendocrine processes in the reproductive cycle of *Anolis carolinensis* and other reptiles. In: Rosemblatt JS, Hinde RA, Beer CG, Busnel MC, eds. *Advances in the Study of Behaviour.* New York, NY: Academic Press;1980:1-74.

2. Edwards TM, Gunderson MP, Milnes MR, Guillette, LJ. Gonadotropin-induced testosterone response in peripubertal male alligators. *Gen Comp Endocrinol.* 2004;175:251-258.

3. Kawazu I, Suzuki M, Maeda K, et al. Ovulation induction with follicle stimulating hormone administration in hawksbill turtles *Eretmochelys imbricata. Curr Herpetol.* 2014;33:88-93.

4. Jones RE, Guillette JR LI, Summers CH, et al. The relashionship among ovarian conditions, steroid hormones and estrous behaviour in *Anolis carolinensis. J Exp Zool.* 1983;227:145-154.

5. Jones RE, Austin HB, Lopez KH, et al. Gonadotropin induced ovulation in a reptiles (*Anolis carolinensis*): histological observations. *Gen Comp Endocrinol.* 1988;72:312-322.

6. Khan UW, Rai U. Role of gonadotropin and Leydig cell-secreted factors in the control of testicular macrophage activities in the wall lizard, *Hemidactylus faviviridis*. *Dev Comp Immunol.* 2008;32:348-355.

7. Khan UW, Rai U. In vitro effect of FSH and testosterone on Sertoli cell nursing function in wall lizard, *Hemidactylus faviviridis* (Rüppell). *Gen Comp Endocrinol.* 2004;136:225-231.

8. Lindzey J, Crews D. Hormonal control of courtship and copulatory behaviour in male *Cnemidophorus inornatus ;* a direct sexual ancestor of a unisexual, parthenogenetic lizard. *Gen Comp Endocrinol.* 1986;64:411.

9. Vetere A, Lubian E, Millefanti M. Follicle-stimulating-hormone-induced mating: behaviour and histologic changes in a pair of ocellated lizard (*Timon lepidus*). *Proc Annu Conf Assoc Reptilian Amphib Vet.* 2016:843.

Pyosalpinx in a Bearded Dragon (*Pogona vitticeps*) Having Previously Undergone Successful Ovariectomy

Heather A. Bjornebo, DVM, DABVP (Reptile and Amphibian Practice), CertAqv

Session #2127

Affiliation: From Arizona Exotic Animal Hospital, 744 N. Center St. #101, Mesa, AZ 85201, USA.

A 2-year-old adult female inland bearded dragon (*Pogona vitticeps*) initially presented with a history of decreased appetite, overall increased activity level, and digging behavior. Physical examination was unremarkable with the exception of significant obesity. Complete blood count and plasma biochemistries revealed an inflammatory leukogram and markedly elevated calcium suggestive of reproductive pathology.[1] An exploratory celiotomy was performed revealing bilaterally enlarged ovaries with numerous adhesions between the ovaries and surrounding tissues. Ovarian follicles were friable and easily ruptured with manipulation, spilling into the coelomic cavity. Samples for culture were collected, the coelom was lavaged, and bilateral ovariectomy was performed. Oviducts appeared normal grossly and were left in place.[1-5] Recovery was uneventful and without obvious complications. No growth occurred on aerobic, anaerobic, and fungal culture. Six months later the patient re-presented with a 1- to 2-week history of decreased appetite. Laboratory results revealed a leukocytosis with toxic heterophils, and the lizard was started on oral compounded (FLAVORx Vet Syrup, FLAVORx Radical Raspberry, FLAVORx Sweetener, FLAVORx Inc, Columbia, MD, USA) 20.4 mg/ml enrofloxacin (Enrofloxacin Flavored Tablets 136 mg, Putney, Inc, Portland, ME, USA) at 10 mg/kg q24h.[1,6] The patient presented again 1 week later with hydrocoelom. Aspiration revealed a clear, yellow fluid, and repeat exploratory celiotomy was recommended.[1] Owners declined surgery and opted for medical management. The patient was started on oral meloxicam (Meloxidyl (meloxicam) 1.5 mg/ml Oral Suspension, Patterson Veterinary, Greely, CO, USA) at 0.5 mg/kg q24h and intramuscular piperacillin/tazobactam (Piperacillin and Tazobactam for Injection, USP, 2.25 mg vial, Mitim S.r.l., Brescia, Italy) at 100 mg/kg q48h.[6] Coelomic fluid was submitted for cytology and aerobic, anaerobic and fungal culture. Cytologic examination revealed a modified inflammatory transudate and culture was negative.[1] The patient's appetite returned and coelomic fluid volume decreased. Within a few weeks the hydrocoelom recurred and owners opted for exploratory surgery. Surgery revealed bilaterally erythematous, distended salpinges containing thick creamy-white exudate. Both salpinges were removed and samples were submitted for histopathology and culture.[1-5] Histopathology revealed diffuse moderate glandular proliferation, cystic dilation with intramural proteinaceous material, and focally extensive bacterial granulomas. No ovarian remnants could be identified grossly during surgery or on histopathology. Culture was positive with a heavy growth of *Staphylococcus* bacteria. The patient recovered well from surgery and no further complications have occurred. Reproductive disease is common in captive female reptiles and often is treated surgically. Some practitioners may elect to perform an ovariectomy instead of an ovariosalpingectomy when pathology does not appear to involve the oviducts, because, in mammalian patients, pyometra is associated with hormonal influence and ovariectomy results in uterine atrophy.[1-5] This case highlights one possible complication even with assumed successful ovariectomy.

References

1. Mader DR. *Reptile Medicine and Surgery.* 2nd edition. St. Louis, MO: Elsevier Saunders;2006.

2. Funk RS. Lizard reproductive medicine and surgery. *Vet Clin North Am Exot Anim Pract.* 2002;5(3); 579-613.

3. Girolamo ND, Mans C. Reptile soft tissue surgery. *Vet Clin North Am Exot Anim Pract.* 2016;19(1):97-131.

4. Lock BA. Reproductive surgery in reptiles. *Vet Clin North Am Exot Anim Pract.* 2000;3(3):733-752.

5. Mader DR, Divers SJ. *Current Therapy in Reptile Medicine & Surgery*. St. Louis, MO: Elsevier Saunders;2014.

6. Klaphake E, Gibbons PM, Sladky KK, Carpenter JW. Reptiles. In: Carpenter JW, ed. *Exotic Animal Formulary*, 5th ed. St. Louis, MO: Elsevier Saunders;2018.

Surgical Revision of Previous Ovariectomies in Turtles: Three Cases

Nicola Di Girolamo, DMV, MSc(EBHC), GPCert(ExAP), PhD, DipECZM (Herpetology)

Session #2203

Affiliation: From the Tai Wai Small Animal and Exotic Hospital, 69-75 Chik Shun Street, Hong Kong.

In the last decade desexing surgery of female chelonians has been increasingly performed thanks to the widespread use of prefemoral coeliotomy.[1] Three previously spayed turtles, 2 red-eared sliders (*Trachemys scripta elegans*) and 1 diamondback terrapin (*Malaclemys terrapin*), presented with variable clinical and biochemical signs consistent with follicular development, including anorexia, increased activity (n = 2), decreased activity (n = 1), increase of calcium, phosphorus, and albumin values. Presence of follicles was confirmed through computed tomography scan (n = 1) or prefemoral ultrasound (n = 2). All the turtles presented variably sized follicles. The turtles underwent surgical removal of the remaining follicles from an approach through the prefemoral fossa. The side where the follicles were more easily visualized through ultrasound was elected for surgery. The turtles were positioned in dorsal recumbency. In order to favor exteriorization of the follicles, the flank opposite to the prefemoral incision was elevated, resulting in a 45 degree angle. The edge of the carapace was positioned at the edge of the table, permitting manoeuvring of the follicles also from a ventral approach. In 2 cases fibrosis and adhesions were observed around the follicles. Removal of all follicles was confirmed by intraoperative endoscopy (n = 3), and intraoperative ultrasound (n = 1). Recovery was uneventful. Ultrasound scan up to 8 months after the second surgery did not show any follicle (n = 2), and biochemistry showed no increase of analytes consistent with folliculogenesis in the following active season (n = 2). Failure to remove ovaries in entirety has been reported in lizards and this report underlines the importance of full removal of the ovaries in chelonians as well.[2]

References

1. Innis CJ, Hernandez-Divers S, Martinez-Jimenez D. Coelioscopic-assisted prefemoral oophorectomy in chelonians. *J Am Vet Med Assoc.* 2007;230:1049-1052.

2. Cruz Cardona JA, Conley KJ, Wellehan JF, et al. Incomplete ovariosalpingectomy and subsequent malignant granulosa cell tumor in a female green iguana (*Iguana iguana*). *J Am Vet Med Assoc.* 2011;239:237-242.

Phallic Intravenous Catheterization and Difficulty in Diagnosing Liver Disease in an Adult Sulcata Tortoise (*Centrochelys sulcata*)

Kyra J. Berg, DVM,
Ryan M. Smith, DVM, DACVECC,
Mathew Stewart, DVM,
Tatiane Terumi Negrão Watanabe, DVM, MS, PhD,
Chiara Hampton, DVM, MS,
Sean M. Perry, DVM,
Virginie Wurlod, Dr vet med, DACVECC, DECVECC,
Nathalie Rademacher, Dr med vet, DACVR, DECVDI, DVM,
Nobuko Wakamatsu DVM, PhD, DACVP, DACPV,
Javier G. Nevarez, DVM, PhD, Dipl ACZM, DipECZM (Herpetology)

Session #2155

Affiliation: From the Department of Veterinary Clinical Sciences (Berg, Smith, Stewart, Hampton, Perry, Wurlod, Rademacher, Nevarez), and the Department of Pathobiological Sciences and Louisiana Animal Disease Diagnostic Laboratory (Watanabe, Wakamatsu), Louisiana State University, School of Veterinary Medicine, Skip Bertman Drive, Baton Rouge, LA 70803, USA.

Abstract: An adult sulcata tortoise presented for endoscopic removal of a linear foreign object. After anesthesia, the patient failed to recover. Intravenous catheters placed in the jugular vein failed. Attempts to catheterize other sites were not successful so it was elected to catheterize vessels in the phallus. Intensive care was provided including mechanical ventilation. Plasma biochemistry, blood gases, and computed tomography were performed. The patient was euthanized and necropsy revealed severe hepatic lipidosis. This case presents a novel catheterization site, mechanical ventilation parameters for this patient, and the difficulty in diagnosing liver disease in reptiles.

Introduction

An adult sulcata tortoise (*Centrochelys sulcata*) was presented for endoscopic removal of a linear foreign object. The object had been ingested just prior to presentation.

Treatment

The tortoise was anesthetized with alfaxalone (10 mg/kg IV; Alfaxan, Jurox Pty Limited, Australia), midazolam (1 mg/kg IV; Midazolam, West-Ward, Eatontown, NJ, USA), and hydromorphone (2 mg/kg IV; Hydromorphone, Baxter, Deerfield, IL, USA). Endoscopic removal of the object was performed without complication.

After anesthesia, the patient failed to recover and resume spontaneous ventilation. The phallus prolapsed post-anesthesia. After failure of all standard intravenous catheter placement sites, a branch of the ventral phallic vein was successfully catheterized utilizing a modified seldinger technique. Intensive care was provided including mechanical ventilation, a norepinephrine constant rate infusion, and isotonic fluids through the phallus catheter.

Plasma biochemistry, repeat blood gases from the phallus vein as well as the brachial vein during mechanical ventilation, and computed tomography (CT) were performed. The plasma biochemistry was within normal reference ranges, while the blood gases demonstrated marked metabolic and respiratory disturbances in spite of modification of mechanical ventilation parameters.

The patient resumed spontaneous ventilation approximately 12 hours after mechanical ventilation. On day 7, the patient declined and was humanely euthanized.

Discussion

Gross necropsy and histopathology demonstrated severe hepatic lipidosis. The lungs had no evidence of barotrauma secondary to mechanical ventilation. Mean CT values of the liver measured between -19 and 32 Hounsfield Units, consistent with hepatic lipidosis. This case presents a novel catheterization site, the difficulty in diagnosing liver disease in reptiles, mechanical ventilation parameters used in this patient, and results of computed tomography which may aid in antemortem diagnosis of hepatopathies in reptiles.[1]

References

1. Marchiori A, da Silva ICC, Bonelli MA, et al. Use of computed tomography for investigation of hepatic lipidosis in captive *Chelonoidis carbonaria* (Spix, 1824). *J Zoo Wildl Med.* 2015;46(2):320-324.

Testosterone and Testicles: Does Size Really Matter?

Sean M. Perry, DVM, and
Mark A. Mitchell, DVM, MS, PhD, DipECZM (Herpetology)

Session #2181

Affiliation: From the Department of Veterinary Clinical Sciences, Louisiana State University, School of Veterinary Medicine, Skip Bertman Drive, Baton Rouge, LA 70803, USA.

Research into the annual reproductive cycles of reptiles is desperately needed to enhance captive propagation efforts and develop assisted reproductive technologies. The objective of this study was to perform a prospective longitudinal study to measure the male annual reproductive cycles of 2 species of chameleons. Ten male veiled chameleons (*Chamaeleo calyptratus*) and 15 male panther chameleons (*Furicifer pardalis*) were used for this study. Testosterone concentrations and testicle size were measured bi-weekly to characterize the reproductive cycles of these lizards under captive conditions. Animals were anesthetized with alfaxalone (10 mg/kg IV) for the procedures. Blood was collected from the jugular vein or ventral tail vein. Ultrasonographic measurements of the testicles were performed utilizing a Sonoscape S8 with a 10-15 mHz Hockey Stick Probe (Universal Ultrasound, Inc Bedford Hills, NY). Gonadal somatic index was estimated based on measurements and individual animal weights. Testosterone concentrations were measured with an automated magnetic immunoassay analyzer TOSOH AIA- 360 (TOSOH Biosiences, Inc, San Francisco, CA) and Arbor Assay Enzyme Immunoassay testosterone kit (Arbor Assays, Ann Arbor, MI). Agreement analysis was used to compare the results between the assays. Testicle size, gonadal somatic index, and testosterone concentrations were correlated to bi-weekly semen collection from both species.

Renal Carcinoma as a Cause of Cloacal Prolapse in Bearded Dragons (*Pogona vitticeps*)

Stacey Leonatti Wilkinson, DVM, DABVP (Reptile and Amphibian Practice)

Session #2228

Affiliation: From Avian and Exotic Animal Hospital of Georgia, 118 Pipemakers Circle Suite 110, Pooler, GA 31322, USA.

Abstract: Three separately owned bearded dragons (*Pogona vitticeps*) presented with marked cloacal swelling or prolapse. All 3 animals were euthanized due to poor prognosis. On necropsy, significant renomegaly was present, which was diagnosed as renal adenocarcinoma. Renal disease, specifically neoplasia, should be a differential diagnosis for cloacal swelling and prolapse in this species.

Introduction

Renal disease and cloacal prolapse are common in lizards, though not typically associated with each other. Because the kidneys lie within the pelvic canal, renomegaly can compress the colon and/or cloaca and cause tenesmus. It can reach such a degree that the kidneys expand outside the pelvis. This report describes 3 female bearded dragons that each presented with cloacal swelling or prolapse secondary to renomegaly that was diagnosed as renal adenocarcinoma. Communication with a pathologist specializing in exotic animals reported this was an unusual presentation for this condition.

Clinical Report

Case 1

A 4-year-old female bearded dragon presented for lethargy, anorexia, tenesmus, and constipation. The patient was thin, the colon was markedly distended with feces, and there was moderate swelling at the cloaca. Euthanasia was elected. Necropsy revealed a markedly distended colon impacted with fecal and bedding material. The kidneys were markedly enlarged with a partially mineralized mass present that corresponded to the cloacal swelling which was compressing the colon. Histopathology revealed renal tubular carcinoma.

Case 2

A 4-year-old female bearded dragon presented for a chronic history of lethargy, anorexia, severe weight loss, and cloacal swelling. The patient was in emaciated condition with a large swelling present at the cloaca. Euthanasia was elected. On necropsy there were developing ovarian follicles. The cloacal swelling was a mineralized mass arising from the kidneys that had expanded caudally out of the pelvis. Histopathology revealed renal adenocarcinoma.

Case 3

A 3-year-old female bearded dragon presented for cloacal prolapse. The tissue prolapsed from the cloaca was approximately 3 cm × 1 cm, firm, and red-black. The cloacal lumen was patent ventral to the prolapsed tissue.

Euthanasia was elected. On necropsy the lumen of the cloaca and colon was confirmed ventral to the abnormal tissue. Dissection of the prolapsed tissue revealed that it was the caudal pole of the left kidney and there was a mass arising from the dorsal aspect of both kidneys. Histopathology revealed renal carcinoma.

Section 33

ARAV Master Classes

Sean Perry, DVM;
Stacey Leonatti Wilkinson, DVM, DABVP (Reptile and Amphibian Practice)
Session Chairs

Diversity of Normal Reptilian Female Reproductive Anatomy and Eggs

Jeanette Wyneken, PhD

Session #2172

Affiliation: From the Department of Biological Sciences, Florida Atlantic University, 777 Glades Road, Boca Raton, FL 33431, USA.

Abstract: An overview of female reproductive systems and eggs is provided. The features shared in common across the Reptilia are covered first, then each major clade is discussed to describe similarities and highlight differences.

Introduction

Reptile reproduction tends to be seasonal consequently reproductive structures (ovaries and oviducts) vary in form in preparation for breeding, during nesting, and then following the season for oviposition. Many reptile species, particularly from tropical regions, breed prior to or during their rainy season, in the warmer spring and summer months or when day lengths are increasing. In temperate species, mating may be in the spring and oviposition for oviparous species or birth for viviparous species follows in weeks or months.

Female reptiles have paired ovaries and the majority of species have paired oviducts that lead to a single cloaca. Current understanding is that both ovaries are active during nesting season, however in species that lay a single egg at a time, it is unknown if eggs come from one ovary, then the next. The oviducts are formed from the embryonic paramesonephic duct (ie, Müllerian duct). Here the term oviduct will include all regions from the infundibulum to the urogenital opening into the cloaca.

Ontogeny of Gonads and Gonadal Ducts

Ovaries form from the genital ridge that has a cortex and an inner medulla. Primordial germ cells (PGC) migrate into the cortex where they become oogonia. Following meiosis the oogonia transform into oocytes and then ova. There is considerable variation in incipient gametes. Some species have oocytes and PGC present in juveniles; in others oocytes and ova are present at hatching. In all reptiles examined to date, (perhaps all taxa), germinal beds may be present throughout life, producing new primordial germ cells, oogonia, oocytes and then follicles.[1] In hatchlings and juveniles, the ovaries have a stratified epithelium overlying many spindly stromal cells that form the medulla. The medulla of the ovary appears somewhat disorganized with connective tissue strands, fibroblasts, and blood vessels present. Gonadal ducts form on each side parallel to the gonad and embryonic kidney and are generally termed paramesonephric ducts. The mesonephric or Wolffian duct and the Müllerian ducts form side by side. Normally, the Wolffian ducts regress in females before hatching or birth. (Wolffian ducts become the vas deferens in males). The Müllerian ducts persist in females and become the oviducts.[2]

As embryos, gonadal sex is determined by the incubation environment (primarily temperature) rather than genetic instructions in most chelonians, crocodilians, *Sphenodon*, and some lizard species. Some lizards, a few turtle species, and all snakes have chromosomally (genetic) directed sex determination.[3] The sexes of hatchlings of most species may be identified prior to the development of secondary sex characteristics by a combination of gross morphologic traits and gonadal histology. In juvenile females, the cortex is multicellular and stratified (in males the cortex is usually one cell layer thick); the gonad appears limp and mobile, and patent paramesonephric (Müllerian) ducts are present on each side.

In juveniles, the cortical cells (oocytes) enlarge and the surface of the ovary becomes grainy as the oocytes and their follicles enlarge. Following puberty, follicles at multiple stages of maturation are present. This follicular variation implies varying degrees of sensitivity to the ovulation-inducing effects of the gonadotropins.[4] Each breeding season's oocytes accumulate yolk that is synthesized in the liver in response to estrogen and vitellogenin production.

Ovaries

The ovaries are elongated or ovoid structures with a granular surface. They appear granular in in even the youngest animals. There normally are 2 ovaries in reptiles. They are suspended from the dorsal coelomic wall by a loose mesentery, the mesovarium on each side of the vertebral column. Ovary formation is directed by incubation temperatures (temperature-dependent sex determination or TSD) in crocodilians, most turtles and tortoises, tuataras (*Sphenodon*), and many lizard species. In snakes, some lizards (Table 1) and a few turtles genera (*Emydura*, *Chelodina*, *Pelodiscus*, *Apalone*, *Claudius*, and *Calemys*), ovary formation from the indifferent gonad is directed by genetic mechanisms (genotypic or genetic sex determination, GSD).[3] In all taxa, sex is determined during the middle-third of development, during organogenesis.[5,6]

There is no default sex in reptiles with TSD. In many turtles and tortoises, there is a cool male-warm female TSD system.[3] In crocodilians and some populations of snapping turtles (*Chelydra serpentina*), females develop when eggs incubate at cool temperatures and at warm temperatures; males result from intermediate incubation temperatures. In *Sphenodon,* alone, there is a warm male-cool female system of sex determination.[7,8]

Lizard ovarian development can be a bit more challenging because genetic sex determination (GSD) and temperature-dependent sex determination (TSD) can co-occur within the same family. To date, members of the Agamidae, some Australian skinks, and some Eublepharid geckos have both systems such that incubation temperature can override genetic sex determining mechanisms, in some cases.[9] Consequently, morphologic sex and chromosomal sex made differ. The literature was surveyed for sex determining mechanisms in squamates and tabulated their findings (Table 1).[10] In some cases the nature of the lizard GSD system (XY or ZW) is not known. All snakes have a ZW GSD system.

In the breeding season, the ovaries increase in size as follicles accumulate yolk (vitellogenesis). It is common for more follicles to undergo vitellogenesis than are ovulated. The scars left in the ovary when mature follicles are ovulated become corpora lutea then corpora albicantia. The presence of corpora albicantia indicate previous ovulation, but may not be an indication of the season or seasons in which they formed.[11] Mature follicles that are not ovulated regress and stromal cells assimilate the yolk.[12] Crocodilians ovaries differ from those other reptiles in that their structure includes well-developed system of lacunae and bundles of smooth muscle around the follicles in all stages of development.[13,14] Tuatara have sacular ovaries with large fluid-filled (presumed to be lymph) spaces in the interior.[15]

Table 1. Taxonomic distribution of particular sex-determining modes across squamate families.

Family	Number of species/family surveyed	Genetic sex determination	Temperature-dependent sex determination
Agamidae (Australian dragon lizard)	25	12 (1 ZW)	13
Chamaeleonidae (Old World chameleon)	4	1 (ZW)	2
Polychrotidae (bush anole)	33	33 (XY)	
Corytophanidae (basilisk, helmetted lizard)	1	1	
Iguanidae (iguana and anole)	1	1	
Crotaphytidae (collared lizard)	1	1	
Tropiduridae (neotropical ground lizard)	6	6 (XY)	
Phrynosomatidae (horned lizard)	43	43 (XY)	
Eublepharidae (eyelid gecko such as leopard gecko)	6	4	2
Diplodactylidae (anole)	5		5
Pygopodidae (Australian limbless gecko)	2	2(XY)	
Gekkonidae (lidless gecko)	32	5 (3 XY, 9 ZW)	15
Dibamidae (blind lizard)	1	1 (XY)	
Teiidae (tegus and racerunner lizard)	3	3 (1 XY)	
Gymnophthalmidae (spectacled lizard)	7	7 (XY)	
Lacertidae (Old World wall lizard, lacerta)	44	44 (ZW)	
Amphisbaenidae (worm lizard, not a lizard)	1	1(ZW)	
Scincidae (skink)	19	15 (2XY)	4 maybe
Anguidae (Northern hemisphere alligator lizard, glass lizard)	1		1 maybe
Varanidae (monitor lizard)	4	4 (ZW)	
Boidae (boa and python)	6	6 (ZW)	
Colubridae (common snake)	63	63(ZW)	
Elapidae (cobra, coral snake, krait, mamba)	18	18 (ZW)	
Hydrophiidae (viviparous sea snake)	58	58(ZW)	
Viperidae (pit viper and Old World viper)	37	37(ZW)	

Source: Adapted from Pokorná M, Kratochvíl L. Phylogeny of sex-determining mechanisms in squamate reptiles: Are sex chromosomes an evolutionary trap? *Zool J Linnean Soc.* 2009;156(1):168-183.

Oviducts

The oviducts of immature females are thin-walled, straight and the diameter quite small. Throughout life, the oviducts are suspended from the coelomic dorsolateral wall by a large mesentery, the mesotubarium. As puberty approaches, the oviducts increase in diameter, length and wall thickness. At puberty, the oviduct walls has 6 basic layers that include an inner epithelium lining oviduct, glands and blood vessels, connective tissue layer, circular muscle layer, longitudinal muscle layer, and outer serosa.[16,17] The entire oviduct has elongated and has differentiated into functionally and structurally into structurally and functionally distinct regions along its length (4, 5, or 6, depending upon taxon and author) summarized in Table 2.[15-22] Girling (2002) organizes the various terms used for similar oviductal structures and discusses the challenges that inconsistent terminology produces.[23] This challenge is obvious in Table 2. She explains what makes up the oviduct as follows:

> "...*Starting anteriorly, the regions include (a) the infundibulum, (b) the uterine tube (also known as the tuba, glandular region, albumen-secreting portion, magnum), (c) the isthmus (aglandular segment, intermediate region), (d) the uterus (shell-forming region), and (e) the vagina (cervix). The different regions are not recognised in all reptilian species, and additional regions may also be included. In descriptions of several squamate species, the infundibulum and uterine*

tube are not differentiated... and the uterus is divided into anterior and posterior regions... In the crocodilians studied to date, the infundibulum and the uterus may each be separated into anterior and posterior portions...."

In reptiles, the oviducts function in fertilization, sperm storage, egg transport, eggshell deposition, maintenance of the early embryo, and expulsion of the egg or fetus.[23,24]

The majority of reptile species have paired oviducts (all crocodilians, all testudines, most snakes and lizards, and *Sphenodon*). Oviduct length may differ between left and right. Oviduct length is staged in most in not all snakes, and some lizards. In few narrow-bodied species (the California legless lizard *Anniella pulchra*, several Australo-Pacific skinks [*Tribolonotus* spp.], Centipede snakes [*Tantilla* spp.]) the left oviduct has been lost or is vestigial.[24]

The histologic features of the oviduct change seasonally.[13] In general, adults have opaque oviduct walls and at least the caudal half is folded in accordion-like pleats.

The regional specializations of the oviducts are most distinct in post-pubescent animals. The most cranial aspect is the thin-walled, funnel-like infundibulum, which is mobile and collects ovulated ova. Ova move into the tuba uterina, which has secretory cells that initiate albumin production, then on caudally into the short ciliated isthmus. Ova then move into the anterior uterus where eggshell membranes start to form. Fertilization is internal in all reptiles. Sperm must transit the length of the oviduct and fertilization is in the more anterior aspects of the oviduct or infundibulum.[19,25] It is common for both oviducts to transport the eggs that form a single clutch.

The Testudinata, Crocodilia, and Sphenodontia all are oviparous and the embryos are lecithotrophic. In turtles, clutches may be large, as in marine turtles that lay multiple clutches in a breeding season. Clutches of relatively small eggs often range from 50-150 and females may produce several hundred eggs per season, depending upon species. At the other end of spectrum are the pancake tortoise (*Malacochersus tornieri*) that lays 1-2 large egg clutches, up to 5 times per season in captivity.[26] The small tortoises produce just a few eggs per season. Squamate clutch sizes vary greatly among species. *Anolis* spp. have clutches of 1 or 2 eggs, *Lacerta vivipara* lay moderate sized clutches of 8-12 eggs, *Iguana iguana* and the grass snake *Natrix natrix* lay moderately larger clutches between 12-50 eggs, and large pythons can lay up to 100 eggs.[27,28] Crocodilians produce a wide range of clutch sizes with the largest species producing the largest clutches.[29] *Sphenodon* produces clutches that average 9-10 eggs (range 1-18).[15]

Recent work on embryonic diapause demonstrated that the distal oviduct of turtles (uterus and vagina) is a hypoxic environment maintains the embryos in the gastrula stage until oviposition.[30,31] Similar studies have not been conducted in other reptiles.

Table 2. Oviduct regional specializations, listed from cranial to caudal. Terminology is not consistent within or among taxa; however, the basic organization of the oviduct regions is consistent.[15-22]

Taxon	Structure and function	Reference
Crocodilia		
Alligator mississippiensis	Infundibulum: uptake of ovulated ova, fertilization Tuba uterina (ie, tube): albumen secretion Anterior uterus: eggshell membrane formation Posterior uterus: calcareous shell formation Vagina: shelled egg retention, muscular egg delivery	Bagwill et al 2009[18]
Crocodilia	Fimbria: collection of ovulated ova Infundibulum: transport of ova / fertilization Magnum: fertilization/ albumen secretion Upper uterine tube: eggshell membrane deposition Distal uterine tube: calcareous shell formation	Richardson et al 2002[19]
Testudinata		
Sea turtles (several cheloniids)	Infundibulum: collection of ovulated ova Aglandular zone Magnum, albumen Shell-forming zone Vagina	Miller 2003[20]
Gopherus polyphemus	Infundibulum: collection of ovulated ova Uterine tube: albumen secretion Isthmus (short aglandular region) Uterus: fibrous and calcareous eggshell formation Vagina: retention of eggs until oviposition	Palmer and Gillette 1988[21]
Squamata		
Lizards		
Crotaphytus collaris	Infundibulum: collection of ovulated ova Anterior uterus: Collagen of eggshell membrane and calcium deposition Posterior uterus	Gillette et al 1989[22]
Eumeces obsoletus	Infundibulum: collection of ovulated ova Anterior uterus: fibers of eggshell membrane Posterior uterus	Gillette et al 1989[22]
Ctenosauria pectinata	Infundibulum tube: Uterus: albumen secretion, fibrous and calcareous eggshell formation Vagina: retention of eggs until oviposition	Uribe et al 1988[16]
Snakes		
Erythrolamprus miliaris	Infundibulum (anterior and posterior infundibulum) Glandular uterus Non-glandular uterus (ie, uterovaginal junction) Pouch (ie, vagina)	Rojas et al 2017[17]
Tuatara		
Sphenodon punctatus	Infundibulum uterine tube: mucus and albumen secretion Uterus: fibrous and calcareous eggshell formation Vagina: muscular connection to the cloaca	Cree 2014[15]

Provisioning of Ova and Embryonic Nutrition

The extent of yolk accumulation by ova is associated with the form of embryonic nutrition and the form of egg. In species with oviparous development, the embryo develops in a cleidoic egg and receives nutrients via lecithotrophy. Lecithrophy refers to the ovum's then embryo's own yolk providing essential nutrients for development. In contrast, embryos of viviparous species, get nutrition via matrotrophy. Matrotrophy is a set of conditions in which maternal provisioning occurs during embryonic development in the reproductive tract.[32] Matrotrophy originated multiple times, independently and take diverse forms, many which are not fully distinct because embryos can transition from one form of nutrition to another. In general, embryos from viviparous species accumulate much smaller quantities of yolk. Matrotrophy includes placentotrophy (nutrition and gas exchange via a placenta formed by fetal membranes in contact with uterine wall) and is common is many skink species (Scindae); Histotrophy (embryos ingest nutrients, typically liquids of maternal origin, that diffuse across tissue spaces to absorptive sites of the embryo); Oophagy is a mechanism by which embryos develop in reproductive tract and are nourished by ingesting sibling yolks or yolk breakdown products), matrophagy occurs when embryos actively eat maternal tissues such as the oviductal mucosa, and embryophagy is defined as developing embryos gaining their nutrition by cannibalize their siblings.[33]

Eggs

Eggshells

Eggs with a thick mineralized calcareous layer overlying the shell membrane are rigid. Eggs with a thin or loosely organize calcareous layer over the shell membrane are pliable (also termed flexible or parchment shelled). Ewert (1985) further separates non-brittle eggs into pliable and hard-expansible eggs.[25] He defines hard-expansible eggs and those that feel brittle when slightly dehydrated and pliable when slightly expanded after absorbing water. Pliable shelled eggs differ in their turgidity with water absorption. These eggs will collapse partially and fold if insufficient moisture is available and rebound when moisture is available unless dehydration is extreme. Rigid shelled eggs remain rigid regardless of moisture. If insufficient water is available, the embryo and its membranes will collapse within the eggshell and not recover.

The eggshell mineral layer is primarily calcium carbonate as aragonite crystals in Testudines, but primarily calcium carbonate as calcite in crocodilians and *Sphenodon*.[12,15] It both protects the developing embryo and is its major calcium source. It is common for the mineral layer to crack and crumble towards the end of development. Ewert (1985) describes the mineral later and its organization in detail for turtles.[25] Eggshell formation in lizards varies in form and organization in oviparous species.[22] Egg membrane form differs with the embryo's nutritional source in viviparous species. Each egg of viviparous squamates is surrounded by a thin eggshell membrane.[34]

The embryo is located on the vitteline membrane, which surrounds the yolk. Albumen surrounds the embryo and its yolks. Extraembryonic tissues provide both protection and support form the embryo. They include the yolk sac (ie, vitelline membrane), chorion, amnion and allantois. The general pattern of development of the extraembryonic membranes is similar among Reptilia.[35,36] In oviparous lizards and snakes, the yolk sac becomes vascular and eventually surrounds the entire yolk mass. In viviparous species the region of the isolated yolk mass contributes to a yolk sac placenta.[37,38]

Cloaca

Each oviduct joins the cloaca via a muscular vagina. There usually is an opening for each oviduct that is a genital pore or slit adjacent to the ureter from the kidney on the same size. In some species, each vagina and ureter open into a urogenital papilla; such is the case for *Sphenodon* and some turtles.[15] Spadola and Insacco (2009) conducted a series of endoscopic studies of the cloaca structures in *Emys trinacris* and described the urogenital papillae as lateral to the urethra.[39] Blackburn (1998) notes that it is typical for the oviducts and ureter to open separately into the cloaca in squamates; however, the oviducts of the southern alligator lizard (*Elgaria multicarinata*) join to form a common tube that is continuous with the posterior urodeum of the cloaca.[24] Some boid snakes have a single urogenital papilla yet have both ureters and both oviducts opening into the cloaca.

Parthenogenic Species

Reptiles include the only truly parthenogenic amniotes. Kearney et al (2009) provide the most up-to-date overview of parthogenetic species.[40] Parthenogenic (unisexual) species or populations have been documented in 8 lizard families and 2 snake families (Table 3). There are 40 known cases of obligate parthenogenesis in the reptiles within 8 families. There are no cases of parthenogenesis in turtles, crocodilian, or *Sphenodon*. All but one natural case of parthenogenesis has been demonstrated to be of hybrid origin.[40]

The morphology of parthenogenic species has been examined in few species. The ovaries and oviducts of parthenogenic geckos show no morphologic differences from their sexual counterparts.[41]

Table 3. Parthenogenic reptile families.[a]

Reptile	Number of verified species	Obligate
Snakes		
Boidae	1	No
Typhlopidae	1	Obligate
Lizards		
Agamidae	3	Obligate
Gekkonidae	7	Obligate
Gymnophthalmidae	2	Obligate
Lacertidae	7	Obligate
Scincidae	1	Obligate
Teiidae	18	Obligate
Varanidae	2	No
Xantusiidae	2	Obligate

[a]Obligate parthenogenic species consist of only females and do not have to mate for embryogenesis to occur. Most parthenogenic species have a hybrid origin. Species that are not obligate parthenogens are those for which asexual reproduction is documented in captivity, but is not typical for the species.

Source: Adapted from Kearney M, Fujita MK, Ridenour J. Lost sex in the reptiles: Constraints and correlations. In: Schon I, Martens K, Dijk P, eds. *Lost Sex: The Evolutionary Biology of Parthenogenesis.* Springer, Dordrecht, The Netherlands: Springer Netherlands;2009:447-474.

References

1. Guraya SS. *Ovarian Follicles in Reptiles and Birds.* Springer Science & Business Media;2013.

2. Kardong KV. *Vertebrates: Comparative Anatomy, Function, and Evolution.* 8th ed. Boston, MA: McGraw-Hill Higher Education; In press.

3. Janzen, FJ, Phillips PC. Exploring the evolution of environmental sex determination, especially in reptiles. *J Evol Biol.* 2006;19(6):1775-1784.

4. Etches RJ, Petitte JN. Reptilian and avian follicular hierarchies: Models for the study of ovarian development. *J Exp Zool.* 1990;4:112-22.

5. Bull JJ, Vogt RC. Temperature-sensitive periods of sex determination in emydid turtles. *J Exp Zool.* 1981;218(3):.435-440.

6. Bull JJ. Temperature-sensitive periods of sex determination in a lizard: Similarities with turtles and crocodilians. *J Exp Zool.* 1987;241(1):143-148.

7. Nelson NJ, Keall SN, Pledger, S, Daugherty CH. Male-biased sex ratio in a small tuatara population. *J Biogeogr.* 2002;29(5-6):633-640.

8. Mitchell NJ, Nelson NJ, Cree A, Pledger S, Keall SN, Daugherty CH. Support for a rare pattern of temperature-dependent sex determination in archaic reptiles: Evidence from two species of tuatara (*Sphenodon*). *Front Zool.* 2006;3(1):9-20.

9. Radder RS, Quinn AE, Georges A, Sarre SD, Shine R. Genetic evidence for co-occurrence of chromosomal and thermal sex-determining systems in a lizard. *Biol Lett.* 2008;4(2):176-178.

10. Pokorná M, Kratochvíl L. Phylogeny of sex-determining mechanisms in squamate reptiles: Are sex chromosomes an evolutionary trap? *Zool J Linnean Soc.* 2009;156(1):168-183.

11. Fox H. The urogenital system of reptiles. In: Gans C, Parsons TS, eds. *Biology of the Reptilia.* London, England: Academic Press;1977:1-157.

12. Ferguson MWJ. Reproductive biology and embryology of the crocodilians. In: Gans C, Billet F, Maderson P, eds. *Biology of the Reptilia*. New York, NY: John Wiley & Sons;1985:329-491.

13. Uribe MCA, Guillette Jr, LJ. Oogenesis and ovarian histology of the American alligator, *Alligator mississippiensis*. *J Morphol.* 2000;245(3):225-240.

14. Calderón M L, De Pérez, GR, Pinilla MR. Morphology of the ovary of *Caiman crocodilus* (Crocodylia: Alligatoridae). *Ann Anat.* 2004;186:13-24.

15. Cree A. *Tuatara: Biology and Conservation of a Venerable Survivor.* Christchurn, NZ: Canterbury University Press;2014.

16. Uribe MCA, Velasco SR, Guillette Jr, LJ, Estrada EF. Oviduct histology of the lizard, *Ctenosaura pectinate.* *Copeia.* 1988;1035-1042.

17. Rojas CA, Barros VA, Almeida-Santos SM. A histological and ultrastructural investigation of the female reproductive system of the water snake (*Erythrolamprus miliaris*): Oviductal cycle and sperm storage. *Acta Zoologica.* 2017;doi.org/10.1111/azo.12234.

18. Bagwill A, Sever DM, Elsey RM, Seasonal variation of the oviduct of the American alligator, *Alligator mississippiensis* (Reptilia: Crocodylia). *J Morphol.* 2009;270(6):702-713.

19. Richardson KC, Webb GJ, Manolis SC. *Crocodiles: Inside Out. A Guide to the Crocodilians and their Functional Morphology.* Surrey Beatty & Sons;2002.

20. Miller JD, Limpus CJ. Ontogeny of marine turtle gonads. In: Lutz P, Musick JA, Wyneken J, eds. *The Biology of Sea Turtles.* Vol 2. Boca Raton, FL: CRC Press;2003:199-224.

21. Palmer BD, Guillette Jr, LJ. Histology and functional morphology of the female reproductive tract of the tortoise *Gopherus polyphemus. Am J Anat.* 1988;183(3):200-211.

22. Guillette LJ, Fox SL, Palmer BD. Oviductal morphology and egg shelling in the oviparous lizards *Crotaphytus collaris* and *Eumeces obsoletus. J Morphol.* 1989;201(2):145-159.

23. Girling JE. The reptilian oviduct: A review of structure and function and directions for future research. *J Exp Zool.* 2002;293(2):141-170.

24. Blackburn DG. Structure, function, and evolution of the oviducts of squamate reptiles, with special reference to viviparity and placentation. *J Exp Zool.* 1998;282(4-5):560-617.

25. Ewert MA. Embryology of turtles. In: Gans C, Billet F, Maderson PFA, eds. *Biology of the Reptilia.* New York, NY: John Wiley & Sons;1985:75-261.

26. Darlington AR, Davis RB. Reproduction of the pancake tortoise, *Malacochersus tornieri*, in captive collections. *Herpetol Rev.* 1990;21:16-18.

27. Hubert J. 1985. Embryology of the squamata. In: *Biology of the Reptilia.* New York, NY: John Wiley & Sons;1985:1-34.

28. Alvarado J, Ibarra L, Suazo I, Rodríguez G, Zamora R. Reproductive characteristics of a green iguana (*Iguana iguana*) population of the west coast of Mexico. *Southwestern Nat.* 1995;40(2):234-237.

29. Greer AE. Clutch size in crocodilians. *J Herpetol.* 1975;9(3):319-322.

30. Rafferty AR, Reina RD. Arrested embryonic development: a review of strategies to delay hatching in egg-laying reptiles. *Proc Biol Sci.* 2012;279(1737):2299-2308.

31. Rafferty AR, RG Evans, TF Scheelings, RD Reina. Limited oxygen availability in utero may constrain the evolution of live birth in reptiles. *Am Nat.* 2013;181(2):245-253.

32. Blackburn DG, Starck JM. Morphological specializations for fetal maintenance in viviparous vertebrates: an introduction and historical retrospective. *J Morphol.* 2015;276(11):E1–E16.

33. Blackburn DG. Evolution of vertebrate viviparity and specializations for fetal nutrition: a quantitative and qualitative analysis. *J Morphol.* 22015;76(8):961-990.

34. Heulin B, Stewart JR, Surget-Groba Y, et al. Development of the uterine shell glands during the preovulatory and early gestation periods in oviparous and viviparous *Lacerta vivipara. J Morphol.* 2005;266(1):80-93.

35. Luckett WP. Ontogeny of amniote membranes and their application to phylogeny. In: Hecht MK, Goody PC, Hecht BM, eds. *Major Patterns in Vertebrate Evolution.* New York, NY: Plenum Press;1977:439-516.

36. Stewart JR. Morphology and evolution of the egg of oviparous amniotes. In: Sumida S, Martin K, eds. *Amniote Origins.* New York, NY: Academic Press;1997:291-326.

37. Blackburn DG, Stewart JR. Viviparity and placentation in snakes. In: Aldridge RD, Sever DM, eds. *Reproductive Biology and Phylogeny of Snakes.* Boca Raton, FL: CRC Press;2011:119-181.

38. Stewart JR, Blackburn DG. Viviparity and placentation in lizards. In: Rheubert J, Seigel D, Trauth S, eds. *Reproductive Biology and Phylogeny of Lizards and Tuatara.* Boca Raton, FL: CRC Press;2015:448-563.

39. Spadola F, Insacco G. Endoscopy of cloaca in 51 *Emys trinacris* (Fritz et al., 2005): Morphological and diagnostic study. *Acta Herpetol.* 2009;4(1):73-81.

40. Kearney M, Fujita MK, Ridenour J. Lost sex in the reptiles: constraints and correlations. In: Schon I, Martens K, Dijk P, eds. *Lost Sex. The Evolutionary Biology of Parthenogenesis.* Springer, Dordrecht, The Netherlands: Springer Netherlands;2009:447-474.

41. Whittier J, Stewart D, Tolley L. Ovarian and oviductal morphology of sexual and parthenogenetic geckos of the *Heteronotia binoei* complex. *Copeia.* 1994;2:484-492.

Reptile Reproduction: Diagnosis and Treatment Options

J. Jill Heatley, DVM, MS, DABVP (Avian Practice), DABVP (Reptile and Amphibian Practice), Dipl ACZM

Session #2147

Abstract: An overview of diagnosis and treatment of disease in the reptilian female reproductive tract and eggs is provided. History, physical examination, imaging, pathology, cytology, among other diagnostics are covered. Medical and husbandry treatment options are outlined.

I. Ovary: Preovulatory Follicular Stasis, Ovarian Neoplasia, Systemic Disease (Metabolic or Infectious), Improper Husbandry

Diagnostics

History

An increase in coelom volume, anorexia, restlessness, and unsuccessful attempts of egg laying may be present.

Physical examination

On abdomen palpation there may be firm egg-like structures or fluid. Digital palpation and noninvasive diagnostic techniques including radiography, ultrasonography, and CT may be used to characterize the ovaries and search for any condition that may cause blockage of the oviduct.

Imaging

Ultrasound is preferred for nonshelled eggs, computed tomography (CT) and magnetic resonance imaging (MRI) may also be useful in species with limited ultrasound windows (chelonian), transillumination limited to thin skinned small creatures. Radiographs include negative contrast air coelomic radiographs, assessment of bone density, soft tissue densities, and endolymphatic mineral reserves. A recent review of reptile reproductive tract imaging suggests that radiographs are not useful for visualizing the ovary of lizards, snakes and chelonians, compared to either ultrasound or CT; however, radiographs are not as useful but may be acceptable for visualization of follicles and eggs with noncalcified shells.

II. Uterus and Shell Gland: Postovulatory Ovostasis, Egg Stasis, Dystocia, Neoplasia, Systemic (Metabolic or Infectious) Disease, Infectious or Inflammatory Disease of the Shell Gland

Diagnostics

History

In amphibians egg retention, or dystocia, may occur when captive gravid females are not provided appropriate environmental parameters or physiologic and natural stimuli for release of eggs.

Clinical signs can be general. They can include increased activity followed by depression.

In snakes initial increase in activity, restlessness, writhing due to discomfort, tail held vertically and cloacal discharge, later exhaustion, depression, snake basking with caudal third upside down, swelling near egg area (caudal third), eggs or young found deposited in cage (partial clutch). As a general rule for snakes, ecdysis should occur ~7-10 (range 3-14) days prior to egg lay and all offspring should be produced over 24 hours.

In lizards, restlessness or digging, anorexia, inactivity, straining with or without cloacal eversion may be observed.

In chelonians, signs include digging motions of the hind limbs, clear cloacal discharge, restlessness, straining, and cloacal prolapse.

Physical examination

Digital palpation and noninvasive diagnostic techniques including radiography, ultrasonography, and CT may be used to characterize the ovaries and search for any condition that may cause blockage of the oviduct. Ballottement may result in "billiard ball sound" or normally formed eggs in tortoises with hard shelled eggs.

Imaging

Dystocia was the cause of imaging consultation for 10.7% of reptile cases in a recent review of 75 lizards, mainly agamids, in Poland. A recent review of reptile reproductive tract imaging suggests that radiographs are comparatively useful to visualize eggs that are either poorly or well calcified of lizards, snakes and chelonians, compared to either ultrasound or CT.

Clinical Pathology

Normocalcemia, proteinemia, lipidemia are expected should eggs appear normally shelled. Hypocalcemia, proteinemia, may require aggressive replacement therapy. Hypercalcemia, proteinemia, lipidemia may be normal if the animal is in a preovulatory state. Abnormal is likely if the animal is in post ovulatory ovostasis. Renal disease and other causes of dehydration should be included in the differential.

Treatment

Husbandry

Provide appropriate nesting, environmental temperature, humidity, and nutrition. Specific light, dark, rain, or aridity are required for nesting or reproduction of some species.

Treat infection, inflammation, and dehydration

Oxytocin, arginine vasopressin

Calcium, other nutritional supplementation

Anesthetics and analgesics

Midazolam, morphine, nalbuphine, meloxicam

Muscle relaxation, analgesia

Prostaglandin induction technique

Lutalyse, prostaglandin F2α plus Sedivet, romifidine, alpha-agonist.

www.inducingturtles.com

III. Coelomitis

Diagnostics

History

Clinical signs can be nebulous and nonspecific. They can include lethargy, anorexia, poor coloration.

Physical examination

Whole body inflammation from sepsis may be evident based on dermatosepticemia, swelling of the inguinal coelomic inlets may be evident, fluid in the coelom may be evident.

Imaging

Ultrasound may be more fruitful than radiographs, which may or may not provide evidence of free gas or of a fluid line. CT/MRI may be more diagnostic in large chelonian based on limited ultrasounds access points.

Clinical Pathology

Evidence of sever inflammation (Band, heterophilia) or sepsis may or may not be present. Coelomic wash or aspiration of fluid for fluid analysis may provide evidence of presence and cause of coelomitis based on protein presences, fluid specific gravity, yolk presence of absence and cell count.

Culture and sensitivity

Culture and sensitivity of fluid is indicated to guide appropriate antibiotic treatment.

Treatment

Nonsurgical options include coelomic lavage. Despite the inflammation that eggs cause within the coelom, the reptile body is made to resorb unused follicles, slowly.

Antibiotics may be indicated should bacterial infection be present. Anti-inflammatories may be indicated should pain be evident; however, this could slow down the absorption process, but also reduce later adhesions of egg yolk peritonitis.

IV. The Egg

Diagnostics

History

The date of lay and dam should be recorded.

Egg examination

Unless egg damage is seen, delay candling for 1-3 weeks to allow embryo development and adhesion. After this point eggs should remain the same side up, marking may facilitate.

Egg viability

Viability can be determined through visual assessment, candling, or auscultation of the embryo heart rate or Egg Buddy (laser HR determination).

Imaging

Generally late in incubation, eggs that become white to light and have dry down based on candling are infertile. Those that remain dark are continuing to develop. Candling, endoscopy, and radiographs are each viable methods of imaging eggs.

Cytology and culture

Egg necropsy, cytology, and culture may be indicated in cases of perinatal or late embryonic death.

Treatment

While egg treatment of birds is well documented (antibiotic treatment of exposed eggs, assisted hatch), early intervention treatments of reptile eggs do not appear well documented. (Audience participation☺) Manual pipping is best reserved for cases of questionable embryo viability and requires sterile technique and magnification.

Recommended Reading

1. Portas TJ. Disorders of the reproductive system. In: Doneley B, Monks D, Johnson R, Carmel B, eds. *Reptile Medicine and Surgery in Clinical Practice.* Ames, IA: John Wiley & Sons;2017:307-321.

2. Pirson S, Geerinckx L, Bosseler L, Simard J, Hellebuyck T. Metastatic granulosa cell tumor in a mangrove snake (*Boiga dendrophila melanota*). *J Exot Pet Med.* 2018;27(3):6-10.

3. Spadola F, Morici M, Oliveri M, Knotek Z. Description of cloacoscopy in the loggerhead sea turtle (*Caretta caretta*). *Acta Veterinaria Brno.* 2017;;85(4):367-370.

4. Monks D, Doneley B. Reptile paediatrics. In: Doneley B, Monks D, Johnson R, Carmel B, eds. *Reptile Medicine and Surgery in Clinical Practice.* Ames, IA: John Wiley & Sons;2017:105-113.

5. Spadola F, Morici M, Santoro M, Oliveri M, Insacco G. Reproductive disorders and perinatology of sea turtles. *Vet Clin North Am Exot Anim Pract.* 2017;20(2):345-370.

6. Keller KA. Reptile perinatology. *Vet Clin North Am Exot Anim Pract.* 2017;20(2):439-454.

7. Cojean O, Vergneau-Grosset C, Masseau I. Ultrasonographic anatomy of reproductive female leopard geckos (*Eublepharis macularius*). *Vet Radiol Ultrasound.* 2018;59(3):333-344.

8. Stahlschmidt Z, Brashears J, Denardo D. The use of ultrasonography to assess reproductive investment and output in pythons. *Biol J Linnean Soc.* 2011;103(4):772-778.

9. Di Giuseppe M, Martínez-Silvestre A, Luparello M, Faraci L. Post-ovulatory dystocia in two small lizards: Leopard gecko (*Eublepharis macularius*) and crested gecko (*Correlophus ciliatus*). *Russian J Herpetol.* 2017;24(2):128-132.

10. Stahlschmidt ZR, Hoffman T, DeNardo DF. Postural shifts during egg-brooding and their impact on egg water balance in Children's pythons (*Antaresia childreni*). *Ethology.* 2008;114(11):1113-1121.

11. Stahlschmidt ZR, Brashears J, DeNardo DF. The role of temperature and humidity in python nest site selection. *Anim Behav.* 2011;81(5):1077-1081.

12. Lourdais O, Hoffman TC, DeNardo DF. Maternal brooding in the Children's python (*Antaresia childreni*) promotes egg water balance. *J Comp Physiol B.* 2007;177(5):569-577.

13. Kophamel S, Lebens M, Gunther P, et al. Lipogranulomatous oophoritis in a Leach's giant gecko (*Rhacodactylus leachianus*). *Wiener Tierarztliche Monatsschrift.* 2018;105(3-4):73-80.

14. Brusch IV GA, Billy G, Blattman JN, DeNardo DF. Reproduction alters hydration state but does not impact the positive effects of dehydration on innate immune function in Children's pythons (*Antaresia childreni*). *Physiol Biochem Zool.* 2017;90(6):646-654.

15. Holtze S, Lukač M, Cizelj I, et al. Monitoring health and reproductive status of olms (*Proteus anguinus*) by ultrasound. *PloS One.* 2017;12(8):e0182209.

16. Nardini G, Girolamo ND. Reptile clinical pathology. *Veterinaria* (Cremona). 2017;31(4):197-205.

17. Stasis E. Disorders of the reproductive system. In: Doneley B, Monks D, Johnson R, Carmel B, eds. *Reptile Medicine and Surgery in Clinical Practice.* Ames, IA: John Wiley & Sons;2017:307.

18. Lamb AD, Watkins-Colwell GJ, Moore JA, et al. Endolymphatic sac use and reproductive activity in the lesser Antilles endemic gecko *Gonatodes antillensis* (Gekkota: Sphaerodactylidae). *Bull Peabody Museum Natl History.* 2017;58(1):17-29.

19. Moeller KT, Demare G, Davies S, DeNardo DF. Dehydration enhances multiple physiological defense mechanisms in a desert lizard, *Heloderma suspectum*. *J Exp Biol.* 2017;220(12):2166-174.

20. Łojszczyk-Szczepaniak AN, Szczepaniak KO, Grzybek M, Lisiak B. Causes of consultations and results of radiological and ultrasound methods in lizard diseases (2006-2014). *Med Weter.* 2018;74(1):65-69.

21. Brischoux F, Dupoué A, Lourdais O, Angelier F. Effects of mild wintering conditions on body mass and corticosterone levels in a temperate reptile, the aspic viper (*Vipera aspis*). *Comp Biochem Physiol A.* 2016;192:52-56.

22. Chai N. Reproductive medicine in amphibians. *Vet Clin North Am Exot Anim Pract.* 2017;20(2):307-325.

23. Perry SM, Mitchell MA. Reproductive medicine in freshwater turtles and land tortoises. *Vet Clin North Am Exot Anim Pract.* 2017;20(2):371-389.

24. Knotek Z, Cermakova E, Oliveri M. Reproductive medicine in lizards. *Vet Clin North Am Exot Anim Pract.* 2017;20(2):411-438.

25. Johnson R, Carmel B. Reproductive strategies in reptiles. In: Doneley B, Monks D, Johnson R, Carmel B, eds. *Reptile Medicine and Surgery in Clinical Practice.* Ames, IA: John Wiley & Sons;201716:91.

26. Ananjeva NB, Uteshev VK, Orlov NL, et al. Comparison of the modern reproductive technologies for amphibians and reptiles. *Russian J Herpetol.* 2017;24(4).

27. Di Girolamo N, Selleri P. Reproductive disorders in snakes. *Vet Clin North Am Exot Anim Pract.* 2017;20(2):391-409.

28. Gumpenberger M. Diagnostic imaging of reproductive tract disorders in reptiles. *Vet Clin North Am Exot Anim Pract.* 2017;20(2):327-343.

Endoscopy and Surgery of the Reptile Reproductive Tract

Stephen J. Divers, BVetMed, DZooMed, DipECZM (Herpetology), DipECZM (Zoo Health Management), Dipl ACZM, FRCVS, and Scott J. Stahl, DVM, DABVP (Avian Practice)

Session #2165

Affiliation: From Department of Small Animal Medicine and Surgery (Zoological Medicine), College of Veterinary Medicine, University of Georgia, 2200 College Station Road, Athens, GA 30602, USA, and Stahl Exotic Animal Veterinary Services, 4105 Rust Road, Fairfax, VA 22030, USA.

Abstract: Reproductive disease remains commonplace in the captive reptile population. Endoscopy can be a valuable for identification of gender in monomorphic juveniles, minimally-invasive sterilization, investigation of reproductive disease and as an aid to correction of surgical issues. When removal of large amounts of tissue is required (eg, dystocia), a traditional surgical approach is still favored.

Introduction

Now that our clients have become better educated regarding captive management and nutrition, reproductive diseases have become increasingly common. This presentation focuses on surgical intervention and specifically diagnostic endoscopic diagnosis, endosurgical resolution and traditional surgical techniques. Given the number of species and varied presentations, an exhaustive review of all possibilities is not possible and therefore examples are provided to highlight specific techniques. For a complete review readers are referred to chapters 64 Diagnostic Endoscopy, 65 Endoscope-Assisted and Endoscopic Surgery, 80 Theriogeniology, and 105 Reproductive surgery in the forthcoming 3rd edition of Mader's Reptile and Amphibian Medicine and Surgery, from which excerpts have been included.[1]

Diagnostic Endoscopy

Introduction

A definitive diagnosis is important to maximize treatment success, but reptile cases are frequently mismanaged simply because of a failure to identify a specific problem and provide appropriate therapy. Definitive diagnosis relies upon the demonstration of a host pathologic response (by histopathology, less reliably cytology, or using paired rising antibody titers), and the causative agent (by microbiology, parasitology, or toxicology). Previously plagued by vague medical histories, non-pathognomonic physical examinations, indistinct diagnostic images, and less than conclusive clinical pathology results, many clinicians have discovered that diagnostic endoscopy offers an unparalleled ability to facilitate ante-mortem diagnosis, and maximize treatment success.

The class Reptilia consists of over 10,000 species, but preference has been given to those species and procedures that are most commonly performed in zoological practice. Prior to the 1990s there were only sporadic reports of reptile endoscopy. The last two decades have seen substantial developments and dedicated literature on the subject.[2,3] In addition, dedicated training is now available for reptile endoscopy (see http://vet.uga.edu/events/events_template/avian-reptile-small-mammal-endoscopy). Given the often small and delicate nature of small reptile species, the continued development of minimally-invasive endoscopy appears assured in these taxa.

Patient selection, evaluation, and contraindications

Most reptiles are presented in a state of advanced, often chronic disease, and therefore medical stabilization is important prior to anesthesia and surgery. The most common endoscopic theriogenology procedures performed by the author include coelioscopy for gender identification and cloacoscopy. In addition to anesthetic complications, small patient size often presents the greatest challenge. Obesity is less of an issue in chelonians and lizards because they store fat in discrete fat bodies rather than diffusely around organs like mammals. Obesity in snakes can still be a significant hindrance during coelioscopy; however, due to their more diffuse coelomic fat reserves.

The benefits of endoscopy are numerous but mostly center on the fact that traditional approaches are more invasive, require longer anesthetic periods, and can carry significantly greater procedural risks (eg, coeliotomy versus coelioscopy). With the ability to examine internally and collect tissue samples, the array of ante-mortem diagnoses has increased considerably with accurate diagnosis resulting in more appropriate targeted therapy, and improved clinical success.

Patient preparation and anesthesia

In most cases, the reptile presents with anorexia of days to months duration, and fasting is usually not a concern; however, when dealing with animals that are still feeding, fasting should be in accordance with body size and feeding strategy. General anesthesia is recommended for all endoscopy procedures (including cloacal examinations), and is required for any invasive procedures (including coelioscopy).[4,5] Conscious examinations may be acceptable in research animals or wildlife under an appropriate experimental license, but in clinical practice, analgesia and anesthesia are mandatory. While some authors have reported using only local anesthesia for chelonian gender identification by coelioscopy, this likely provides inadequate analgesia.[6] In a recent comparison of local versus general anesthesia for chelonian gender identification by coelioscopy, struggling was significantly reduced for procedures conducted under general anesthesia compared to local lidocaine alone.[5]

Operating room preparation, ergonomics and equipment handling

Some consideration should be given to room set-up and positioning which will greatly facilitate procedural success. Procedural and species anatomy are key factors in animal positioning and are dealt with under each individual procedure. The endoscopy tower and especially the video monitor should be positioned directly in front of the endoscopist. Trying to observe a monitor to one side, or even worse over a shoulder, creates confusion, frustration and discomfort. For right-handed surgeons, the instrument table should be placed on the right (and vice versa for those who are left handed).

The fact that most reptiles weigh < 2 kg necessitates careful control of the telescope with the eye-piece and camera supported using the superior hand, and the distal shaft of the telescope held by thumb and forefinger of the inferior hand. Handling the telescope in this fashion provides fine control without tremor; however, for a single surgeon to be able to use instruments a modified hold is required. In order to insert and manipulate instruments through the operating sheath it is necessary to change from a two-handed to a one-handed technique. The usual thumb and forefinger support of the telescope shaft is now adjusted so that the inferior hand takes the entire weight of the sheath-telescope-camera system. The thumb is slid up the shaft of the sheath, and the fingers are curled over the top to encircle the sheath. Now that the sheath is grasped in a fist-like grip with the sheath further supported by the thumb to prevent rotation, the superior hand can be removed to manipulate an instrument down the operating channel. This can only be performed using a correctly sheathed telescope - damage will otherwise occur to the telescope.

Cloacoscopy

While snakes can also be positioned in sternal or dorsal recumbency, chelonians and lizards are generally easier in dorsal. Endoscopy using warm saline irrigation promotes unparalleled examination of the proctodeum, urodeum, and coprodeum. Detailed appraisal of the distal colon, cloacal mucosa, urogenital papillae, oviductal openings in females, and (when present) the urethral opening and bladder. With practice, the endoscope can be directed through the urethra into the thin-walled bladder in chelonians, and some lizards. It is also possible to enter the distal oviducts of reproductively active females. Cloacoscopy has been used to remove shell and egg material from the distal oviducts and bladder, and as a guide for the exteriorization of partially prolapsed tissue and resection. Mucosal biopsies should be taken with care especially from the colon due to its thin nature, and the risk of perforation. More recently, a technique for neonate gender identification that centered on observing the gonads through the bladder wall via a cloacoscopic approach has been described in Hermanns tortoises (*Testudo hermanni*).[7] This technique has the advantage of avoiding entry into the coelomic cavity but, because examination is through a membrane, may not be as sensitive as direct coelioscopic visualization. In addition, research at UGA involving sliders (*Trachemys*) indicated that perforation of the urethra and bladder were common and the technique could not be recommended (Proenca, unpublished data). There may be species-specific safety limitations for this cystoscopic technique.

Coelioscopy

As reptiles lack a true abdomen, the term coelioscopy is preferred over laparoscopy. For a detailed discussion of generic laparoscopy methodologies the reader is referred to the dedicated literature as only reptile specifics will be highlighted here.[8] Laparoscopy has been shown to offer significant advantages over traditional surgical options, both in human and veterinary medicine. In particular, laparoscopy is, with practice, faster, less traumatic, causes less post-operative pain, with a more rapid return to normal function.[9-20] Until the advent of 2- and 3-mm human pediatric equipment, laparoscopy was limited to a single-entry system using the sheathed telescope; however, multiple entry theriogenologic techniques are now possible and practical for animals over 500 g.[21]

Precise positioning for coelioscopy depends upon the reptile's body shape and confirmation, the structure(s) of interest, and the preferences of the endoscopist.[22] The entry site is aseptically prepared and draped, and adhesive clear plastic drapes permit better visualization of the patient. Insufflation is essential for lizards and snakes and, although the shell prevents coelomic collapse, can still be helpful in many chelonians. Typically, CO_2 insufflation pressures of 3-5 mm Hg are used, which are lower than those recommended for mammals (8-12 mm Hg). In some situations, saline infusion may be preferred over gas. In the author's experience this is especially true when dealing with neonates, juveniles, and other very small reptiles, or aquatic species, where residual gas may have a negative effect on post-operative buoyancy.

Lizards

For access to the dorsal gonads, a lateral (paralumbar) approach is preferred. While anatomic differences exist between families, the green iguana serves as a useful model for saurian coelioscopy.[23,24] In the iguana, there is no significant difference in visualization of lung, liver, pancreas, small intestine, large intestine, ovary, oviduct, testis, epididymis, vas deferens, bladder, fat body, or kidney from either left or right approaches.[24] Right-handed surgeons will find that a left paralumbar approach is easier while left-handed surgeon will prefer a right paralumbar approach. Unless physical examination, diagnostic imaging or clinicopathology suggest an asymmetric lesion, the endoscopist's preference can prevail. The iguana is positioned in lateral recumbency, with the left pelvic limb taped caudad against the tail base. The entry area is bordered craniad by the ribs, dorsad by the spine, and caudad by the pelvic limb; however, in larger lizards an intercostal approach is also practical. Aseptically, a small skin incision is made in the center of the defined area. The skin and underlying muscle is grasped and elevated away from the coelomic viscera and the sheath (with obturator), aimed parallel to the table and towards the head, are gently forced through the coelomic musculature and into the coelomic cavity. It is wise to temporarily cease artificial ventilation until the telescope has been introduced into the coelom, thereby reducing the possibility of damage to an inflated lung or viscus. The obturator is replaced by the telescope and the insufflation line attached to one of the sheath ports. By making a small skin incision and breach in the muscle, the sheath will be tight fitting and insufflation gas leakage will be prevented. If there is any gas leakage, a mattress suture can be placed through the skin and around the sheath. Once a pneumocoelom has been created, it may be necessary to gently touch the tip of the telescope against a coelomic membrane to clean the terminal lens of condensation or tissue fluid. If there is fat or blood on the lens it is often necessary to remove the telescope from the sheath, clean with gauze moistened with sterile saline, and replace. It is unwise to continue with poor visualization.

Upon entry via a left paralumbar approach in the iguanid lizard, the first organ to note is the large brown liver lying in the mid-ventral coelom. There are no diaphragmatic, post-pulmonary, or longitudinal membranes in iguanas; however, these membranes do exist to a greater or lesser extent in tegus, monitors, and certain other saurians. Minor perforation of these membranes by the telescope will not cause any harm but care is required to avoid perforating the underlying viscera. Lung ventilation will be substantially reduced during CO_2 insufflation, and careful communication with the anesthetist is required to balance ventilation and coelomic pressures. The gonads are located just caudodorsal to the stomach and spleen, on either side of dorsal midline. Rigid endoscopy can be utilized to determine the gender of reptiles, even at a very early age. This is extremely useful in monomorphic species, or when involved in an endangered species breeding program that requires gender determination in neonates or juveniles.[5] Endoscopy also provides feedback on reproductive activity and disease. The testis is usually ovoid and smooth, and size may vary dramatically with season. The immature or inactive ovary appears as a cluster of small, fluid-filled follicles. As the ovary matures some of the follicles will enlarge and appear yellow to orange in color as they fill with yolk. Some species of lizards have thin post-hepatic and post-pulmonary membranes that may be pigmented and obscure the gonads. The membrane can be breached to allow better visualization by grasping and tearing with retrieval forceps, or incising using scissors. Care must be taken not to damage the follicles as leakage results in coelomitis. Dorsal to the gonads are the adrenal glands which lie adjacent to the renal veins. The vas deferens of males and oviducts of females are also visible and can be followed caudad to the pelvic inlet. The kidneys of most lizards are situated in the caudodorsal coelom; however, those of the iguana are intrapelvic but can still be visualized. The kidneys of male squamates may show significant color and size changes associated with sexual segment hypertrophy, and should not be confused with pathology.

Snakes

Snakes are less commonly subjected to coelomic endoscopy because (i) they have more diffuse coelomic fat, (ii) it is impossible to examine all the major organs via a single-entry point, (iii) insufflation is generally more difficult, (iv); however, targeted coeliotomy can be readily performed to evaluate the gonads, or via a separate approach the vas deferens or oviducts. If a targeted endoscopic approach to a single coelomic region is required, lateral entry between the ribs in larger snakes, and ventrolateral entry just medial to the ribs in smaller snakes are practical options.[22]

Chelonians

With the chelonian supported in lateral recumbency, the most useful coelioscopic approach is via the left or right prefemoral fossa.[22] Right-handed surgeons will generally find it easier to enter the left prefemoral fossa, and left-handed surgeons the right. Again, unless unilateral pathology exists, the surgeon's preference can prevail. A large bladder can hinder coelioscopy and chelonians should be encouraged to urinate prior to surgery. This can be achieved by bathing the animal, or by gentle digital stimulation of the cloaca prior to anesthetic induction.

The pelvic limb is retracted and taped caudad to expose the prefemoral fossa. In chelonians with a well-developed plastron hinge (eg, box turtles) it is wise to place a wedge between the caudal plastron and carapace to maintain exposure. Following aseptic preparation, a small (2-4 mm) craniocaudal skin incision is made in the center of the prefemoral fossa. The subcutaneous fat and connective tissues are bluntly dissected using hemostats down to the coelomic aponeurosis (formed by the broad tendinous portions of the transverse and oblique abdominal muscles) being careful to remain cranial to the sartorius muscle, and ventral to the iliacus muscle that lies between the femur and ventral surface of the ilium. The coelomic aponeurosis is penetrated by advancing hemostats (or the sheath with obturator) towards the head (ie, parallel, not perpendicular, to the table). Some force is required to breach of the coleomic membrane, and gain access to the coelomic cavity. An objective evaluation of coelioscopy in freshwater turtles (*Trachemys scripta* spp) has been published.[25,26]

The gonads and oviducts or vas deferens can be viewed from either side. The bladder is situated in the most dependent area, while the ovaries and oviducts of reproductively active females can occupy much of the coelom. The male testes (cream, yellow or brown in color) are situated in the caudodorsal coelom, and closely associated with the vasa deferentia, epididymides, adrenal glands, and retrocoelomic kidneys. Recently, coelioscopy using saline-infusion has been shown to be accurate for identifying gender in hatchling and neonate chelonians. A study in Chinese box turtles (*Cuora flavomarginata*) concluded that gonads could be visualized and sex easily identified in animals as small as 10 g.[5] Over 300 individual juveniles of various chelonian species have been subjected to early gender identification by coelioscopy (Divers, unpublished data). Thus far, this technique has been found to be accurate and safe and has become established as a significant clinical service to breeders, and curators of head-start conservation programs.

Pathology and endoscopic biopsy

Tissue samples can be easily harvested from most organs, and, in general, any abnormal soft tissue structure. Care if required when sampling gonads, but larger follicles or any fluid filled structures should not be biopsied because of the real dangers of yolk-associated coelomitis. Endoscopic biopsies are delicate and measure approximately 1 mm^3 when collected using 1.7-mm forceps. Handling and other histologic artifacts are reduced by gently dislodging the tissue from the biopsy cups into a small volume of sterile saline, before then decanting into a biopsy cassette or bag, and submitting in 10% neutral buffered formalin.[27] Picking the biopsy out of the instrument using a needle will cause severe damage, and even moistened cotton-tipped applicators have been shown to cause tissue alteration. Biopsies for microbiology are best submitted in sterile saline for immediate processing. Alternatively, if mailing to a laboratory, they should be submitted in appropriate transport media. For the submission of samples for toxicology or parasitology, it is wise to consult with the laboratory prior to sample collection.

Endoscope-Assisted Procedures

An endoscope-assisted procedure is a hybrid that combines elements of endoscopy with traditional surgery. In its simplest form, an endoscope-assisted procedure uses endoscopy to identify and grasp an internal structure which is then exteriorized through a miniature coeliotomy incision for completion of the procedure outside the body cavity. The advantages of endoscope-assisted procedures are that they decrease the size of the coeliotomy incision and exposure of internal viscera to the environment. An assortment of endoscope-assisted coelomic procedures including ovariectomy, cystotomy, salpingotomy, and salpingohysterectomy are now possible. These techniques rely on the ability to exteriorize the tissue of interest out through a small, targeted coeliotomy incision. If the tissue cannot be exteriorized and the procedure cannot be completed externally, then a true endosurgical or traditional open technique is required.

Endoscope-assisted procedures have been most commonly utilized in chelonians, where a soft tissue prefemoral approach has now superseded the need for transplastron coeliotomy in many situations. Of particular practical relevance is endoscope-assisted, prefemoral ovariectomy (or oophorectomy) of mature female chelonians that negates the need for a transplastron coeliotomy.[28] By virtue of the extensive mesovarian found in mature females, an endoscope-assisted procedure is practical in animals weighing a few hundred grams to 100 kg or more.[28,29] With the chelonian in dorsal recumbency a unilateral surgical approach to the coelom is made via one prefemoral fossa. The telescope is inserted into the coelom and used to locate the ipsilateral ovary. Then, using the telescope to guide atraumatic forceps (or even long hemostats in small chelonians), the interfollicular tissue is grasped and gently elevated to the skin incision. Once the first follicle is exteriorized, the telescope is no longer required, as the ovary is carefully manipulated out of the coelom. Once the entire ovary is exteriorized, oophorectomy proceeds using traditional techniques. The same procedure is repeated for the contralateral ovary, but it is usually not necessary nor recommended to remove the oviducts unless diseased. In most species, both ovaries can be removed from the same incision; however, in some terrestrial tortoises with particularly restrictive prefemoral fossae or voluminous large intestinal tracts, it may be necessary to perform a bilateral procedure. The prefemoral approach is closed routinely. This technique has several advantages over traditional transplastron approaches, namely, less invasive with faster healing.

Similar endoscope-assisted procedures have also been accomplished for the removal of retained eggs from bladders. The general technique and endoscopy skills are very similar. First, diagnostic imaging is utilized to determine which side is favored for the approach. Then following a prefemoral coeliotomy, the telescope is used to identify the structure of interest and guide atraumatic tissue forceps. While any tissue that is due to be resected and removed can be grasped using Kelly or Babcock forceps (or similar), any tissue that is to ultimately be replaced should ideally be handled using atraumatic or fenestrated forceps. This is especially important when handling the bladder or reproductive tracts.

Endosurgery

To date, endosurgery has been more commonly performed in chelonians, although endosurgical orchiectomy of iguanids has also been described.[3] While established endosurgical dogma dictates that telescopes are often placed through the umbilicus or linea alba, and instrument ports are placed far enough apart to facilitate triangulation, reptile anatomy often necessitates a modified approach.

Endosurgical procedures usually commence with the placement of the telescope cannula, and there are two main techniques commonly used. Either the telescope cannula is surgically placed via a mini-coeliotomy incision, after which the CO_2 line is attached and the coelom insufflated, or a veress needle can be used. Placing the telescope cannula surgically, as preferred by the author, ensures intracoelomic placement, reduces the risks of iatrogenic trauma, but does require a mattress suture around the cannula to close the incision and prevent gas leakage.

Chelonians

Due to the constraints of the chelonian shell, it is seldom possible to place the telescope and ports where established protocols would suggest. There are two major approaches, depending upon the size of the animal. For larger chelonians, the telescope and instruments may all be inserted through one prefemoral fossa, while in smaller animals the telescope and instruments may be placed through one or both fossae. There are advantages and disadvantages to both approaches. Placing all instrumentation through a single fossa enables the surgeon to control both instruments even in large animals, but instrument triangulation and coordinated use can be more difficult due to the acute angles involved (often $\leq 20^o$) as instruments run more parallel. Nevertheless, compared to a bilateral fossa technique, a unilateral approach is less invasive with only a single surgical approach required, and in addition enables the animal to be positioned in lateral recumbency which may aid organ visualization. A single surgeon can utilize both prefemoral fossa in smaller animals but here the angle of triangulation is often obtuse (90-180°), again making coordinated instrument manipulation more challenging. If the target structure is cranial to the bladder and the chelonian is less than 20 kg it is often preferable for one instrument and telescope to be inserted through one prefemoral fossa, while the second instrument is inserted through the contralateral fossa. For such procedures, the animal is in dorsal recumbency, and enjoys the advantage that instruments can be more easily triangulated within the cranial to mid coelom; however, in larger chelonians it can be difficult to impossible for a single surgeon to operate through both prefemoral fossa simultaneously, and in such cases a second surgeon is required to operate through one fossa. The endosurgical approach to caudal structures (eg, bladder, kidney, adrenal, or gonad) is more problematic as it is often difficult to access the same structure simultaneously from both fossae. Therefore, where a caudal structure is the focus of the procedure, insertion of telescope and instruments through a single prefemoral fossa is often preferable. In such cases, the chelonian is generally positioned in dorsal or lateral recumbency which provides greater accessibility as the intestines and bladder fall away into the more dependent areas of the coelom. This approach is easier in those animals with a large prefemoral fossa, but even so the close placement and parallel nature of instrumentation can make triangulation more difficult. In addition, it is often easier (especially in smaller animals) to make a single large prefemoral incision for the insertion of all instruments, rather than place three cannulae adjacent to one another. Insufflation is seldom essential due to the non-collapsible nature of the chelonian shell, and therefore endoscopic devices can still be used through an incision without the need for cannulae.

Oophorectomy of immature females

Intracorporeal oophorectomy has been described in immature, Galapagos tortoises (*Geochelone nigra*) using 5 mm instrumentation;[29] however, there is no reason why this same technique could not be applied to smaller chelonians with suitably sized (3 mm) instrumentation. In addition, this technique provides a model for other possible endosurgical procedures. The paired, immature ovaries reside within the caudodorsal coelom, and in close association with the retrocoelomic kidney. The bladder obscures the contralateral ovary and so a unilateral endosurgical approach must be repeated on each side. With the animal in lateral recumbency, a prefemoral approach to the coelom is made in order to accommodate the telescope and two instruments via a single incision. Cannulae are generally are not helpful because the instruments are so closely aligned and insufflation is not required. The telescope is positioned and held by a surgical assistant in order to maintain the view of the ovary. The immature ovary is elevated away from the serosal surface of the caudodorsal shell using Babcock or Kelly forceps, and monopolar scissors are used to dissect the ovary free from its mesovarian attachments. The prefemoral incision is closed routinely, and the procedure repeated on the opposite side.

Orchiectomy

Researchers originally developed a similar approach for safe and effective orchiectomy, using the red-eared slider (*Trachemys scripta elegans*) as a model.[30] The animal is again positioned in lateral recumbency and, following prefemoral coeliotomy, a surgical assistant identifies and maintains a clear view of the testis. The endosurgeon then uses Babcock or Kelly forceps to grasp and elevate the testis away from the kidney and epididymis to expose the mesorchium and associated vessels. Both monopolar radiosurgery and vascular clips have been used to ensure hemostasis before scissors are used to dissect the testis free. At least in *Trachemys* it is possible to remove both testes via a unilateral prefemoral approach.

Further procedural development was required to permit orchiectomy in larger tortoises where the mesorchial attachments are typically more substantive and require extensive dissection to free the testis. Using the desert tortoise (*Gopherus agassizii*) as a model, a single-surgeon technique was developed that employed a mechanical support arm for the telescope, freeing the surgeon to utilize a combination of endoscopic vascular clips and radiosurgical dissection to dissect each testis free.[31]

Lizards

It is particularly important to remember that the standardized placement of cannulae through the linea alba cannot be applied to most lizards due to the presence of the midline abdominal vein; however, placement of a midline cannula, caudal to the umbilicus is certainly practical and often preferred, particularly in iguanids.

Orchiectomy

With the iguana in right lateral recumbency, the telescope and protection sheath are inserted, just lateral to the ventral midline, close to the anastomosis of the left and right pelvic veins as they form the ventral abdominal vein. The first cannula is placed cranial to the telescope, just caudal to the last rib, while the second cannula is placed caudal to the anastomosis of the pelvic veins in the ventral midline. Grasping forceps are inserted through one cannula and used to elevate the testis, while scissors (attached to monopolar radiosurgery) are inserted through the other cannula and used to coagulate and cut through the mesorchium and associated vessels. In larger iguanas, bipolar forceps are used to coagulate blood vessels prior to sharp dissection. The testis is extracted through the forceps cannula hole, which can be enlarged if necessary. The cannula holes are closed with a single suture or tissue adhesive. The iguana is rotated into left lateral recumbency, and using the same entry sites, the procedure is repeated for the second testis.

Traditional Male Reproductive Surgery

In cases where a large amount of tissue has to be surgically removed (eg, ovariosalpingectomy to resolve dystocia), it makes little sense to adopt a minimally-invasive approach. For these cases, a traditional coeliotomy is still favored.[32,33]

Orchiectomy

Indications for castration in lizards and chelonians are to reduce intra and interspecies aggression, inhibit the ability to reproduce (population control), research, and to eliminate testicular pathology such as neoplasia or orchitis (abscess, granuloma). Indications for orchiectomy in snakes and crocodilians are less common and are primarily to manage pathologic change.

Lizards

For castration in most lizards a ventral (paramedian or midline) coeliotomy approach is recommended; however, with laterally compressed species such as old-world chameleons a flank (lateral) coeliotomy approach can be utilized. Once the coelom is safely entered the testes are identified. Generally, the testes are paired (right testis slightly more cranial) and found dorsally in the mid to caudal coelom just cranial to the kidneys. Exposure can be improved with the use of retractors such as the lone star retractor. The testes are encased in a fibrous, mesorchium capsule which provides some stability for manipulation; however, if the capsule ruptures the testis is friable and difficult to manage. When possible, the testis is gently elevated to visualize vessels for ligation. Surgically important anatomic differences exist between the right and left testes in lizards. The right testis is attached directly to the vena cava with short vessels (approximately 1-2 mm) with the right adrenal gland located on the *opposite* side of the vena cava. In this location accidental damage or removal of the right adrenal gland is unlikely; however, the left testis has its own distinct testicular artery and vein and the adrenal gland is located *adjacent* to the left testis utilizing the same vessels. The elongated pink-tan granular adrenal gland should not be removed or damaged during ligation of the left testis. Once each testis is isolated the vascular supply is identified the vessels are ligated with vascular clips or absorbable monofilament suture. In some species (eg, bearded dragons, *Pogona vitticeps*) the mesorchium is pigmented and the vasculature difficult to visualize. Magnification, focal lighting and microsurgical instruments can be useful in these situations. After ligation the mesorchial pedicles are observed for bleeding prior to routine closure.

For laterally compressed lizards such as old-world chameleons a flank coeliotomy approach can be utilized. With this lateral coelomic approach the testes are easily identified upon entry as they are typically positioned at the entry site. The same careful ligation of the testicular vessels is performed and both testes can usually be removed from one incision unless there is associated pathology. Orchiectomy is less commonly performed on snakes, but the methodology is similar.

Chelonians

The least invasive and preferred technique for castration in chelonians is the prefemoral coelomic approach instead of the ventral transplastron coelomic approach. The testes in most chelonians are accessible from a prefemoral approach (especially with endoscopic assistance) and the trauma of cutting through plastron bone is avoided; however, with testicular pathology or neoplasia, or in some chelonian species a transplastron coeliotomy may be necessary.

Once the coelom is entered, with either approach, the bladder, intestinal tract and lungs are gently manipulated to allow access to the testes. The testes of chelonians are paired and found dorsally in the caudal coelom cranial to the kidneys. With a prefemoral approach the testes is identified and grasped with forceps or hemostats and gently manipulated up to the level of the prefemoral incision to allow visualization of the mesorchium and the testicular vessels. It is difficult, if not impossible, to exteriorize the testes. With the transplastron approach the testis is manipulated up towards the ventral plastronotomy incision, but again, the surgeon typically has to work within the coelom. When the testis is globoid a single vascular clip may be sufficient, otherwise multiple vascular clips, suture or radiosurgery or laser may be needed. Both testes are accessible from a transplastron coeliotomy; however, with a prefemoral approach, the contralateral testis can be difficult to identify and remove from the ipsilateral approach without endoscopy, and a second prefemoral coeliotomy incision is often necessary.

Testicular biopsy

In some situations, the collection of histopathologic and microbiologic samples from testes may be required to determine underlying disease and dysfunction or infertility. Indications for testicular biopsy include neoplasia, orchitis (abscess, granuloma) and abnormal sperm production (infertility). In most reptiles endoscopic testicular evaluation and biopsy is effective, minimally invasive, and preferred to open surgical approaches; however, an open coeliotomy approach may be necessary if endoscopy equipment is unavailable, or when evaluating the testes of snakes which are inherently more challenging to endoscope.

Phallectomy

Phallectomy has been utilized to control reproduction in chelonian conservation projects. It is a less invasive and safer procedure than castration in chelonians and allows normal (testosterone driven) social behaviors without insemination of females. Phallectomy (chelonians) and hemipenal amputation (snakes and lizards) are frequently performed procedures as prolapses are common. One important, unique consideration with reptile phallus surgery is, unlike mammals, amputation does not affect the urinary system as these structures do not contain a urethra. Additionally, for squamates, the removal of one hemipenis will still allow copulation with the remaining hemipenis.

Copulatory organs are vascular and transfixing sutures are required. For hemipenal amputation the hemipenis is clamped close to its base with a hemostat and one or two absorbable monofilament transfixing sutures are placed below the hemostat depending on the patient size. When possible, the remaining tissue should be short enough to allow the sutured stump to recess into the caudal vent or hemipenal sulcus to avoid postoperative exposure trauma. In chelonians, the phallus is retracted caudally and clamped at the base with a hemostat or tissue clamp and a minimum of two absorbable monofilament transfixing and encircling sutures are placed, each approximately half way across the phallus. In larger chelonians the blood supply to each distinct longitudinal ridge (basically dividing the phallus in half longitudinally) must be identified and each body (corpus cavernosa) and blood supply individually clamped and double ligated with absorbable monofilament sutures. Bleeding can be substantial if a vessel is damaged in placing a transfixing suture and additional sutures may be required. Sutures should be placed close to the base for a good "hold" but minimize leaving any functional phallus. The remaining phallus stump can be closed with a simple continuous pattern using absorbable monofilament suture.

Traditional Female Reproductive Surgery

The most common indications for ovariectomy and ovariosalpingectomy in lizards, chelonians and less commonly snakes is to manage reproductive follicular activity and dystocia. Preventive (elective) ovariectomy can be performed in lizard and chelonian species that are prone to reproductive related issues.

Preventive (elective) and postovulatory ovariectomy

For ovariectomy in most lizards a ventral coeliotomy (midline or paramedian) approach is recommended, but a lateral approach is preferred in old world chameleons. In chelonians a ventral plastron approach may be necessary, while prefemoral approaches are more difficult without an endoscope. Once the coelom is entered the ovaries are identified as paired dorsal structures in the mid to caudal coelom, often just cranial to the kidneys. In cases of dystocia, the oviducts are filled with eggs or feti and need to be exteriorized and removed first to allow exposure of the ovaries. With a preventive (immature or seasonally inactive) ovariectomy or a postovulatory ovariectomy the ovaries are generally small and appears as a small cluster of fluid filled follicles located deep in the coelom. Each ovary is identified and elevated, but can rarely be exteriorized. The surgeon may have to work within the coelom and so the coeliotomy incision should be sufficiently large.

In lizards, the right ovary is attached directly to the vena cava with short vessels (1-2 mm in length) and the right adrenal gland is located on the *opposite* side of the vena cava and thus unlikely to be damaged due to its location; however, the left ovary has its own distinct ovarian artery and vein, and the adrenal gland is located *adjacent* to the left ovary utilizing the same vessels. The elongated, pink-tan, granular adrenal gland must not be removed or damaged during ligation of the left ovary. Hemostasis is achieved with vascular clips (easier, recommended) or absorbable monofilament suture (more difficult). In some species (eg, bearded dragons,) the mesovarium may be pigmented and the vasculature difficult to see. Also, in bearded dragons the right liver lobe is closely associated with the vasculature of the right ovary and the surgeon must avoid damaging the liver during ligation of the right ovarian vessels. Magnification and focal lighting and microsurgical instruments can be useful in these situations. Ovaries are easily identified upon entry as they are typically positioned near the entry site.

Ovariectomy (pre-ovulatory follicular stasis)

The surgical approach is similar; however, the ovaries are grossly enlarged and possess numerous large follicles. Following coelomic entry, the greatly enlarged ovaries which appear as a cluster of yolk filled follicles are obvious. The mesovarium is more elastic and will usually stretch to allow manipulation of each ovary out of the incision; however, the ovary is often larger than the incision and the follicles may be friable and begin to coalesce (especially if diseased) and can rupture with handling. Gentle manipulation and patience are required. Once each ovary is isolated and the vascular supply identified, the vessels are ligated with clips or absorbable monofilament suture. Particularly large vessels should be double ligated. The ovary is transected from its mesovarial attachments and removed. The process is repeated for the second ovary. If normal in appearance the oviducts can be left in situ. Also, if any ovarian remnant remains, the oviducts can prevent egg yolk coelomitis. This is a legitimate concern especially in species such as bearded dragons where the mesovaria are short and pigmented, and the liver is closely associated with the right ovarian vasculature. It can be difficult to be certain that all the ovarian tissue is removed in this species, and clients should be warned of this possibility.

Salpingoctomy

A salpingectomy is the first procedure performed in cases of surgical dystocia, and upon entry into the coelom the oviducts filled with eggs or feti will usually be the first structures identified. The oviducts are paired and found adjacent but not connected to the ovaries, and when full of eggs/feti may occupy most of the coelom. Once identified, the oviducts are gently exteriorized from the cranial (fimbria) to caudal (junction with urodeum) to allow visualization of the vascular supply. The fimbria is identified, clamped with a hemostat, and ligated with absorbable monofilament suture or vascular clips, and transected. In many cases, the fimbria are thin and poorly vascular, such that simple radiosurgical dissection (without ligation) may be all that is required. Starting cranially and working caudally, the associated vessels are ligated individually or in groups using vascular clips, absorbable monofilament suture, or radiosurgery or laser. The distal oviduct is then clamped with a hemostat at its juncture with the urodeum and double ligated with circumferential and transfixing absorbable monofilament sutures, before being transected allowing the entire oviduct and contained structures to be removed *en bloc*. This procedure is repeated on the opposite oviduct. Following the removal of both oviducts, ovariectomy is performed as described above. Failure to completely remove both ovaries will likely result in fatal yolk coelomitis in the future.

Salpingotomy

If a lizard is presented with a dystocia and maintaining future breeding capability is important then salpingotomy is recommended. This is easiest when just 1-2 eggs/feti have to be removed. Following identification and exteriorization of the oviducts, the number and location of the retained eggs/feti are evaluated and the initial incision can be determined. To improve future reproductive success, trauma to the oviduct must be minimized, and the initial incision is made longitudinally along the antemesenteric border with a scalpel blade to ensure primary intention healing. The location of the incision should be strategically determined to allow the removal of as many eggs or feti as possible through the same incision. Diagnostic samples including oviductal biopsy for histopathology, and microbiology should be collected at this time. If necessary, the incision can be extended with a scalpel, using radiosurgery for pin-point hemostasis. Often with dystocia there are fibrin adhesions and secondary inflammation inhibiting the eggs or feti from moving freely through the oviduct. If the eggs or feti are adhered to the mucosa, the oviduct can be packed off with moistened lap sponges, and warm saline infused into the lumen. The saline, along with gentle manipulation, may allow more eggs or feti to be freed, and increase the number that can be manipulated out of a single incision. Nevertheless, several incisions may be required to successfully remove all eggs or feti. The oviduct incisions are closed with a simple continuous or inverting pattern using fine monofilament absorbable suture. The coelom is lavaged with warm sterile saline, and routinely closed.

Egg yolk coelomitis, ectopic follicles and eggs

Egg yolk coelomitis is not an uncommon sequalae to reproductive disease in reptiles. It is important to thoroughly evaluate the coelom during any reproductive procedure to ensure that such pathology is not missed. Diagnostic samples taken for histopathology (or cytology) and microbiology should be taken before the coelom is copiously and repeatedly lavaged with warm sterile saline to remove as much yolk material as possible. In more chronic cases, a fibrin-like yolk material may be adhered to serosal surfaces and will require gentle debridement to remove. A Gram stain can be performed to help direct immediate antimicrobial treatment, and subsequently adjusted based on future sensitivity results.

References

1. Divers SJ, Stahl SJ. *Mader's Reptile and Amphibian Medicine and Surgery*. 3rd ed. St. Louis, MO: Elsevier Saunders;2018.

2. Divers SJ. Diagnostic endoscopy In: Mader DR, Divers SJ, eds. *Current Therapy in Reptile Medicine and Surgery*. St Louis, MO: Elsevier Saunders;2014:154-178.

3. Divers SJ. Endoscope-assisted and endoscopic surgery In: Mader DR, Divers SJ, eds. *Current Therapy in Reptile Medicine and Surgery*. St Louis, MO: Elsevier Saunders;2014:179-196.

4. Schumacher J, Yellen T. Anesthesia and analgesia In: Mader DR, ed. *Reptile Medicine and Surgery*. 2nd ed. St Louis, MO: Elsevier Saunders;2006:442-452.

5. Hernandez-Divers SJ, Stahl SJ, Farrell R. Endoscopic gender identification of hatchling Chinese box turtles (*Cuora flavomarginata*) under local and general anesthesia. *J Am Vet Med Assoc.* 2009;234:800-804.

6. Rostal D, Grumbles J, Lance V, et al. Non-lethal sexing techniques for hatchling and immature desert tortoises (*Gopherus agassizii*). *Herp Mono.* 1994;8:103-116.

7. Selleri P, Di Girolamo N, Melidone R. Cystoscopic sex identification of posthatchling chelonians. *J Am Vet Med Assoc.* 2013;242:1744-1750.

8. Tams TR, Rawlings CA. *Small Animal Endoscopy*. 3rd ed. St Louis, MO: Elsevier Mosby;2011.

9. Falcone RE, Wanamaker SR, Barnes F, et al. Laparoscopic vs. open wedge biopsy of the liver. *J Laparoendosc Surg.* 1993;3:325-329.

10. Golditch IM. Laparoscopy: Advances and advantages. *Fertil Steril.* 1971;22:306-310.

11. Grauer GF, Twedt DC, Mero KN. Evaluation of laparoscopy for obtaining renal biopsy specimens from dogs and cats. *J Am Vet Med Assoc.* 1983;183:677-679.

12. Hancock RB, Lanz OI, Waldron DR, et al. Comparison of postoperative pain after ovariohysterectomy by harmonic scalpel-assisted laparoscopy compared with median celiotomy and ligation in dogs. *Vet Surg.* 2005;34:273-282.

13. Kehlet H. Surgical stress response: Does endoscopic surgery confer an advantage? *World J Surg.* 1999;23: 801-807.

14. Lagares-Garcia JA, Bansidhar B, Moore RA. Benefits of laparoscopy in middle-aged patients. *Surg Endosc.* 2003;17:68-72.

15. Orlando R, Lirussi F, Okolicsanyi L. Laparoscopy and liver biopsy: further evidence that the two procedures improve the diagnosis of liver cirrhosis. A retrospective study of 1,003 consecutive examinations. *J Clin Gastroenterol.* 1990;12:47-52.

16. Rawlings CA, Diamond H, Howerth EW, et al. Diagnostic quality of percutaneous kidney biopsy specimens obtained with laparoscopy versus ultrasound guidance in dogs. *J Am Vet Med Assoc.* 2003;223:317-321.

17. Reissman P, Gofrit O, Rivkind A. Exploratory laparoscopy: a crucial advantage of laparoscopic over standard appendectomy. *South Med J.* 1994;87:576.

18. Vander Velpen GC, Shimi SM, Cuschieri A. Diagnostic yield and management benefit of laparoscopy: a prospective audit. *Gut.* 1994;35:1617-1621.

19. Weickert U, Buttmann A, Jakobs R, et al. Diagnosis of liver cirrhosis: A comparison of modified ultrasound and laparoscopy in 100 consecutive patients. *J Clin Gastroenterol.* 2005;39:529-532.

20. Yu SY, Chiu JH, Loong CC, et al. Diagnostic laparoscopy: indication and benefit. *Zhonghua Yi Xue Za Zhi (Taipei).* 1997;59:158-163.

21. Hernandez-Divers SJ. Surgery: Principles and techniques In: Raiti P,Girling S, eds. *Manual of Reptiles.* 2nd ed. Cheltenham, England: British Small Animal Veterinary Association;2004:147-167.

22. Hernandez-Divers SJ, Hernandez-Divers SM, Wilson GH, et al. A review of reptile diagnostic coelioscopy. *J Herpetol Med Surg.* 2005;15:16-31.

23. Divers SJ. Lizard endoscopic techniques with particular regard to the green iguana (*Iguana iguana*). *Semin Avian Exot Pet Med.* 1999;8:122-129.

24. Hernandez-Divers SJ, Stahl S, Hernandez-Divers SM, et al. Coelomic endoscopy of the green iguana (*Iguana iguana*). *J Herpetol Med Surg.* 2004;14:10-18.

25. Divers SJ, Stahl SJ, Camus A. Evaluation of diagnostic coelioscopy including liver and kidney biopsy in freshwater turtles (*Trachemys scripta*). *J Zoo Wildl Med.* 2010;41:677-687.

26. Hernandez-Divers SJ. Endoscopic renal evaluation and biopsy of Chelonia. *Vet Rec.* 2004;154:73-80.

27. Stahl SJ, Hernandez-Divers SJ, Cooper TL, et al. Evaluation of transcutaneous pulmonoscopy for examination and biopsy of the lungs of ball pythons and determination of preferred biopsy specimen handling and fixation procedures. *J Am Vet Med Assoc.* 2008;233:440-445.

28. Innis C, Hernandez-Divers SJ, Martinez-Jimenez D. Coelioscopic-assisted prefemoral oophorectomy in chelonians. *J Am Vet Med Assoc.* 2007;230:1049-1052.

29. Knafo SE, Divers SJ, Rivera S, et al. Sterilization of hybrid Galapagos tortoises (*Geochelone nigra*) for island restoration: Part 1 female endoscopic oophorectomy under ketamine-medetomidine anaesthesia. *Vet Rec.* 2010;168:47-51.

30. Innis CJ, Feinsod R, Hanlon J, et al. Coelioscopic orchiectomy can be effectively and safely accomplished in chelonians. *Vet Rec.* 2013;172:526.

31. Proenca LM, Fowler S, Kleine S, et al. Single surgeon coelioscopic orchiectomy of desert tortoises (*Gopherus agassizii*) for population management. *Vet Rec.* 2014;175:404.

32. Mader DR, Bennett RA, Funk RS, et al. Surgery In: Mader DR, ed. *Reptile Medicine and Surgery*. 2nd ed. St. Louis, MO: Elsevier;2006:581-630.

33. Stahl SJ. Reptile production medicine. *Semin Avian Exot Pet Med.* 2001;10:140-150.

Reproductive Tract Diseases of Reptiles and Amphibians

Drury Reavill, DVM, DABVP (Avian Practice), DABVP (Reptile and Amphibian Practice), DACVP

Session #2156

Affiliation: From Zoo/Exotic Pathology Service, 6020 Rutland Drive #14, Carmichael, CA 95608-0515, USA.

Abstract: This paper will cover the diseases of the reproductive system in both reptiles and amphibians. It is not a comprehensive review. More in-depth disease descriptions can be found in the articles and reference books listed in the literature cited.

Introduction

After a brief summary of the normal gross appearance, the paper is divided by etiologic disease categories. A short discussion of the disease conditions of reptiles and amphibians are found in each section. For the purposes of this paper, chelonian will refer to all of the shelled reptiles. The term "tortoise" will be used to describe the large land animals that have stout elephantine feet. Few references to marine turtles, caecilian, and crocodilian species are included, as these are uncommonly seen in most clinical practices.

Normal Gross Anatomy

It will be important to recognize the normal gross appearance of the reproductive tract in order to determine if there are abnormalities that require further examination, either by surgical biopsies or histologic examination during a necropsy procedure.

Males of some lizard species, such as the geckos, skinks, chameleons, and iguanids, possess a sexual segment of the renal tubule. The sexual segment can give a portion of the kidneys a more swollen appearance. In chameleons, these sexual segments are generally located in clusters situated throughout the kidneys. The clusters are typically wedge shaped and may be confused with disseminated disease.[1] The sexual segment of lizards and snakes is anatomically the last part of the nephron before it enters the collecting duct.[2]

Reproductive organs in lizards are paired and located craniomedial to the kidneys. The oval testes are yellow to off-white in most species, and the ovaries are clusters of variably developed follicles that are yellow to pink. Males have paired hemipenes that are inverted in the ventral tail distal to the vent. These blind-ended organs mechanically transfer sperm via the seminal groove and are used for reproduction, not urination.[3]

Males of many snake species have a sexual segment that when they are reproductively active, will result in enlargement of the kidney. The sexual segment of the kidney in male snakes will produce a secretion rich in proteins and lipids, which is used as a copulatory plug.[4] In snakes, the sexual segments are typically situated at the surface of the kidneys. Histologically these segments are comprised of enlarged tubular epithelial cells that contain abundant eosinophilic granules. When they are active, producing the secretions, they can be grossly confused with tubules dilated with urates or even gout.[2]

The testes in snakes are elongated, lightly colored, and located between the pancreas and the kidneys. They will enlarge during breeding season as do the testes of any of the reptilian and amphibian species. The ovaries are located near the pancreas. Male snakes have paired hemipenes that are also inverted into the ventral tail base. They have no connection with the urinary system. In snakes there is a cloacal or musk gland that lies parallel to the inverted hemipenes and opens into the dorsal cloaca.[4]

The ovaries and testes of chelonians are paired and are anterior to the kidneys. Male chelonians do not have hemipenes, like other reptiles. They have a dark purple single erectile copulatory organ that lies on the floor of the proctodeum. There is a seminal groove, which aids in sperm transport.[5,6]

The ovaries in reptiles will vary in appearance depending on the stage of oogenesis. Inactive and/or immature ovaries will be small and granular. An active ovary will have multiple large, spherical vitellogenic follicles. Vitellogenesis is the accumulation of yolk. No true uterus exists in lizards and snakes. In chelonia, the uterus is where the outer egg shell is deposited.[7]

There is some variation in the gross anatomy of different amphibian species. In pregnant females, the follicles or ova will fill the coelomic cavity.

Developmental Disorders

Hermaphrodites

The toads of *Bufo*, *Pedostibes*, *Pseudobufo*, *Atelopus*, *Leptophryne*, *Dendrophryiscus*, and some *Nectophryne* have a structure called the Bidder's organ.[8] This is a wedge-shaped band or cap-like mass of pigmented oocytes that are closely associated with the cranial pole of the testes. This is a normal structure for the listed toads; however, the organ can be easily misclassified as the ovotestes of hermaphrodites. Hermaphroditism has been recognized in a Madagascar tomato frog (*Dyscophus antongilii*).[8] This animal had both testes that exhibited active spermatozoa formation as well as masses of heavily pigmented yolked ovarian follicles. Ovotestis (intersex) is reported in a horned marsupial frog (*Gastrotheca cornuata*).[9] Hermaphroditism has also been reported in a phenotypic male common green iguana (*Iguana iguana*).[10]

Deposition Disorders

Cholesterol

A deposition of cholesterol crystals and associated inflammation (xanthomatosis) has been described in round island geckos (*Phelsuma geuntheri*), leopard gecko (*Eublepharis macularius*), Northern and Leaf-tail geckos.[11-13] All have been females and the xanthomatous inflammation was speculated to be due to abnormalities in cholesterol metabolism possibility associated with folliculogenesis. The cholesterol crystals are acicular clefts associated with chronic and acute inflammation (xanthomas) and identified in the kidney, ovary, coelom, brain, and liver.

There are two reports in Cuban tree frogs (*Osteopilus septentrionalis*) of xanthomatosis involving many organs including the kidneys and ovaries.[14,15] The condition is one in captive amphibians having a high serum or plasma cholesterol and concomitant lesions of corneal lipidosis. The underlying cause of this disorder is unknown, but it is likely that a diet high in cholesterol plays a role.

Nutrition-Related Disorders

A study was conducted to compare thiamine concentrations in American alligators (*Alligator mississippiensis*) and Florida largemouth bass (*Micropterus salmoides floridanus*). Results suggest that thiamine deficiency might be playing an important role in alligator embryo survival. These results suggest that thiamine deficiency might be playing an important role in alligator embryo survival but not in reproductive failure and recruitment of largemouth bass. The cause(s) of this thiamine deficiency are unknown but might be related to differences in the nutritional value of prey items across the sites studied and/or to the presence of high concentration of contaminants in eggs.[16]

In amphibians, hypovitaminosis A is associated with squamous metaplasia of the oviduct and cloaca, as well as the oral cavity, nasal cavity, pharynx, esophagus, stomach, skin, urinary bladder, and ureter. It is recommended that the tongue and urinary bladder are routinely examined in amphibians and to collect the left lobe of the liver for tissue vitamin A analysis.[17]

Agents of Disease

Inflammation of the oviduct/uterus and testes can be due to infectious agents, or in the case of females, secondary to retained yolk with impaction. The inflammation in the latter case is secondary to irritant tissue effects of the yolk protein. In general, systemic infections can also involve the reproductive tract. There are few disease agents specifically associated with the reproductive tract.

Venereal transmission

Metazoan

There is an interesting life cycle of the atractid nematode *Cyrtosomum penneri* (Cosmocercoidea: Atractidae) which is transmitted venereally in *Anolis sagrei*. In these lizards, it is passed through the cloaca of the infected lizard to the cloaca of the other during mating.[18] The nematode is a common gastrointestinal parasite of *Anolis sagrei*.

Virus

The eosinophilic globular intracytoplasmic inclusion bodies within the epithelial cells of many tissues is the lesion associated with inclusion body disease (IBD) of Boid snakes and Palm vipers (*Bothriechis marchi*).[19,20,21] The etiological agent of IBD is an arenavirus.[22,23] This disease can be responsible for clinical signs including chronic regurgitation, incoordination, loss of righting reflexes, paresis, and an increased incidence of secondary infections such as stomatitis and pneumonia. Clinical signs referable to the central nervous system are more prominent in the Pythoninae subfamily. At this time, the definitive modes of transmission is unknown; however, the expected routes are fecal/oral contamination, airborne via respiratory discharges, venereally, or by the snake mite as a vector.

Systemic infections with lesions in the reproductive tracts

Microsporidia

Microsporidia are small, intracellular protozoans. They form an infective thick-walled spore containing the sporoplasm which once ingested, is discharged and migrates to its target organ. Microsporidian protozoa include *Encephalitozoon* sp., *Pleistophora* sp., *Nosema* sp., and other species. The reports in reptiles are primarily of systemic fatal infections in bearded dragons (*Pogona vitticeps*).[24] The primary lesion is a severe hepatic necrosis with clusters of light basophilic intracytoplasmic microorganisms packed within distended hepatocytes as well as being free in areas of necrosis. Similar microorganisms can be found in cytoplasmic vacuoles of renal epithelial cells, pulmonary epithelial cells, gastric mucosal epithelial cells, enterocytes, capillary endothelial cells, ventricular ependymal cells in the brain, and in macrophages within granulomatous inflammation of the colon, adrenal glands, and ovaries. They are oval, approximately 1µm in length, and stain faintly on H&E sections. The spores are acid-fast positive and stain variably gram positive.

Fungus

Chromomycosis is primarily a systemic fungal infection by pigmented fungi of the family Dematiaceae.[25] These fungi are saprophytes that live in soil and decaying vegetable matter, predominantly in warmer climates of tropical and subtropical countries.[26] Some identified members include *Cladosporium herbarum, Scolecobasidium humicola, Fonsecaea* spp., *Phialophora* spp., *Rhinocladiella* spp., *Hormodendrum* spp., *Curvularia* spp., and *Drechslera* spp.[27-32] The fungal organisms produce a characteristic amber-brown, thick-walled, septate structures known as sclerotic bodies.[32] Chromomycosis infection is most commonly described in amphibians but has also been reported in mammals, a radiated tortoise, pythons, boas, and a mangrove snake with a concomitant fibrosarcoma. Transmission is suspected to be by the alimentary tract and skin abrasions. Skin trauma incurred during feeding may be the most likely portal of entry (especially in colonies of animals or with vigorous eaters). The stress of captivity or other debilitating disease processes can also contribute to the development of the disease.[33,25]

Chromomycosis generally presents first as an ulcerated or granulomatous skin lesion. These lesions can be nodules, grey-black papules and/or irregularly shaped grey-black ulcers. The most common sites for external lesions are the dorsal and ventral skin surfaces. Over a period of time (which can be greater than six months) the animal succumbs to disseminated disease.[25] On gross examination multiple organs may contain multiple coalescing grey-black nodules. These granulomas are commonly described in the liver and kidney, but also occur in the spleen, lung, heart, adrenal, gastrointestinal tract (serosal and mucosal surfaces), bone, reproductive tract, and central nervous system.

Bacteria

Any number of bacterial infections can involve the reproductive system. The majority of bacterial oophoritis and salpingitis in both reptiles and amphibians are gram-negative bacteria and primarily Salmonella, Pseudomonas, Aeromonas, and Citrobacter species.[34] In addition, Salmonella species have been recovered from eggs suggesting transovarian transmission in some reptile species.[35,36] There may also be a peripheral heterophilia in response to a bacteremia and a coelomitis.

Mycobacterial infections, which are generally systemic diseases, can be identified within the reproductive tract.

Although all reptiles are suspected to be susceptible, the reports are rare. The granulomas typically have central necrosis supporting large numbers of acid-fast rod-shaped bacteria and surrounded by multinucleate giant cells as well as macrophages. There may be a fibrotic capsule, depending on the chronicity of the lesions.

Mycobacterial species resulting in systemic infections of amphibians include *Mycobacterium ranae* (previously reported as *M. fortuitum*), *Mycobacterium xenopi*, *Mycobacterium lifiandii*, and *Mycobacterium ranicla*.[37-40] These infections are generally opportunistic and are present in the environment. They can be cultured from aquariums as well as from the skin of amphibians.[37] *Mycobacteria* spp. generally cause disease opportunistically through skin or mucosal abrasions in debilitated amphibians. Clinical signs and gross lesions include: sluggish behavior and long periods of motionless in water, multiple abscesses on feet with cellulitis extending into the legs and skin, nodular cutaneous lesions that may be ulcerated, splenomegaly and numerous small nodules in subcapsular locations in the testicle, ovary, oviduct, liver, kidneys, and spleen. Histologically these nodules are granulomas comprised of macrophages surrounded by a rim of epithelioid cells and several layers of fibroblasts. Nodules may contain numerous pleomorphic acid fast bacilli.

Inflammatory Conditions and Trauma

Scent gland and hemipenile impactions

These conditions are primarily identified in the common green iguana as well as some commonly kept snakes. There will be casts of the inspissated exudate that accumulates within the lumen of the invaginated hemipenis or both hemipenes and occasionally involves the adjacent scent gland or glands.[41] This will be composed of keratin and other secretory material commonly supporting secondary bacterial and fungal infections. Hemipenile pouch impactions or infections can lead to hemipenile prolapse.

Hemipenile or penile prolapse

The majority of these conditions are due to trauma during eversion of the copulatory organ in breeding season. The trauma results in extensive swelling, preventing retraction and leads to further trauma. Secondary infections are common.[42] Shell deformities or straining due to impactions or urolithiasis can result in penile prolapse of chelonians.

Extra-oviductal eggs and ectopic egg

In tortoises and turtles, a phenomenon recognized in females is of fully developed eggs being retropulsed into the urinary bladder.[7] The eggs are generally unable to be passed back out and will develop into a calculus.

Coelomic extra-oviductal shelled eggs have been described in a variety of turtles.[43] These eggs which may be supporting variable shell membranes can elicit an inflammatory reaction and be found attached to internal organs or free-floating. It is not uncommon to isolate bacteria from the albumen and yolk. The proposed mechanism is by retropulsion from the oviduct to release by rents in the oviduct (trauma, pressure necrosis).

Salpingitis

Inflammation of the oviduct or uterus and testes can be due to infectious agents, or in the case of females, secondary to retained yolk with impaction. The inflammation in the latter case is secondary to irritant tissue effects of the yolk protein.

Torsion

Ovarian torsion is associated with ovarian disease such as oophoritis and egg retention.[44] Torsions in lizards may also involve the adrenal gland given the close position of these organs.

Uterine rupture has been reported in Rio Cauca caecilian or rubber eel (*Typhlonectes natans*) a viviparous amphibian.[45] Malposition of the embryo after intrauterine hatching was found associated with tears through the uterine wall and hemorrhage.

Yolk coelomitis

Yolk coelomitis is a common sequela of pre-ovulatory follicular stasis and rupture of the ova. Normally in animals that fail to ovulate, the follicles should be reabsorbed. The inflammation is secondary to irritant tissue effects of the yolk protein. The condition is more commonly identified in captive lizards and chelonians. It is also reported to be a major cause of death in captive sexually mature female Fiji Island banded iguanas (*Brachylophus fasciatus*).[46] The clinical signs are generally fairly non-specific; lethargy, cutaneous discoloration, coelomic effusion. Gross examination find the coelomic cavity containing various amounts of yellow-white friable material adhered to most serosal surfaces and opaque yellow to red-tinged fluid. There may be ovarian follicles in various stages of development including atretic follicles and perhaps collapsed and/or ruptured vitallogenic follicles. Atresia of large vitellogenic follicles seems to be a common finding and it may be these follicles are the ones that may burst/leak. Inappropriate husbandry or poor body condition resulting in follicular or ovostasis is generally associated with cases of yolk coelomitis. Septic oophoritis and coelomitis are common sequelae.[47]

Dystocia

Pre-ovulatory stasis (follicular stasis) occurs when there is interruption in follicular development, and post-ovulatory stasis (dystocia) occurs after ovulation, when the egg has developed (deposition of albumin and formation of shell membranes) but is unable to move along the oviduct. Differentiation between pre and post can occasionally be determined by shape and location; ovarian follicles tend to be dorsal and spherical, and less mobile than oviductal eggs, which are usually oval and more ventral and caudal within the abdomen.[48] Snakes tend to develop post-ovulatory stasis more frequently than pre-ovulatory. Follicular stasis is common in captive chelonians.[49]

In both oviparous amphibians and reptiles, pre-ovulatory stasis is generally related to inappropriate clues for egg-laying (nesting substrate, temperature, humidity, pheromones, lighting) or poor health of the female. It is unknown if the cause of death in some gravid females is due to metabolic derangements or that other disease conditions became more significant because of the gravid condition.[50,51] sources claim preovulatory stasis is not seen in wild populations.

Post-ovulatory stasis (dystocia) can be divided into obstructive and non-obstructive. Obstructive can include defects in the egg itself; malpositioning, irregular surfaces, abnormal shape and/or size, rupture. Other causes may be of pelvic masses, oviductal lesions (strictures, inflammatory or tumor masses), or oviductal torsion.

Dystocia and subsequent surgical intervention has been described in Rio Cauca caecilians or rubber eel (*Typh-*

lonectes natans) a viviparous amphibian. Briefly the embryo development is divided into 34 stages; intrauterine hatching occurs between stages 25 and 26, metamorphosis from stage 30 to 32, and birth at stage 34. Anorexia in the female and increasing time between passing of live and some dead young triggered the decision for cesarean surgery. Specific causes for the dystocia was not identified.[45]

Tumors

Seminomas are reported in snake, lizard species, and one report in a spur-thighed tortoise (*Testudo graeca*).[52] The seminoma was unilateral and large enough to compress the liver.

Teratomas are composed of tissues derived from at least two of the three embryonic germ layers (endoderm, mesoderm, and ectoderm). Typically, they arise from the gonads. The ovaries are common sites for teratoma development in the common green iguana (*Iguana iguana*) [53] and other unspecified saurians [54], as well as a chelonian.[55] Clinical signs are the result of a space-occupying mass and may include digestive and respiratory dysfunction as well as coelomic cavity distention. Differentials in females include neoplasia or granuloma of the ovary or oviduct, cystic ovary, segmental pyosalpinx, and neoplasm or granuloma of the adrenal gland or spleen. In the few described cases the tumors had a rapid growth rate and a potential for malignancy. Early surgical removal is recommended.

On the basis of the morphologic and immunohistochemical results, a diagnosis of ovarian undifferentiated carcinoma was made in a corn snake (*Elaphe guttata guttata*). Histologic specimens consisted of a highly cellular mass composed of pleomorphic, spindle-shaped cells and, occasionally, round to polygonal cells arranged in irregular fascicles. The neoplastic cells were immunoreactive for cytokeratin (AE1/AE3), smooth muscle actin, and skeletal muscle actin, but did not stain for vimentin or desmin.[56]

A European viper with a cloaca and oviduct adenocarcinoma died after photodynamic therapy with chloroaluminum sulfonated phthalocyanine.[57]

MIscellaneous

Testicular cysts

Testicular cysts have been previously described in a veiled chameleon and an emerald tree boa, but documentation is nonexistent in other reptilian or amphibian species.[42] These cysts can become very large and result in a mass effect on coelomic organs. In other species (poultry) testicular cysts can be a sequel of sodium intoxication in hatchlings or exposure to mycotoxins.

References and Suggested Reading

1. Zwart P. Renal pathology in reptiles. *Vet Clin North Am Exot Anim Pract.* 2006;9(1):129-159.

2. Bishop JE. A histological and histochemical study of the kidney tubule of the common garter snake, Thamnophis sirtalis, with special reference to the sexual segment in the male. *J Morphol.* 1959;104:307-357.

3. Hanley CS, Hernandez Divers S. Practical Gross Pathology of Reptiles. *Semin Avian Exot Pet Med.* 2003;12(2):71-80.

4. O'Malley B. Snakes. In: *Clinical Anatomy and Physiology of Exotic Species.* Philadelphia, PA: Elsevier Saunders;2005:77-96.

5. O'Malley B. Tortoises and turtles. In: *Clinical Anatomy and Physiology of Exotic Species.* Philadelphia, PA: Elsevier Saunders;2005:41-56.

6. Rosskopf WJ, Shindo MK. Syndromes and conditions of commonly kept tortoise and turtle species. *Semin Avian Exot Pet Med.* 2003;12(3):149-161.

7. Perry SM, Mitchell MA. Reproductive medicine in freshwater turtles and land tortoises. *Vet Clin North Am Exot Anim Pract.* 2017;20(2):371-389.

8. Frye FL, Miller H. Hermaphroditism in a tomato frog (*Dyscophus antongilii*). *J Zoo Wildl Med.* 1994;25(1);154-157.

9. Pessier AP, Baitchman EJ, Crump P, et al. Causes of mortality in anuran amphibians from an ex situ survival assurance colony in Panama. *Zoo Biol.* 2014;33(6):516-26.

10. Miller SM, Serrano MA, Garner MM. Ectopic ovum in a hermaphrodite common green iguana, *Iguana iguana*. *J Herpetol Med Surg.* 2003;13(4):23-26.

11. Cooper JE, Bloxam QMC, Tonge SJ. Pathology of round island geckos, *Phelsuma geuntheri*: Some unexpected findings. Dodo. *J Wildl Preserv Trusts.* 1998;34:153-158.

12. Garner MM, Lung NP, Murray S. Xanthomatosis in geckos: Five cases. *J Zoo Wildl Med.* 1999;30(3): 443-447.

13. Reed SD, Reed FM, Castleman WL. Successful surgical management of advanced xanthomatosis in a leopard gecko, *Eublepharis macularius*. *J Herpetol Med Surg.* 2007;17(1):19-21.

14. Carpenter JL, Bachrach A Jr, Albert DM, et al. 1986 Xanthomatous keratitis, disseminated xanthomatosis, and atherosclerosis in Cuban tree frogs. *Vet Pathol.* 23(3):337-339.

15. Russell WC, Edwards EL, Stair EL, et al. Cornal lipidosis, disseminated xanthomatosis, and hypercholesterolemia in Cuban treefrogs (*Osteopilus septentrionalis*). *J Zoo Wildl Med.* 1990;21(1):99-104.

16. Sepulveda MS, Wiebe JJ, Honeyfield DC, et al. Organochlorine pesticides and thiamine in eggs of largemouth bass and American alligators and their relationship with early life-stage mortality. *J Wildl Dis.* 2004;40(4):782-786.

17. Rodriguez CE, Pessier AP. Pathologic changes associated with suspected hypovitaminosis A in amphibians under managed care. *Zoo Biol.* 2014;33(6):508-515.

18. Langford GJ, Willobee BA, Isidoro LF. Transmission, host specificity, and seasonal occurrence of *Cyrtosomum penneri* (Nematoda: Atractidae) in lizards from Florida. *J Parasitol.* 2013;99(2):241-246.

19. Jacobson ER, Oros J, Tucker SJ, et al. Partial characterization of retroviruses from boid snakes with inclusion body disease. *Am J Vet Res.* 2001;62(2):217-224.

20. Raymond JT, Garner MM, Nordhausen RW, Jacobson ER. A disease resembling inclusion body disease of boid snakes in captive palm vipers (*Bothriechis marchi*). *J Vet Diagn Invest.* 2001;13(1):82-86.

21. Vancraeynest D, Pasmans F, Martel A, et al. Inclusion body disease in snakes: A review and description of three cases in boa constrictors in Belgium. *Vet Rec.* 2006;158(22):757-760.

22. Stenglein MD, Jacobson ER, Chang L-W, et al. Widespread recombination, reassortment, and transmission of unbalanced compound viral genotypes in natural arenavirus infections. *PLoS Pathog.* 2015;11(5): e1004900.

23. Stenglein MD, Sanders C, Kistler AL, et al. Identification, characterization, and in vitro culture of highly divergent arenaviruses from boa constrictors and annulated tree boas: Candidate etiological agents for snake inclusion body disease. *mBio.* 2012:3(4) e00180-12.

24. Jacobson ER, Green DE, Undeen AH, et al. Systemic microsporidiosis in inland bearded dragons (*Pogona vitticeps*). *J Zoo Wildl Med.* 1998;29(3):315-323.

25. Suedmeyer Wk, Gillespie DS, Pace L. Chromomycosis in a marine toad, *Bufo marinus*. *Bull Assoc Reptilian Amphib Vet.* 1997;7(3):13-15.

26. Miller EA, Montali RJ, Ramsay EC, Rideout BA. Disseminated chromoblastomycosis in a colony of ornate-horned frogs (*Ceratophrys ornata*). *J Zoo Wildl Med.* 1992;23(4):433-438.

27. Anver MR. Diagnostic exercise: Systemic chromomycosis in frogs. *Lab Anim Sci.* 1980;30:165-166.

28. Bube A, Burkhardt E, Weiss R. Spontaneous chromomycosis in the marine toad (*Bufo marinus*). *J Comp Pathol.* 1992;106(1):73-77.

29. Cicmanec JL, Ringler DH, Beneke ES. Spontaneous occurrence and experimental transmission of the fungus, *Fonsecaea pedrosoi*, in the marine toad, *Bufo marinus*. *Lab Anim Sci.* 1973;23(1):43-47.

30. Elkan E, Philpot CM. Mycotic infections in frogs due to a *Phialophora*-like fungus. *Sabouraudia.* 1973;11: 99-105.

31. Rush HG, Anver MR, Beneke ES. Systemic chromomycosis in *Rana pipiens*. *Lab Anim Sci.* 1974;24(4): 646-655.

32. Tell L, Nichols DK, Bush M. Clinical challenge. *J Zoo Wildl Med.* 1994;25(1):173-175.

33. Schmidt RE, Hartfiel DA. Chromomycosis in amphibians: Review and case report. *J Zoo Anim Med.* 1977;8:26-28.

34. Arnbjerg J, Ruelokke, Aalbaek MLB. Infectious salpingitis and coelomitis in a spiny-tailed monitor lizard. *Exot DVM.* 2002;4(2):17-19.

35. Kourany M, Telford SR. *Salmonella* and *Arizona* infections of alimentary and reproductive tracts of Panamanian lizards. *Infect Immun.* 1982;36(1):432-434.

36. Kaufman AF, Morrison ZL. An epidemiologic study of salmonellosis in turtles. *Am J Epidemiol.* 1966;84: 364-370.

37. Done LB, Willard-Mack CL, Ruble G, Cranfield M. Diagnostic exercise: ulcerative dermatitis and cellulitis in American toads. *Lab Anim Sci.* 1993;43(6):619-621.

38. Joiner GN, Abrams GD. Experimental tuberculosis in the leopard frog. *J Am Vet Med Assoc.* 1967;151(7): 942-949.

39. Joiner GN, Abrams GD. Tuberculosis in the leopard frog (*Rana pipiens*). *J Infect Dis.* 1970;122(1):96-99.

40. Fremont-Rahl JJ, Ek C, Williamson HR, et al. *Mycobacterium liflandii* outbreak in a research colony of *Xenopus* (Silurana) *tropicalis* frogs. *Vet Pathol.* 2011;48(4):856-867.

41. Johnson JD. Urogenital system. In Girling SJ, Raiti P, eds. *BSAVA Manual of Reptiles.* Gloucester, England: Br Small Anim Vet Assoc;2004:261-272.

42. Knotek Z, Cermakova E, Oliveri M. Reproductive medicine in lizards. *Vet Clin North Am Exot Anim Pract.* 2017;20(2):411-438.

43. Jackson, Jr. CG, Landry MM, Jackson MM. Reproductive tract anomaly in a box turtle. *J Wildl Dis.* 1971;7(3):175-177.

44. Juan-Sallés C, Garner MM, Monreal T, Burgos-Rodriguez AG. Ovarian torsion in a green iguana, *Iguana iguana*, and a rhinoceros, *Cyclura cornuta*. *J Herpetol Med Surg.* 2008;18(1):14-17.

45. Barbon,AR, Goetz M, López J, Routh A. Uterine rupture and cesarean surgery in three Rio Cauca caecilians (*Typhlonectes natans*). *J Zoo Wildl Med.* 2017;48(1):164-170.

46. Stacy BA, Howard L, Kinkaid J, Vidal JD, Papendick R. Yolk coelomitis in Fiji Island banded iguanas (*Brachylophus fasciatus*). *J Zoo Wildl Med.* 2008;39(2):161-169.

47. Gardner BR, Barrows MG. Yolk coelomitis in a white-throated monitor lizard (*Varanus albigularis*). *J S Afr Vet Assoc.* 2010;81(2):121-122.

48. Stahl SJ. Pet lizard conditions and syndromes. *Semin Avian Exot Pet Med.* 2003;12(3):162-182.

49. Takami Y. Single-incision, prefemoral bilateral oophorosalpingectomy without coelioscopy in an Indian star tortoise (*Geochelone elegans*) with follicular stasis. *J Vet Med Sci.* 2017;79(10):1675-1677.

50. Stetter MD. Noninfectious medical disorders of amphibians. *Semin Avian Exot Pet Med.* 1995;4(1):49-55.

51. Melidone R, Knoll JS, Parry N. Preovulatory stasis and dystocia in oviparous lizards. *Vet Med.* 2008;103(11):595-598.

52. Pees M, Ludewig E, Plenz B, Schmidt V. Imaging diagnosis-seminoma causing liver compression in a spur-thighed tortoise (*Testudo graeca*). *Vet Radiol Ultrasound.* 2015;56(2):E21-E24.

53. Anderson NL, Williams J, Sagartz JE, Barnewall R. Ovarian teratoma in a green iguana (*Iguana iguana*). *J Zoo Wildl Med.* 1996;27(1):90-95.

54. Hernandez-Divers SM, Garner MM. Neoplasia of reptiles with an emphasis on lizards. *Vet Clin North Am Exot Anim Pract.* 2003;6:251-273.

55. Newman SJ, Brown CJ, Patnaik AK. Malignant ovarian teratoma in a red-eared slider (*Trachemys scripta elegans*). *J Vet Diagn Invest.* 2003;15(1):77-81.

56. Petterino C, Bedin M, Podest G, Ratto A. Undifferentiated tumor in the ovary of a corn snake (*Elaphe guttata guttata*). *Vet Clin Pathol.* 2006;35(1):95-100.

57. Roberts WG, Klein MK, Loomis M, Weldy S, Berns MW. Photodynamic therapy of spontaneous cancers in felines, canines, and snakes with chloro-aluminum sulfonated phthalocyanine. *J Natl Cancer Inst.* 1991;83(1):18-23.

Managing Cloacal Prolapses in Reptiles

Christoph Mans, Dr med vet, Dipl ACZM, DipECZM (Zoo Health Management)

Session #2249

Affiliation: From the School of Veterinary Medicine, University of Wisconsin-Madison, 2015 Linden Drive, Madison, WI 53706, USA.

Abstract: Cloacal prolapses are commonly diagnosed in reptiles. Correct identification of prolapsed tissue can be challenging, but it is critical to determine an appropriate approach to treatment. Diagnostics include whole body radiographs, computed tomography, routine blood work, fecal parasitology, and saline-infusion cloacoscopy. Treatment can include repositioning, resection and anastomosis. Besides identifying the prolapsed tissue itself and providing immediate care to prevent further tissue trauma, it is important to investigate the primary underlying cause of the cloacal prolapse. Cloacal eversion, penile phallus, hemipenile, oviductal, colon, rectal, and urinary bladder prolapses can occur in reptiles.

Introduction

Cloacal prolapses are common diagnoses in reptiles. In a retrospective study the incidence of cloacal prolapses in reptile patients was 1.9%.[1] Prolapses were more frequent in lizards and chelonians than in snakes and the most common prolapsed organs were the penis, phallus, and hemipenis.[1] The gastrointestinal (GI), urinary, and reproductive tract all terminate in the cloaca in reptiles and therefore oviduct, urinary bladder, penis, phallus (turtles and tortoises) or hemipenes (lizards and snakes), and the intestine can prolapse through the cloacal opening. Furthermore, the cloaca itself may evert and prolapse (ie, true cloacal prolapse).[1]

Correct identification of the prolapsed tissue can be challenging, but is critical, in order to determine the correct treatment approach. In particular, if the tissue had been prolapsed for several hours to days, then changes its appearance due to tissue necrosis can make identification challenging. The tissue should be carefully manipulated and cleaned to aid identification.

Diagnostics

Whole-body radiographs or computed tomography should be performed in most reptiles presenting with a cloacal prolapse. Radiographs will help to rule out the presence of mineralized egg, which may be associated with increased straining, and potentially leading to cloacal prolapse. Radiographs also can rule out the presence of uroliths, which can be either located in the urinary bladder or have migrated from the urinary bladder into the cloaca. Snakes and some lizard species (eg, bearded dragons) do not have a urinary bladder, and this should be considered when evaluating these species. In addition, the GI tract can be assessed by radiographs for presence of impactions (eg, ingesta, sand) or tympany. Poor bone quality is most likely indicative of nutritional secondary hyperparathyroidism and the patient may be hypocalcemic, which may lead to cloacal prolapse. Additional imaging techniques, such as computed tomography are very useful to evaluate coelomic organs and evaluate for coelomic effusion (eg, due to yolk coelomitis).

Routine blood work is indicated in cases where a metabolic or systemic underlying disease process is suspected. A plasma biochemistry panel will allow rule out of hypocalcemia, impaired renal function, amongs other abnormalities. Increased calcium levels in female reptiles, are indicative of reproductive activity and therefore the cloacal prolapse may be associated with reproductive tract activity or disease. A complete blood count (CBC) will allow rule out of anemia (eg, due to acute blood loss from the prolapsed cloacal tissue) or chronic disease, as well as help to evaluate for a systemic inflammatory response. An increase in the total white blood cell count as well as morphologic changes in the granulocytes support the diagnosis of an inflammatory process; however, without baseline values for each individual patient obtained during a period when no underlying disease process is present, it will be challenging to interpret a single CBC result obtained during a disease process.

Fecal parasitology should be performed in most reptiles diagnosed with cloacal prolapses, in particular in animals with intestinal prolapse. Heavy parasite burden can lead to enteritis, diarrhea, straining, and ultimately cloacal prolapse. Fecal samples should be evaluated as a wet-mount preparation as well as following flotation. If a fecal sample is not readily available, then an enema may promote defecation. Consider empiric antiparasitic treatment if a fecal sample cannot be obtained, but endoparasites are a likely an underlying cause.

Saline-infusion cloacoscopy under sedation or general anesthesia is the most valuable diagnostic tool available to evaluate cloacal prolapse cases. For this procedure a rigid endoscope housed in a working sheath is used. An intravenous infusion set is connected to a fluid bag and to one of the ports of the working sheath. The infusion of the cloaca with 0.9 % saline solution or a balanced electrolyte solution will distend the cloaca and allow for evaluation of the normal cloacal structures and the origin of the cloacal prolapse, which allows to diagnosis the cloacal prolapse by exclusion. If the urinary bladder and rectum both can be identified during saline infusion cloacoscopy, then it can be ruled out that these organs have prolapsed. Manipulation of the cloacal prolapse and following the prolapsed tissue to its base, will help to identify the origin of the prolapsed tissue.

Treatment

Treatment of cloacal prolapses should only be performed once the tissue has been identified. If the tissue is viable, repositioning of the prolapse under sedation, intrathecal anesthesia, or general anesthesia can be attempted if it has been determined that reposition is the preferred treatment method. Repositioning is performed by means of gentle pressure with cotton-tipped applicators after the application of hyperosmotic solution or sugar on the tissue. Alternatively, if the prolapsed organ has a lumen (ie, intestine, cloaca) it can be infused with saline, which will aid in repositioning. Once repositioned, 2 sutures are placed at the side of the vent to reduce the chances of further prolapse. The use of purse-string suture, although widely used in the past, is not recommended due to the risk of damage to the cloacal sphincter.[2] In lizards and snakes percutaneous tagging sutures also can be used if the cloaca itself is prolapsed. If these techniques fail to prevent reoccurrence of cloacal tissue prolapse, then cloacopexy or colonpexy should be considered.[2]

If a hollow organ prolapses (eg, colon, oviduct) and the tissue has suffered consistent trauma, a resection and anastomosis is performed. With the reptile anesthetized in dorsal recumbency the prolapsed tissue is cleaned. A cylindrical stent (eg, a syringe case) is placed in the lumen and the tissue is transfixed by orthogonal needles. The nonviable tissue is excised and mattress sutures are placed all around the organ. With this approach continuity of the hollow organ is maintained. In cases of oviductal prolapse usually substantial oviductal disease and/or coelomitis is present, and these cases often benefit from coeliotomy and a more thorough assessment. Oviductal prolapse usually results in trauma to the oviductal vasculature, which can lead to substantial hemorrhage and hemocoelom.

Besides identifying the prolapsed tissue itself and providing immediate care to prevent further tissue trauma, it is important to investigate the primary underlying cause of the cloacal prolapse. Without identification and correction of the primary underlying cause the prognosis for resolution of most cloacal prolapses is poor. While the underlying or concurrent diseases are being diagnosed, it is important to prevent further trauma to the prolapsed cloacal tissue.

Cloacal Eversion

The cloaca itself may evert and prolapse without any other organ system being involved.[1] Causes for cloacal eversion (true cloacal prolapses) include metabolic causes, straining, increased intracoelomic pressure, obesity, dystocia, inflammation/infection of the cloaca, amongst others. The everted and prolapsed cloaca may be mistaken as the urinary bladder, but the urinary bladder wall is thinner and has no lumen. The prolapsed tissue, if viable, should be replaced as soon as possible and stay sutures placed to reduce the cloacal opening.

Penile, Phallus, and Hemipenile Prolapse

Prolapse of the penis (syn. phallus) in chelonians is common and can occur secondary to behavioral reasons (eg, fighting males, agitated males during breeding season) or due to underlying disease (eg, shell deformation leading to increased coelomic pressure, neurologic causes, trauma). Penile prolapse secondary to cystic or cloacal calculi has been reported and therefore radiographs always should be performed to rule out this condition.[3] Penile trauma can be a cause for severe and life-threatening hemorrhage, and therefore the penis should be carefully examined for very small traumatic lesions. In lizards and snakes the hemipenes can prolapse due to irritation from probing or from iatrogenic protrusion of the hemipenes performed for sexing. Other causes include smegma accumulation or inflammation of the hemipenes. In many cases the underlying cause remains unknown. The author recommends surgical amputation for all penile prolapses in chelonians, due to the high risk of reoccurrence, unless breeding ability needs to be maintained. In lizards and snakes the hemipenes can sometimes be carefully replaced and stay sutures placed; however, if the hemipenes are enlarged or the tissue is non-viable or very inflamed, surgical amputation is recommended. Breeding ability can be maintained, if only one hemipenis is amputated. Amputation may be performed in toto, as penis and hemipenes do not contain the urethra.[2] In small individuals, one to two transfixing ligatures are placed at the base of the penis/hemipenis. The tissue is excised distal to the ligatures. In larger chelonians, each half of the phallus may be individually double ligated with encircling absorbable sutures. The penile stump can be closed with a simple continuous or a purse-string suture, in particular in larger species.[2] Tortoises have two cartilaginous cords within the penis, making transection and ligation more challenging, compared to turtles.

Oviductal Prolapse

Prolapse of the oviduct is frequently associated with reproductive activity and occurs most common in chelonians and lizards. The prolapsed oviduct often appears plicated, thin-walled and has a lumen; however, with prolonged duration of the prolapse, the tissue is usually swollen, necrotic or covered by a fibrinous scab. Diagnostics should always include diagnostic imaging to assess the reproductive tract. Causes include reproductive disease (oviductal impaction, salpingitis, yolk-coelomitis, dystocia) or increased intracoelomic pressure. Resection of the prolapsed tissue at the level of the entry into the urodeum of the cloaca has been reported as a treatment option, but it does not address the underlying disease, and therefore a coeliotomy is usually necessary to treat the underlying disease and a ovariosalpingectomy is frequently performed. Intracoelomic bleeding is a common complication of oviductal prolapses due to trauma of the oviductal vessels.[2]

Colon and Rectal Prolapse

Prolapsed rectum and colon have a lumen and a thick wall compared to other prolapsed organs. Usually substantial swelling if present as the prolapse persists. Causes include endoparasitism, straining, constipation, increased coelomic pressure, or metabolic disease (eg, hypocalcemia). If viable the tissue may be replaced. If necrosis is evident resection and end-to-end anastomosis is necessary.

Urinary Bladder Prolapse

Prolapsed urinary bladders have a very thin wall, which is easily traumatized. The prolapsed tissue has no lumen and often the prolapsed tissue is of substantial size. Underlying causes are similar as for other cloacal prolapses. Cystitis, urolithiasis and increased coelomic pressure are common causes. Viable bladder tissue can be carefully replaced by infusing the organ with saline. If the tissue is necrotic amputation should be performed. The ureters in reptiles do not enter into the urinary bladder, but into the urodeum of the cloaca, and therefore trauma or ligation of the ureters is not a concern during cystectomy. The procedure can be performed transcloacally or intracoelomically, in particular in lizards.

References

1. Hedley J, Eatwell K. Cloacal prolapses in reptiles: A retrospective study of 56 cases. *J Small Anim Pract.* 2014;55:265-268.

2. Di Girolamo N, Mans C. Reptile soft tissue surgery. *Vet Clin North Am Exot Anim Pract.* 2016;19:97-131.

3. Huang J, Eshar D, Andrews G, et al. Diagnostic Challenge. *J Exot Pet Med.* 2014;23:418-420.

Scute Pyramiding in Tortoises: Prevention and Resolution of the Condition

Mark E. Burgess, DVM

Session #2135

Affiliation: From Southwest Animal Hospital Exotic Animal Practice, Beaverton, OR 97005, USA.

Abstract: Pyramiding or doming of the carapacial scutes in captive tortoises is common, especially in large, fast-growing species. Despite the prevalence of this condition, relatively little consensus has been reached in the literature as to the etiology. Various factors, including diet, ultraviolet light levels, and even humidity, have been suggested as underlying factors. In this discussion, the author reviews the literature, and adds his own observations over 32 years of clinical practice. This paper associates pyramiding with osteodystrophy due to low ultraviolet light exposure and/or improper dietary calcium:phosphorus ratios, and describes how to successfully prevent the disease, as well as halting it in patients already manifesting clinical signs.

Introduction

Shell growth anomalies are commonly observed in captive turtles and tortoises. The clinical appearance varies between turtle groups, with many aquatic species and North American box turtles manifesting different growth defects than tortoises in the author's experience. This discussion focuses on tortoises, in particular large, fast-growing species such as the African spurred tortoise, *Geochelone sulcata* (*Centrochelys sulcata*) and the leopard tortoise (*Stigmochelys pardalis*). These commonly-kept species manifest shell anomalies readily, likely due to their aggressive growth rates which place more demand on optimal husbandry. A primary visible anomaly observed by the author (and others) is pyramiding or doming of the carapace scutes.[1-5]

The scientific literature regarding this condition is surprisingly sparse. Although the condition has been widely recognized in veterinary and breeder discussions, few scientific articles have been published. Much discussion has centered around metabolic bone disease as a result of improper dietary calcium:phosphorus ratio (Ca:P) and/or low ultraviolet-B light levels.[1,6-8] Some authors have mentioned dietary protein and environmental humidity as factors influencing scute pyramiding in hatchling African Spurred tortoises.[1,9,10] Another author hypothesized multiple factors influencing shell growth, including calorie intake and output, body temperature, acid-base balance, hydration, and environment.[3]

The author's observations and clinical results have yielded insights into factors which appear to be the most important in the clinical setting. These will be discussed along with the cited literature.

Literature Review and Clinical Observations

Much of the previously reported discussion of scute pyramiding has been based on hypothetical considerations of dietary or environmental factors. Only one paper has detailed a scientific investigation into factors influencing scute pyramiding.[9] This article noted a small positive association of pyramiding with increased dietary protein levels. This was not unexpected, as high dietary energy content (especially protein) has been proposed by others as a possible cause of excessively rapid or abnormal shell growth in herbivorous chelonians.[1,2,10]

More surprising was the association the article made between lower humidity levels and increased scute pyramiding, whereas high humidity levels (up to 99% relative humidity) caused reduced pyramiding. Interestingly, the lower humidity parameters included relative humidity levels up to nearly 75%. Very high humidity levels reduced pyramiding in the study animals, but did not eliminate it. In that study, serum calcium and phosphorus levels were reported as within normal limits in all the tortoises.

When considering pyramiding, recognition of current versus prior growth anomalies is important, as it helps elucidate the etiology of the condition. Tortoises with active/current scute pyramiding will manifest a vertical scute growth pattern (on the carapace) that continues into the most recently laid-down growth rings (those at the periphery of each scute.) Those tortoises with inactive pyramiding (wherein the growth anomaly has resolved) will show verticality of the central portion of the dorsal scutes, but the more recent peripheral rings will show a more flattened (normal) growth pattern. Detailed discussion of the reptile's husbandry history often reveals which dietary or environmental changes led to the improved shell conformation.

In the author's experience, nearly every tortoise with active pyramiding also manifests a softened, flexible plastron (palpated at the ventral midline). This indicates that osteodystrophy is associated with the scute growth anomaly. This was the most logical prediction, as shell shape should be determined mostly by the conformation of the underlying bone, not the superficial keratin layer. Confusion may arise when tortoises with pyramiding lack detectable shell softness, but these cases may no longer have active osteodystrophy. As the prior growth deformities are not reversible, some central doming of the scutes will persist lifelong. These pyramids will become proportionately smaller (compared to the overall shell size) if the animal continues to grow normally thereafter.

In the author's experience, ambient humidity does not appear to be a primary factor in tortoise scute pyramiding. The author has observed that African spurred tortoises raised in the arid American southwest often manifest excellent shell conformation. The warm climate allows these animals to spend much time outdoors, obtaining ultraviolet (UV) light exposure and consuming grass and leafy plants. Their nutritional profiles and ultraviolet exposures are thereby optimized. By contrast, animals raised in the more humid Pacific Northwest climate zone (where the author practices) almost always manifest pyramiding. The cool climate necessitates animals being housed indoors for much of the year, and cloudy conditions often reduce UV exposure when outdoors. Unless the tortoise's owners follow strict habitat and dietary guidelines, pyramiding invariably develops. Those guidelines include close attention to dietary protein and Ca:P ratios, and provision of adequate UV light throughout the *entire* habitat.

A close inspection of the prior study reveals possible explanations for the reported findings.[9] Regarding the normal serum calcium levels in the study animals: the subjects were hatchlings, and were presumably normal at hatch time. The study followed their growth for only five months. This is long enough for diet imbalances or low UV-B levels to induce pyramiding, but possibly not long enough to completely deplete the bony calcium reserves. In conditions of calcium deficiency, the body will attempt to maintain serum calcium levels within normal parameters, as hypocalcemia will interfere with neuromuscular function and can lead to weakness, tremors or even full seizures. Calcium will be pulled from the bones until those reserves are depleted, so serum calcium levels may not drop until osteodystrophy is very advanced.

The correlation between very high humidity levels and reduced degree of pyramiding is interesting. African Spurred Tortoises naturally live in arid regions with low humidity levels, including the southern Sahara desert, Ethiopia, Sudan, Nigeria and Senegal (Wikipedia. Available at https://en.wikipedia.org. Accessed March 28, 2018.) They do shelter in burrows during the hottest portions of the day, but the annual average rainfall along the southern Sahara ranges from 100 millimeters (4 in) to 250 millimeters (10 in).[11] The evaporation rates are extraordinarily high, making for very dry conditions.[12] Nowhere else on Earth has air been found as dry and evaporative as in the Sahara region. Therefore, it appears erroneous to conclude that lower humidity levels are a primary cause of scute pyramiding.

The author's own observations in clinical practice support this conclusion. The question remains as to why very high humidity levels were reported to reduce the degree of scute doming. With osteodystrophy, the underlying bone layer in pyramided scutes will be softer and more malleable than normal. In this situation, high moisture content in the air may allow for softening of the surface keratin, leading to less rigidity of the surface, and flattening of the pyramided scutes. This would reduce (but not eliminate) the pyramiding. Photos of subject animals in the prior study showed visible pyramiding at all humidity levels. Higher humidity reduced, but did not eliminate, the deformity.

In normal animals, the keratin stiffness would not be as likely to influence the shell conformation. This would make humidity a lessor factor in pyramid development, in animals with normal bone density. Low humidity would therefore only be a (secondary) factor in the expression of pyramiding in tortoises with pre-existing osteodystrophy.

Other than humidity, what factors could have caused the pyramiding in the study animals? In the author's practice, close attention to dietary energy content and Ca:P ratios, combined with provision of strong UV-B levels (280-320 nm spectrum) to the patient for at least 8-10 hours daily, has proven highly effective in preventing scute pyramiding in all tortoise species. Proper husbandry also reliably halts the disease in animals already afflicted. The author has never needed to address humidity as part of the protocol in managing this pathology. Using these same guidelines, an inspection of the husbandry parameters in the prior study protocol reveals potential problem areas.

The study diet was comprised of a pelleted feed with varying protein content, from13.7-30.7% protein.[9] This was fed along with endive at a 1:1 ratio until day 105, then at a 2:1 ratio from day 106 to day 150. The Ca:P ratio of the diet varied from 3.7:1 to 4:1. As the pelleted feed comprised much of the study diet, its impact on growth and health would be significant. Without a more detailed nutrient analysis (which the paper didn't provide), it is difficult to assess what other imbalances may have been present. The fresh vegetable intake was limited to only endive, which meant that the provision of balanced nutrition would rely on the pellets primarily. One obvious concern would be the high protein content of the richer pellets (which was deliberate, to assess the effect on scute pyramiding.)

The Ca:P ratio was quite high in the study diet, far exceeding the usual recommendations for normal bone development and maintenance in humans and animals of 1:1 to 2:1 in favor of calcium.[1,13,14] The effect of this relative calcium excess is uncertain; it may be of little consequence, as one study suggested,[15] or in theory could stimulate more rapid bone growth and perhaps interfere with normal mineralization. High energy intake and excessive mineral supplementation have been associated with osseous developmental anomalies (such as osteochondrosis) in fast growing dogs.[16,17]

The habitat setup in the published study was also concerning, and in the author's experience this could predispose to metabolic bone disease. The housing consisted of a terrarium 100 cm (39.4 inches) in length, and 80 cm (31.5 inches) wide. A single ultraviolet-producing fluorescent tube was utilized to provide UV exposure; the product utilized was reliable (Reptisun 5.0, ZooMed Laboratories Inc., San Luis Obispo, CA, USA); however, fluorescent UV sources have significant range limitations,[18] with smaller bulbs being more concerning due to lower overall energy output. The length and wattage of the light source was not listed; if the bulb wasn't large enough (91cm [36 inches] long, at a minimum) to properly cover the enclosure's full length, then ambient UV levels may have been inadequate in portions of the habitat. Similarly, if the light was positioned toward one end of the cage, then the opposite end may have experienced low UV levels.

In the study, 150 watt bright lights provided adequate day length, but the ultraviolet source was only active for 4.5 hours daily. In the author's experience, reptiles usually need 8-10 hours minimum exposure to the common UV fluorescent tubes to assure adequate UV exposure. Studies have shown that commercial UV lamps (both fluorescents and mercury vapor bulbs) were less effective at elevating plasma 25-hydroxyvitamin D (D_3) levels compared to natural sunlight.[18-20] This would necessitate longer exposure times, compared to outdoor sunlight, to achieve adequate vitamin D_3 levels. One study indicated that 2 hours of exposure to a strong artificial UV source raised serum D_3 levels significantly in bearded dragons,[21] but this effect could vary greatly depending on the type and age of the light bulb, and distance between light source and animal. Another bearded dragon study showed that 12-14 hours of daily exposure to desert compact UV coil bulbs (various brands), positioned at approximately 25 cm (9.8 inches) above the basking site, failed to stimulate synthesis of plasma 25(OH)D_3 to levels similar to free-living animals.[18]

Lastly, the study habitat included hiding caves, which if utilized by the animals during the daytime, would have prevented proper ultraviolet exposure.[9] This is especially concerning given the short window of UV exposure provided daily. Taking all the parameters into account, it is likely that the tortoises in the study did not consistently receive adequate UV-B levels. This could be a primary factor in creating the pyramiding noted in the study.

Clinical Management

Diet

Dietary management of osteodystrophy primarily relies on provision of vegetable dietary items with low protein content and proper Ca:P ratio. A Ca:P ratio ranging from 1:1 up to 2:1 has been considered necessary for proper bone development and maintenance in most vertebrate species.[22,23] The author prefers to utilize food sources with Ca:P ratios between 1.5:1 and 2:1. Vegetables that meet these parameters are primarily leafy greens, although a few fruits (blackberries, raspberries, papaya) have strong Ca:P ratios. A few other fruits (grapes, apples, mangos) have Ca:P ratios that fall into the lower end of the preferred range (ranging from 1:1 to 1.4:1).[24] Most other fruits, and non-leafy plant portions such as carrots, peas and corn, have moderately to severely inverted mineral ratios (phosphorus higher than calcium), and would therefore promote osteodystrophy.

Some leafy greens with adequate Ca:P ratios may still be weak calcium sources, due to heavy oxalate content which binds the calcium. Vegetable calcium oxalate is poorly absorbed and biologically less unavailable.[25,26] Notable offenders with disproportionately high oxalate content (compared to total calcium) include spinach, parsley, and beet greens (Vegan Health: Evidence-Based Nutrient Recommendations, https://veganhealth.org, accessed 3-28-2018). Predictably, the literature is also inconsistent with regards to this subject. Some authors have stated that kale is high in oxalates and a poor calcium source,[1,10] but another study showed that kale is low in oxalate content and an excellent source of calcium.[26] The author utilizes a table of mineral contents of common fruits and vegetables to help educate reptile owners. Ideally the diet should be comprised mainly of Ca:P balanced leafy greens and grasses, including as much variety as possible.

Reliable provision of other trace nutrients (vitamins and minerals) is also challenging. Utilizing a good variety of vegetable sources will increase the likelihood of balancing the diet, but any homemade mixed-ingredient diet contains a degree of nutritional uncertainty. Supplement powders and liquids marketed for reptiles are generally unproven, without feeding trials and growth/performance studies to verify safety or efficacy. In addition, powdered concentrates typically come without reliable dosing directions, and most manufacturers in truth don't know the appropriate dose for a given body mass. Add in the variability between products, and the consumer is left to guess what the supplement dose should be. With concentrated products, the potential for long-term over dosage (or under dosage) is a potential concern. One study noted that two different supplements yielded different results, with metabolic bone disease manifesting with one product, but not the other.[27] Another study noted that the recommended dose for a marketed supplement failed to yield optimal bony calcification; doses three times the label dose yielded good bone density and growth, but higher doses caused metastatic calcification.[28] These studies touched upon only a fraction of the vitamin-mineral supplements available in today's pet trade, but they adequately demonstrate the potential risks with such products.

The author's solution is to utilize an appropriately formulated tortoise food as a small portion of the diet. This provides trace nutrients in a safer (non-concentrated) format with minimal risk of over dosage. Ratios of calcium and phosphorus in commercial diets are precisely balanced, and protein levels of appropriately-formulated products are low (typically 8-12% protein in dry kibble diets). The author prefers dry commercial foods, as they are more stable in storage than canned products, and can be fed dry, moistened, or ground into a powder (using a mortar and pestle, or a coffee bean grinder) and sprinkled onto dampened greens. Some brands have performed feeding trials prior to marketing their foods, generating more evidence to support their product compared to typical vitamin supplements (Michael Massie, president of Pretty Bird International, Stacy, MN, USA; oral communication, April 2018). These foods appear to provide adequate trace nutrient content (based on clinical health) when used long term with a vegetable-based diet. In the author's experience, most species of reptiles are maintained in good health status, with very good longevity, with no concentrated vitamin or mineral supplement utilized. Successful feeding protocols in the author's practice have utilized tortoise kibbles as 5-15% of the overall diet.

Lighting

The other half of the pyramiding equation involves adequate provision of ultraviolet light, especially "UV-B" at the 280-320 nm wavelengths. Ultraviolet light is needed to activate vitamin D and facilitate calcium absorption from the gut. In low sunlight regions or cool climates where the animal cannot be housed outdoors, UV deficiency is a significant concern. Artificial UV sources commonly fall into two general categories: fluorescent bulbs and mercury vapor bulbs. One study showed efficacy of an LED bulb in emitting UV-B light, achieving serum 25-hydroxycholecalciferol levels equal to or greater than a fluorescent UV-B bulb.[29]

A primary concern with artificial UV sources is low intensity of light production (compared to sunlight) and limited effective range from the light source.[18-20] In one study, a variety of brands of 'desert species' compact UV coil lamps were tested, and at 120 days of usage, the maximum UV-B output (at 25 cm distance) for any brand was 57 uW/cm^2.[18] By contrast, the ambient outdoor UV-B levels in the animals' natural habitat ranged as high as 506 uW/cm^2 at 14.00 (2pm). UV levels were assessed with a Solarmeter 6.5 UV Index meter and a Solarmeter 6.2 broadband UVb irradiance meter (Solartech Inc., Harrison Township, MI, USA). The small size of the compact bulb would be disadvantageous in the author's experience, both in terms of overall energy output, and its inability to stretch the length of the cage; animals may easily move beyond the bulb's effective range within the habitat.

In the author's experience, fluorescent bulbs can provide adequate lighting in smaller habitats, provided that the light is large enough to cover the entire cage length, and that the tortoise doesn't hide in a shaded shelter. The light must also be changed regularly (typically every six months) as UV output fades rapidly with bulb usage. One study showed that over 120 days, UV output from various fluorescent light sources decreased from 10% to 27%.[18]

Larger or taller habitats will challenge the limits of fluorescent technology, even the largest UV tubes. Mercury vapor bulbs may be better suited for these situations; their higher cost (and energy usage) is offset by their larger effective ranges and longer lifespans compared to fluorescents. These bulbs are very hot, and produce a "spotlight" pattern of light, making them less ideal in small cages. Large habitats allow for the bulb to be elevated, minimizing burn risk. A higher bulb position also allows the light pattern to spread out and cover a larger floor area.

Regardless of the type of UV source, owners may still find it difficult to provide adequate UV light for larger tortoises in cool/cloudy climates. The very large sizes of the adult animals necessitates extensive habitats. Artificial UV sources require that the animal remain within the effective range of the bulbs for extended periods each day. Multiple powerful light sources may be required to provide adequate UV-B as the animal ranges through its habitat. Zoos with large reptile habitats often use high output mercury vapor bulbs, positioned safely far above the animals, to provide light over extended areas (UV Guide UK, available at www.uvguide.co.uk, accessed 3-25-2018).

Even adequate UV sources do not guarantee success. Solid materials (even transparent) will block UV transmission to varying degrees; UV sources should be positioned to allow direct light exposure to the animal.[30] Light fixture covers should not be used, nor should lights be positioned outside of a glass cage top or wall. Hiding caves are not ideal if the animal shelters in them during daylight hours. On the other hand, tortoises often are shy, and appear restless when they are constantly exposed in the open. In the outdoor setting, an animal may hide within tall grass, rocks or brush, while still finding sunlight to bask in. In a terrarium, the pet may be faced with 2 choices: sit exposed with no shelter (and clear glass walls all around), or hide in a cave. Neither option is ideal. Cage habitats need to provide areas of visual security, to minimize stress and the need to bury or hide. Transparent cage walls can be masked with decorative paper at one end, so the animal can feel hidden when sitting in that area. Tall rocks, artificial plants, or logs can provide a shelter to hide behind (not under), with the light positioned to allow UV exposure while hiding. A box with an entry at one end to crawl into may be a useful shelter for some tortoises; the box top is open, allowing UV exposure while the tortoise feels hidden within.

Summary

Scute pyramiding is a commonly encountered problem in captive tortoises. Despite the wide range of published hypotheses regarding etiology and prevention, the author has found that close attention to details of lighting and diet produce consistently good clinical outcomes. Normal juvenile tortoises can be maintained in good health without pyramiding; those that present with pre-existing shell deformity can be treated, and their pathology halted. Utilizing dietary items with low energy content (in particular, protein) and well-balanced Ca:P ratios is critical. Leafy greens and grasses should comprise the bulk of the diet, including as much variety as possible. Dietary vitamin/mineral supplements should probably be avoided, in favor of a well-balanced tortoise food used in moderation. Ultraviolet light sources need to be large and powerful enough for the habitat being utilized, and changed out regularly to keep UV output at adequate levels. Lights need to have direct exposure to the animals, without solid blocking materials obstructing the UV light. Tortoises need to spend many hours daily within effective range of their light sources.

References

1. Kirchgessner M, Mitchell M. Chelonians: Nutritional disorders. In: Mitchell A, Tully T, eds. *Manual of Exotic Pet Practice.* St. Louis, MO: Elsevier Saunders;2009:234.

2. Frye FL. *Biomedical and Surgical Aspects of Captive Reptile Husbandry.* Malabar, FL: Krieger;1991:400.

3. Donaghue S. Nutrition. In: Mader D, ed. *Reptile Medicine and Surgery.* 2nd ed. St. Louis, MO: Elsevier Saunders;2006:284.

4. Mader D, ed. *Reptile Medicine and Surgery.* Philadelphia, PA: Saunders;1996:389.

5. Beyon P, Lawton PC, Cooper JC. *Kompendium der Reptilienkrankheiten: Haltung, Diagnostik, Therapie.* Hannover: Schluttersche;1997:158.

6. Obst FJ, Meusel W. *Die Landschildkroten Europas.* Wittenberg: A. Ziemsen Verlag;1972:65.

7. McArthur S. *Veterinary Management of Tortoises and Turtles.* Oxford, UK: Blackwell Publishing;1996:40.

8. Highfield AC. *Practical Encyclopedia of Keeping and Breeding Tortoises and Freshwater Turtles.* London, UK: Carapace Press;1996:97.

9. Wiesner CS, Iben C. Influence of environmental humidity and dietary protein on pyramidal growth of carapaces in African spurred tortoises (*Geochelone sulcata*). *J Anim Physiol Anim Nutr (Berl).* 2003;87:66-74.

10. McArthur S. Problem-solving approach to common diseases of terrestrial and semi-aquatic chelonians. In: McArthur S, Wilkinson R, Meyer J, eds. *Medicine and Surgery of Tortoises and Turtles.* Ames, IA: Blackwell Publishing;2004.

11. Houérou HN. *Bioclimatology and Biogeography of Africa.* Berlin, Germany: Springer-Verlag;2009.

12. Brown GW, ed. *Desert Biology: Special Topics on the Physical and Biological Aspects of Arid Regions.* New York, NY: Academic Press Harcourt;1974.

13. Loughrill E, Wray D, Christides T, Zand N. Calcium to phosphorus ratio, essential elements and vitamin D content of infant foods in the UK: Possible implications for bone health. *Matern Child Nutr*. 2017;13:e12368.

14. Wise MB, Ordoveza AL, Barrick ER. Influence of variations in dietary calcium:phosphorus ratio on performance and blood constituents of calves. *J Nutr.* 1963;79:79-84.

15. Stancel CF, Dierenfeld ES, Schoknecht PA. Calcium and phosphorus supplementation decreases growth, but does not induce pyramiding, in young red-eared sliders, *Trachemys scripta elegans*. *Zoo Biol.* 1998;17:17.

16. Goedegebuure SA, Hazewinkel HA. Morphological findings in young dogs chronically fed a diet containing excess calcium. *Vet Pathol.* 1986;23:594-605.

17. Richardson DC, Zentek J. Nutrition and Osteochondrosis. 1998;28:115-135.

18. Diehl JJE, Baines FM, Heijboer AC, et al. A comparison of UVb compact lamps in enabling cutaneous vitamin D synthesis in growing bearded dragons. *J Anim Physiol Anim Nutr (Berl)*. 2018;102:308-316.

19. Selleri P, Di Girolamo N. Plasma 25-hydroxyvitamin D(3) concentrations in Hermann's tortoises (*Testudo hermanni*) exposed to natural sunlight and two artificial ultraviolet radiation sources. *Am J Vet Res.* 2012;73:1781-1786.

20. Laing CJ, Trube A, Shea GM, Fraser DR. The requirement for natural sunlight to prevent vitamin D deficiency in iguanian lizards. *J Zoo Wildl Med.* 2001;32:342-348.

21. Oonincx DG, Stevens Y, van den Borne JJ, et al. Effects of vitamin D_3 supplementation and UVb exposure on the growth and plasma concentration of vitamin D_3 metabolites in juvenile bearded dragons (*Pogona vitticeps*). *Comp Biochem Physiol B Biochem Mol Biol.* 2010;156:122-128.

22. Wise MB, Ordoveza AL, Barrick ER. Influence of variations in dietary calcium:phosphorus ratio on performance and blood constituents of calves. *J Nutr.* 1963;79:79-84.

23. Loughrill E, Wray D, Christides T, Zand N. Calcium to phosphorus ratio, essential elements and vitamin D content of infant foods in the UK: Possible implications for bone health. *Matern Child Nutr.* 2017;13:e12368.

24. Bowes A, Pennington J. *Food Values of Portions Commonly Used.* 15th ed. New York, NY: Harper Collins;1989.

25. Heaney RP, Weaver CM. Oxalate: effect on calcium absorbability. *Am J Clin Nutr.* 1989;50:830-832.

26. Heaney RP, Weaver CM. Calcium absorption from kale. *Am J Clin Nutr.* 1990;51:656-657.

27. Hetényi N, Sátorhelyi T, Kovács S, Hullár I. Effects of two dietary vitamin and mineral supplements on the growth and health of Hermann's tortoise (*Testudo hermanni*). *Berl Munch Tierarztl Wochenschr.* 2014;127:251-256.

28. Fledelius B, Jørgensen GW, Jensen HE, Brimer L. Influence of the calcium content of the diet offered to leopard tortoises (*Geochelone pardalis*). *Vet Rec.* 2005;156:831-835.

29. Cusack L, Rivera S, Lock B, et al. Effects of a light-emitting diode on the production of cholecalciferol and associated blood parameters in the bearded dragon (*Pogona vitticeps*). *J Zoo Wildl Med.* 2017;48:1120-1126.

30. Acierno MJ, Mitchell MA, Roundtree MK, Zachariah TT. Effects of ultraviolet radiation on 25-hydroxyvitamin D_3 synthesis in red-eared slider turtles (*Trachemys scripta elegans*). *Am J Vet Res.* 2006;67:2046-2049.

Correlating Pathology and Clinical Disease in Amphibians

Robert J. Ossiboff, DVM, PhD, DACVP

Session #2144

Affiliation: From Aquatic, Amphibian, and Reptile Pathology Program, Department of Comparative, Diagnostic, and Population Medicine, College of Veterinary Medicine, University of Florida, 2015 SW 16th Ave, Gainesville, FL 32610-0123, USA.

Abstract: While amphibians, particularly frogs and to a lesser degree salamanders, are commonly maintained in captivity, they are less commonly presented to clinicians for disease management. And often when ill amphibians are brought in for veterinary care, it is in the late stages of the disease process. In such cases, it can be very important for clinicians to recognize the gross appearance of common amphibian disease processes to diagnose potential infectious or metabolic diseases that can serve as indicators of overall collection health. During this master class, participants will be introduced to fundamental aspects of amphibian, anatomy, and disease by way of case presentations of commonly encountered clinical amphibian diseases. Clinical histories and gross presentations will be correlated with pathologic and microbiologic or molecular findings. As many disease manifestations are cutaneous in amphibians, and the skin is a crucial organ for amphibian survival, a particular emphasis will be placed on amphibian skin.

Introduction

Amphibians are commonly maintained in captivity. Frogs, salamanders, and to much lesser degree caecilians, can be found in private, zoological or aquarium, and academic collections. Though there are nearly 8000 species of amphibians, in many ways, the manifestation of disease in amphibian species is quite conserved. For clinicians, recognizing the common presentations of disease in amphibians can be very useful for establishing lists of differential diagnoses. Furthermore, the unique physiology and anatomy of amphibians makes their collective response to certain types of insults or diseases quite different from that of other vertebrates. And their unique physiology may provide important considerations for disease treatment in individuals or disease management in collections. During this master class, common amphibian diseases will be presented with an emphasis on correlating gross and histologic pathology, anatomy, and physiology with diagnoses.

Cutaneous Physiology and Disease

Amphibian skin plays a crucial role in numerous fundamental physiologic process in amphibians, including osmoregulation, maintenance of electrolytes, and respiration. In each of these processes, the thickness of the amphibian epidermis, either as a measure of the total thickness or the thickness of its parts, is a crucial determinant of normal function. Mitochondrial-rich (MR cells) are specialized epidermal cells that facilitate electrolyte transport across the epidermis.[1] They lie immediately deep to the stratum corneum. Cutaneous osmoregulation is achieved passively as a result of MR cell-dependent electrolyte transport. Cutaneous respiration is also a passive process; carbon dioxide diffuses out of the blood stream and oxygen is bound by circulating erythrocytes. This process is dependent upon capillaries present within the very superficial dermis, or in some species, within the epidermis itself.

The response of amphibian skin to irritation, whether it be from environmental irritants (chemicals, abrasion) or infectious organisms (fungus, oomycetes, bacteria) includes both epidermal hyperplasia and hyperkeratosis. While in other species these non-specific responses would be of limited clinical significance, in an amphibian, they can be fatal. Increasing the thickness of the stratum corneum increases the depth of the MR cells from the skin-environment interface, and can substantially affect plasma electrolyte levels.[2] Increasing the thickness of the epidermis as a whole can also substantially affect cutaneous respiration. Changes to both of these processes can cause the host to change normal behavior (excessive soaking, air gulping in obligate aquatic species). Many amphibian diseases have alterations to these fundamental process as part of their pathogenesis, and for some diseases, such as chytridiomycosis, they are the primary cause of death.

Cutaneous Manifestations of Disease

Many important infectious diseases of amphibians have cutaneous manifestations. Classical amphibian chytridiomycosis, caused by the fungus *Batrachochytrium dendrobatidis*, is characterized by both epidermal hyperplasia and hyperkeratosis.[3] While in some cases, particularly in susceptible species, this may not be accompanied by overall changes in appearance, in other species it can be visible as dullness to the skin or an overall dampening of color. In salamanders with *Batrachochytrium salamandrivorans* infections, gross lesions are evident as raised to ulcerated, pale grey foci with discoloration.[4] Both Gram-negative bacterial septicemia and ranavirus infections can manifest as cutaneous hemorrhage, necrosis, and congestion. *Mycobacterium* infections in some frogs, particularly dendrobatids, present as ulcerated, gray, nodular cutaneous swellings of the extremities and body wall. In wild larval anurans, Perkinsea-like infections, an emerging pathogen of amphibians, can be evident as areas of cutaneous pallor. Cutaneous nematode migration can also be an important cause of epidermal hyperplasia, ulceration, and inflammation in many species.[5]

Non-infectious diseases can also be evident cutaneously. Expansion of the subcutaneous lymphatic sacs by fluid (edema) is a non-specific sign in amphibians that can be quite dramatic. The finding can be the result of one or more of a number of disease processes, including chronic renal disease, gastrointestinal disease, and bacterial infections.[6] While hypovitaminosis A most commonly presents as weight loss and decreased prehension of prey in anurans, in caudates it can present as cutaneous swellings around the cloaca and eyelids secondary to glandular squamous metaplasia.

Summary

As amphibians often do not show clinical signs or gross lesions of clinical disease until late in the disease process, recognition of common manifestations of disease in amphibians can be critical for disease mitigation and treatment in amphibian collections. While not all amphibian diseases present consistently, and not every individual will present with the lesions shared during this master class, I hope that this introduction to what is of course a very broad, detailed, and variable topic will prove useful to you as a clinician.

References

1. Guo P, Hillyard SD, Fu BM. A two-barrier compartment model for volume flow across amphibian skin. *Am J Physiol Regul Integr Comp Physiol*. 2003;285:R1384-R1394.

2. Voyles J, Young S, Berger L, et al. Pathogenesis of chytridiomycosis, a cause of catastrophic amphibian declines. *Science*. 2009;326:582-585.

3. Baitchman EJ, Pessier AP. Pathogenesis, diagnosis, and treatment of amphibian chytridiomycosis. *Vet Clin North Am Exot Anim Pract.* 2013;16:669-685.

4. Blooi M, Pasmans F, Rouffaer L, et al. Successful treatment of *Batrachochytrium salamandrivorans* infections in salamanders requires synergy between voriconazole, polymyxin E and temperature. *Sci Rep.* 2015;5:11788.

5. Cunningham AA, Sainsbury AW, Cooper JE. Diagnosis and treatment of a parasitic dermatitis in a laboratory colony of African clawed frogs (*Xenopus laevis*). *Vet Rec.* 1996;138:640-642.

6. Clancy MM, Clayton LA, Hadfield CA. Hydrocoelom and lymphedema in dendrobatid frogs at National Aquarium, Baltimore: 2003-2011. *J Zoo Wildl Med.* 2015;46:18-26.

Clinical Approach to Reptile Neoplasia

Tom Hellebuyck, Dr med vet, DipECZM (Herpetology), and
Jules Simard, med vet

Session #2258

Affiliation: From the Clinic for Poultry, Exotic Companion Animals, Wildlife and Experimental Animals, Department of Pathology, Bacteriology and Avian Diseases, Faculty of Veterinary Medicine, Ghent University, Salisburylaan 133, B-9820, Merelbeke, Belgium.

Abstract: Neoplasia is commonly seen in the practice of reptile medicine. This masterclass provides an overview of neoplastic disorders in reptiles that may be encountered by the clinical practitioner. The clinical presentation and behavior of different neoplastic disorders, tumor staging as well as the establishment of appropriate diagnostic and therapeutic approaches will be discussed. Novel insights towards the etiology, predisposing factors and the pathogenesis of neoplastic diseases in the reptile patient will be emphasized.

Introduction

Although it was once thought of as uncommon, an increasing prevalence of neoplasia seems to be noted in captive reptiles.[1-6] The latter may at least be partly attributed to the fact that reptile pet owners more readily seek veterinary advice as well as the expanding availability and use of appropriate diagnostic tools in the practice of reptile medicine.[4,7] The increasing lifespan of reptile pets as well as predisposing environmental factors that may promote the development of neoplastic disorders in a wide variety of reptilian species on the other hand, may also contribute to this increasing prevalence. Nevertheless, the diagnosis of neoplasia in the reptile patient may be easily missed at initial presentation as associated clinical signs are often non-specific.[1,4] Continuous efforts should be made to improve our understanding of the pathogenesis of neoplastic disorders in reptiles and to enhance appropriate diagnostic and therapeutic procedures. As in other animals and humans, the prognosis of any neoplastic disease in reptiles mainly depends on an early and correct diagnosis. This masterclass aims to provide a clinical approach to neoplastic disorders that may be encountered in the reptile patient.

Reported prevalence of neoplasia in reptiles based on data of specialist diagnostic services show considerable variation; however, snakes and lizards seem to be more commonly affected then chelonians in general.[1,2,4,6,8] Neoplastic disorders seem to be relatively rare in crocodilian species.[5] In snakes, hematopoietic and lymphoid tumors, epithelial neoplasia, cutaneous chromatophoroma, squamous cell carcinoma (SCC), ovarian adenocarcinomas and renal adenocarcinoma have been most frequently reported. In lizards, the cutaneous, hematopoietic and hepatic system are most commonly affected.[1,4-6,9,10] Turtles and tortoises seem to show a lower incidence of neoplasia and less tendency to develop metastasis in comparison to squamates and the integument seems to be most frequently affected organ system. In addition to cutaneous SCC, fibromas, adenocarcinomas and mast cell tumors are regularly reported in chelonians.[4,6,11-13] Endocrine tumors seem to occur more frequently in chelonians than in any other order. Thyroid adenocarcinoma is the most commonly reported single tumor of endocrine origin.[14] Although some evidence has been delivered for the predisposition of certain reptile species for the development of specific types of neoplasia, it should be considered that they may be over-represented because of their popularity as pets.

Primary Etiologies and Predisposing Factors

Although most neoplasias in reptiles seem to develop spontaneously, several viruses have been shown to possess oncogenic potential. The association of fibropapillomatosis and a herpesvirus in green sea turtles (*Chelonia mydas*) and papovaviruses and papillomatosis in European green lizards (*Lacerta viridis*) presumably remain the most notorious examples,[15-21] but papillomatosis in Bollivian side-neck turtles (*Platemys platycephala*), a carpet python (*Morelia spilota*) and a tegu (*Tupinambis teguexin*) was also associated with presumed papovavirus and poxvirus infection respectively.[22,23] Intracytoplasmic eosinophilic inclusions characteristic for boid inclusion body disease (bIBD) were demonstrated in a peripheral odontogenic fibromyxoma in a boa constrictor.[24] Although the tumoral tissue tested positive for arenavirus, the virus was considered to take advantage of the rapidly dividing neoplastic cells to replicate and disseminate rather than possessing oncogenic properties.[24] As neoplastic disorders are frequently seen in bIBD positive snakes, immunosuppression due to arenavirus infection may predispose to the development of neoplastic disease.[24-27] Taking into account the large number of asymptomatically bIBD positive/arenavirus infected snakes, especially boa constrictors, the association of neoplasms and other diseases should be made cautiously.[24]

Although age, trauma, scar tissue and immunosuppression are recognized as important factors that promote the development of cutaneous basal cell carcinoma (BCC) and SCC in humans, exposure to ultraviolet (UV) radiation is known to be the major cause of these neoplasms.[28-30] In contrast to cutaneous SCC, BCC seems to be associated with intermittent rather than continuous exposure to UV radiation during childhood and adolescence.[28,31] In addition to SCC and BCC, chromatophoromas and hemangiomas are also considered to be potentially associated with exposure to UV light irradiation in captive reptiles. Although providing adequate sources of UV radiation is an essential part of the appropriate captive care of many of these species, the role of artificial UV radiation as an etiologic factor in the development of neoplasms in captive reptiles remains to be further investigated.[32,33] Furthermore, the role of genetic predisposition and hormone profile alterations towards the development of neoplasias in reptiles need to be further elucidated.[34-36]

Tumor Types by System

Cutaneous neoplasia

Skin neoplasms are presumably one of the most commonly diagnosed neoplastic disorders in the practice of reptile medicine. Nevertheless, this may be partly attributed to the ease of visual inspection of the skin, shell and adnexa of the integument by the owner as well as the clinical practitioner.[7,31] Cutaneous SCC is frequently seen in lizards, snakes and chelonians[1,2,4-6,32,37,38] and mucocutaneous junctions seem to be the most important predilection sites.[1,32] Some lizard species such as inland bearded dragons (*Pogona vitticeps*), monitors and chameleons seem to be predisposed to the development of cutaneous SCC and the skin of the eyelids and skin of the back seem to be frequently affected sites.[1,4,5,32,39] In snakes the cloacal region seems to be the most frequently affected and SCCs may originate from the pericloacal skin, scent glands or hemipene.[1,4] Although less often seen in chelonians, SCC may occur in the skin of eyelids, intermandibular space and is regularly observed to affect the shell.[38] Typically, SCCs are locally invasive, but vascular invasion is rare as well as metastasis.[7] Especially in lizards cutaneous SCC may show a multicentric appearance.

In Greek tortoises (*Testudo hermanni*), basal cell carcinoma has recently been described as a neoplastic entity that primarily arises from the shell but oral BCCs have been observed in tortoises and lizards.[40] The clinical presentation and performance of BCC largely corresponds to what is seen in SCC. In end stage cases of BCC, multiple metastatic implants were found in the liver, lungs and myocardium of tortoises.[40] Despite the overall low metastatic potential, SCC and BCC may become locally invasive processes that invade underlying tissues and may show necrosis and become secondarily infected. For this reason these processes may be confused with chronic abscesses or granulomas.

Chromatophoromas are commonly seen in snakes and lizards.[5,8,41-46] These tumors are further classified based on type of pigment production in melanophoromas, iridophoromas and xanthophoromas and differentiation relies on immunohistochemical staining.[41,45,47] These tumors are typically seen on the trunk and eyelids. In most cases chromatophoromas are highly invasive, difficult to excise in toto and show a relatively high tendency to metastasize.[4,2,6,41,42] Melanophoromas often present as single, white cutaneous masses in snakes. Based on a retrospective study, malignancy seemed higher for melanophoromas than for iridophoromas in snakes, lizards and tortoises. The poor prognosis of mucinous melanophoroma is mainly attributed highly invasive nature in the surrounding tissues rather than the occurrence of metastasis.[41,48-49]

Cutaneous fibrosarcomas have been described in lizards and snakes but seem to be rare in crocodilians and chelonians. These tumors may also occur in the muscles, mainly epaxial muscles of the back, or as intracoelomic masses for which the primary tumor site cannot always be determined.[5,6,8,50] Although, fibrosarcomas are highly infiltrative and readily reoccur following resection, metastasis seems to be uncommon.[51]

Soft tissue sarcomas constitute a heterogeneous group of neoplasias that originate from the mesenchymal tissue and may be related to muscle, fat, neurovascular or fibrous tissues.[51] Differentiation relies on IHC staining, but the latter is largely unexplored in reptiles. These tumors are frequently observed in squamates .[4] Although generally considered to have low metastatic potential, metastasis has been reported in reptiles, mainly for visceral soft tissue sarcomas.[2,4-6]

Cutaneous papillomas have been described in lizards, crocodilians and turtles.[15-21] Morphologic features include papilliform proliferation, but without a mesenchymal component.[4] Fibropapillomatosis is a well-recognized condition of marine turtles and a herpesvirus is the putative causative agent with fibropapillomas occurring solitary or multicentric in a wide variety of locations. Concurrent visceral fibromas are often concurrent and multicentric (especially in the kidney).[19-21,52,53]

Dermal lipomas are relatively uncommon in reptiles and mostly show a subcutaneous localization.[4] Corn snakes seem to be predisposed to the development of subcutaneous lipomas and these are predominantly located at the lateral area of the base of the tail.[4,50,54] In most cases lipomas remain restricted to the subcutis and do not necessarily require surgical intervention, although infiltration in adjacent muscle tissue may be observed.

Hemangiomas have been observed in the skin of lizards, snakes and the shell of a turtle.[55] These benign tumors originating from the vascular endothelium may associated with exposure to UV radiation, as previously documented in dogs.[55]

Alimentary tract tumors

Various neoplastic processes arising from the oral cavity or esophagus have been described in reptiles, including SCC, BCC, fibromyxoma, fibrosarcoma, melanoma and epulis.[1,4,24,40] Due to the anatomic localization, these oral neoplasms are readily detected and especially in early stages in toto excision may be possible. For those cases that show infiltration of neoplastic cells within surrounding soft tissue or when bony involvement is present, surgical treatment options may be limited. Adenocarcinomas of the stomach, small intestines, colon or cloaca have also been reported predominately in snakes although cases exist in lizards and chelonians.[4] Often adenocarcinomas are highly invasive and their presence and related clinical signs are often left unrevealed until end stage disease occurs. The latter clinical signs, such as regurgitation, vomiting, tenesmus, or ileus often result from the partial or total occlusion of the gastric or intestinal lumen caused by the intramural/intraluminal growth of the neoplastic mass. Similar clinical signs are seen in intramural colonic lymphomas, for example.[4,5]

Neuroendocrine gastric carcinomas (NGC), have particularly been reported in relatively young inland bearded dragons.[56,57] Often metastatic tissue are found in the parenchyma of major organs or may arise from serosal surfaces or the coelomic membrane.[56-58] Indicative signs in affected bearded dragons may consist of vomiting, cachexia, or clinical signs resulting from metastatic disease. Chronic non-regenerative anemia and hyperglycemia are clinical pathologic findings.[58] The cryptic location of the relatively fast growing NGC often results in late detection in the stage where metastasis is present and normal gastric function is severely impaired.[58]

Miscellaneous enteric tumors and polyps are occasionally seen and often show ulceration and secondary bacterial infection. Although these processes grossly mimic neoplasms, they are histologically benign and complete excision is generally curative.[4]

While pancreatic acinar cell carcinomas have been documented in squamates and turtles, acinar cell adenoma seems to be uncommon and shows a relatively low tendency to metastasize.[1]

Hepatic tumors

Biliary adenocarcinomas (cholangiocarcinoma, bile duct carcinoma) and adenomas (biliary cystadenoma, cholangioma) have been mainly documented in squamates.[2,5,8,17,59] Although often multicentric within the liver, metastasis of adenocarcinomas is uncommon and clinical sings are often not observed until the neoplasia has achieved a considerable size.[60] Biliary cystadenoma shows secretory activity resulting in the development of cysts with variable size.[60] Hepatocellular carcinomas are most commonly observed in snakes.[2,4,5,8] Although metastasis is uncommon, surgical removal is often impossible due to tumor dimensions combined with the typically elongated aspect of the liver in snakes and the involvement of the vena cava. Hepatomas are frequently seen in snakes and similar to hepatocellular carcinomas they also achieve a large size, and may be solitary or multicentric.[4] Well-differentiated hepatomas are often difficult to differentiate from normal hepatocytes or nodular hyperplasia in cirrhotic livers. The absence of portal areas and hepatic veins as well as the dimensions and compression of the surrounding hepatic tissue may be considered indicative from hepatoma.[6,61] Cytologic preparations should be interpreted cautiously as far as it comes to differentiating these tumors from hepatocellular proliferative disease.

Renal tumors

Renal adenocarcinoma is mainly seen in snakes. In most cases these tumors present as large, solitary and unilateral tumors that may eventually replace a large part of the healthy renal parenchyma.[4] Because of their predominant unilateral occurrence, low mitotic index and locally invasive nature, nephrectomy may be a useful therapeutic option for some cases.[62] Renal tubular adenoma is seen in snakes and lizards, but often constitutes an incidental finding.[4]

Reproductive tract tumors

Various neoplastic disorders of the reproductive tract have been reported in reptiles.[4,63] In general, neoplasms of the reproductive tract are more common in lizards than in snakes, except for GCTs, and are relatively rare in chelonians and crocodilians.[1,4,5] Ovarian neoplasms reported include hemangioma, fibroma, carcinoma, sarcoma and granulosa cell tumor (GCT).[4,6,65-66] GCTs in snakes show a predominant slow growth and overall benign behavior,[4,65] but metastatic disease of GCTs have been documented several cases.[63] Call-exner bodies may be observed but are not necessarily seen in reptilian GCTs.[67] Garter snakes (*Thamnophis* spp.) seem to be predisposed to the development of GCT.[2,65] GCTs are rarely identified in lizards and are mostly benign.[4,64] A single case of a malignant GCT has been reported in a green iguana (*Iguana iguana*) following incomplete ovariosalpingectomy with metastasis to multiple organs, serosal surfaces and the coelomic membrane.[36]

Ovarian adenocarcinoma seems to show a low metastatic behavior and is frequently seen in lizards, especially in iguanas and snakes.[68-71] As for other ovarian tumors, detection often relies on diagnosis trough imaging to reveal the cause of localized or generalized coelomic distention or the presence of a demarcated mass during coelomic palpation.

Tumors of the male reproductive tract are relatively rare in reptile species but a seminoma has been reported in a crocodile, lizards and snakes.[1,4] Although the mitosis index is often high, seminomas tend to be well encapsulated a show low tendency to be locally invasive or metastasize. Benign Leydig cell (interstitial cell) tumors and luteoma have been reported in various reptile species but seem to be rather incidental findings.[1,4,5] Teratomas have been described and may consist of tissues arising from ecto-, meso- and endoderm.[71-73]

Endocrine tumors

In comparison to mammals, paraneoplastic syndrome is not often evidenced in reptiles with endocrine tumors.[74] Islet tumors rarely occur in reptiles. Pheochromocytomas in snakes are presumably the most frequently occurring adrenal tumors in reptiles. Cortical adenoma has been anecdotally reported in an iguana and a snake.[4] Thyroid (cyst)adenomas, thyroid carcinomas and parathyroid adenoma have been reported in reptiles[4,14] and intracoelomic endocrine carcinomas of undetermined origin are sporadically observed in reptiles.[1] In a few cases the primary tumor was identified as a NGC in bearded dragons.[58]

Soft tissue carcinomas

The differentiation of anaplastic mesenchymal tumors in reptiles is often challenging.[4] Mostly tumors are identified as soft tissue sarcomas based on the predominant presence of mesenchymal spindle cells in tumors for which the tissue of origin cannot be determined. These may include neurofibrosarcoma, fibrosarcoma, nerve sheath tumors, myxosarcoma, leiomyosarcoma, rhabdomyosarcoma and amelanotic melanoma.[1,4,5] Mainly the lack of reliable immunohistochemistry protocols to determine the character of the infiltrating cells render the definitive diagnoses of these tumors difficult. Soft tissue sarcomas are most prevalent in crotalids, iguanas and geckos.[1] Often primary sarcomas originate from the dermis and may metastasize to the viscera.[1,4,5]

Musculoskeletal tumors

Rhabdomyosarcoma is rare but has been diagnosed in lizards, especially in geckos, and snakes originating from the body wall, myocardium and the oral cavity.[8,50,77,78]

Ossifying fibromas, chondrosarcomas and osteosarcomas are relatively rare in reptiles.[9,35,79-81] Osteosarcomas are most often observed in snakes and to a lesser degree in lizards. In snakes they often arise from the vertebrae and ribs.[4,5,9] Although metastases are not often seen, osteosarcomas tend to be locally invasive and destructive with osteolysis and new bone formation.[9] Affected individuals often present in the stage where the tumor has invaded the muscles or subcutis and sometimes breaches trough to overlying epidermis. Chondrosarcomas are relatively rare, but most frequently occur in colubrids and crotalids and quite often metastasize.[4,9,81] Pathologic vertebral fractures are seen in snake with chondrosarcomas of the vertebral articulations.[4] Leiomyomas and leiomyosarcomas are relatively uncommon in reptiles, and have been associated with a variety of organs such as the intestine, cloaca, oviduct and testicles in reptiles and show a varying tendency to metastasize.[4,50,75,76]

Hematopoietic and lymphoid tumors

Lymphoid tumors with or without leukemia are presumably the most frequently occurring malignancies in reptiles and are mainly diagnosed in snakes and lizards.[1,3-5,8,25] Typically they show a multicentric appearance and a predominant blast morphology. Lymphoblastic lymphoma is one of the most commonly reported single neoplasia in lizards.[1,3,4,82,83] A plasmacytoid and histiocytic morphology is seen infrequently. Other myeloproliferative neoplastic conditions are relatively uncommon in reptiles. A staging system has not been developed for reptiles.[85] Myeloid leukemia has been diagnosed in snakes and lizards.[3,25]

Diagnostic Approach

As generally accepted in other animal species, an early and correct diagnosis of neoplastic disorders in reptiles results in the most favorable prognosis. Tumor type, location and size are the main factors that determine clinical signs at initial presentation.[14] Initial signs of many types of neoplasia in reptiles often are vague. Even as disease progresses, the associated clinical signs may remain non-specific, such as anorexia, lethargy, dyspnea, coelomic distention, constipation and/or paresis or paralysis.[4,5,14] In most cases, neoplastic disease will be diagnosed based on the presence of visible or palpable masses or indicative findings obtained from a multidirectional diagnostic approach that is applied because of the presence of generalized signs of disease. The latter often results from systemic effects caused by a primary neoplastic process or metastatic disease. In addition, several neoplastic disorders may resemble other disease processes that are accompanied by growths and swellings with secondary deformation and tissue destruction.[5] Cutaneous neoplasms for example may be confused with abscesses, chronic granulomas or degenerative skin disorders, especially when necrosis and/or secondary infection of the affected tissue is present. The importance of coelomic palpation as a part of the clinical exam in any reptile patient cannot be overly emphasized in order to reveal abnormal shape, size or consistency of internal organs, the presence of asymmetric processes or masses for which the origin cannot be determined.

Signalment, history, husbandry, and physical examination combined with known predispositions related to age, gender, management issues, species or concurrent disease can lead to a presumptive diagnosis of neoplastic disease.[4] Although many neoplastic disorders more readily occur in aged or geriatric reptile patients, several neoplasms are typically seen in relatively young animals (eg, NGC in bearded dragons).[1,4,58]

A complete blood count and plasma biochemistry can aid in revealing systemic manifestations of the disease and especially for paraneoplastic syndromes and some hematopoietic tumors an inflammatory leukogram and severe lymphocytosis may be observed respectively.[10,84-88] Although far from pathognomonic, hypercalcemia, hypo- or hyperglycemia, hyperphosphatemia and increased creatine kinase and alanine transaminase levels may indicate neoplastic disease in combination with other laboratory testing.[14]

Imaging techniques may offer valuable information about tumor type and prognosis based on tumor size and the presence or absence of detectable metastasis. Especially for those types of neoplasia that present as large intracoelomic masses or that show progressive invasion and destruction of surrounding tissues, medical imaging techniques are mandatory to assess tumor dimensions and borders. Radiology provides a rapid tool to assess obvious changes such the presence of masses, organ enlargement, mass effect as well as the relationship between internal organs and may allow to reveal the presence of metastatic tumors with a considerable size.[5,88] In comparison to radiography, ultrasonographic imaging is superior when it comes to assessing tissue architecture, vascularization and infiltration as well as the differentiation between soft tissue and fluid.[89,90] The collection of ultrasonographically guided fine-needle aspirates of suspected neoplastic processes should be considered as an important part of the diagnostic plan.[1,4,90,91] Computed tomography and magnetic resonance imaging are becoming more widely available and allow the collection of three-dimensional volumetric images, enhanced organ detail and the detection of relatively small and focal neoplasms or metastasis.[14,91] Moreover, the use of contrast may enhance the visualization of internal organs and vascularization. These and other advanced imaging techniques can largely contribute to facilitate early detection of tumors as well as staging. [14,88,91] Endoscopy allows to directly visualize internal masses or lesions that were identified using other imaging techniques and allows the collection of tissue biopsies.[14] A definitive antemortem diagnosis of neoplastic diseases in reptiles often requires microscopic examination of fine-needle aspirates, imprints and/or histologic sections of surgically removed tissues or biopsies. Cytologic evaluation of general cellular morphology and cellular characteristics, cohesiveness and cytoplasmic and nuclear features can provide information towards assessing malignancy.[14,93] The examination of histologic sections from surgically removed tissue or biopsies remains superior when it comes to tumor identification and evaluation of tissue architecture and assessing tumor margins and vascular infiltration. Immunohistochemistry is a highly valuable tool to determine and differentiate tumor types. Unfortunately, few IHC or tissue markers of neoplasia are validated or are readily available for the use in reptiles.[1,4,31,58]

Therapeutic Approach

Treatment options for neoplastic disorders in reptiles largely correspond to those used in other animals. In general, local neoplasms require local therapy and systemic disease should be treated with systemic therapies.[5] Treatment plans should be molded at the level of the individual patient and should be based on species, age, tumor staging and grading. If a conclusive diagnosis is obtained, the chosen treatment option should aim for a maximal response in the tumor while minimizing the risk of unacceptable complications and toxicity in the patient.[5,14] The process of staging refers to the assessment of tumor burden that is present in a cancer patient. This process commonly includes routine imaging, bone marrow evaluation, and routine hematologic and biochemistry tests. Unfortunately, little is known about the significance of tumor stage and prognosis in reptilian cancer patients.[5]

Before embarking on any specific treatment of a neoplastic disorder, supportive care, pain management, hydration and the provision of an optimized hospitalization environment should be realized whenever needed.

Depending on the tumor type involved, complete surgical resection of a neoplastic mass should be attempted or should be used in combination with other treatments. Incomplete resection and the presence of metastatic disease are common reasons for therapeutic failure.[2,3,5] If complete surgical resection is not possible, some palliative form of therapy should be considered to enhance local tumor control.[5] Local or systemic chemotherapy may be considered but methods of application and dosages are often empirical. Accordingly, chemotherapy neither guarantees extension nor improved quality of life.[14] Intralesional chemotherapy, has been described for several neoplastic disorders in reptiles with a variable success rate.[5,69,94,95] In addition to the lack of chemotherapeutic efficacy and appropriate dosage regimens, the risk of potential human toxicity of used agents, limited vascular access for continued administration of intravenous chemotherapeutics, the development of immunosuppression and other secondary effects should be always considered and should be monitored.[5,94,95] Continuous hematologic assessment and complete blood counts of the reptile patient should reveal the onset of leukopenia. The latter may be highly variable and depends on the involved patient and the used protocol.[5]

Electrochemotherapy has been used with success for the treatment of fibropapillomas in green sea turtles and seems to be a treatment option showing few side effects for benign tumors or tumors with low tendency to metastasize. Analgesia, sedation and/or anesthesia is required as the electrical pulses are considered painful.[14,95,96]

Radiation oncology may be considered as a primary curative or palliative treatment or may be combined with other chemotherapeutic treatment modalities, for example for the treatment of residual tumor tissue after surgery due to incomplete margins by the surgeon.[98-100] In reptiles, the use of radiation therapy currently stays limited to a few cases with non-resectable tumors. Local as well as systemic side effects seem to be common.[97,98]

Depending on the tumor type, photodynamic therapy, cryosurgery and surgical removal of neoplasias using laser therapy are promising local treatment alternatives, but require further investigation for the reptile patient and often the availability of specialized equipment and training.[5,14] In any case, a definitive diagnosis based on histopathologic examination of biopsies is mandatory as some severe disruption of tissue architecture may interfere with obtaining an accurate histologic diagnosis.[5]

References

1. Hernandez-Divers SM, Garner, MM. Neoplasia of reptiles with an emphasis on lizards. *Vet Clin North Am Exot Anim Pract.* 2003;6:251-273.

2. Ramsay EC, Munson L, Lowenstine L., Fowler ME. A retrospective study of neoplasia in a collection of captive snakes. *J Zoo Wildl Med.* 1996;27(1):28-34.

3. Gyimesi ZS, Garner, MM, Burns RB, et al. High incidence of lymphoid neoplasia in a colony of Egyptian spiny-tailed lizard (*Uromastyx aegyptius*). *J Zoo Wildl Med.* 2005;36(1):103-110.

4. Garner MM, Hernandez-Divers SM, Raymond JT. Reptile neoplasia: a retrospective studie of case submissions to a specialty diagnostic service. *Vet Clin North Am Exotic Anim Pract.* 2004;7:653-671.

5. Mauldin GN, Done LB. Oncology. In: Mader DR, ed. *Reptile Medicine and Surgery.* 2nd ed. St. Louis, MO: Elsevier Saunders;2006:299-322.

6. Sykes JM, Trupkiewiez JG. Reptile neoplasia at the Philadelphia Zoological Garden, 1901-2002. *J Zoo Wildl Med.* 2006;37(1):11-19.

7. Hellebuyck T, Pasmans F, Haesebrouck F, Martel A. Dermatological diseases in lizards. *Vet J.* 2012;193: 38-45.

8. Catao-Dias JL, Nichols DK. Neoplasia in snakes at the National Zoological Park, Washington, DC (1978-1997). *J Comp Pathol.* 1999;120:89-95.

9. Dietz J, Heckers KO, Pees M, Aupeperle H. Knochentumoren bei Echsen und Schlangen, ein seltener klinischer Befund in der Reptilienpraxis. *Tierärztl Prax.* 2015;43:31-39.

10. Hernandez-Divers SM, Orcutt CJ, Stahl SJ, et al. Lymphoma in lizards: Three case reports. *J Herpetol Med Surg.* 2003;13:14-22.

11. Gonzales-Viera O, Bauer G, Aguair LS, et al. Cutaneous fibroma in a common snapping turtle (*Chelydra serpentina*). *J Comp Pathol.* 2008;138:574-576.

12. Santoro M, Stacy BA, Morales JA, et al. Mast cell tumor in a giant Galapagos tortoise (*Geochelone nigra vicina*). *J Comp Pathol.* 2008;138:156-159.

13. Cladaugh K, Haag KM, Hanley CS, et al. Undifferentiated sarcoma resolved by forelimb amputation and prosthesis in a radiate tortoise (*Geochelone radiata*). *J Zoo Wildl Med.* 2005;36(1):117-120.

14. Christman J, Devau M, Wilson-Robles H, et al. Oncology of reptiles. Diseases, diagnosis, treatment. *Vet Clin North Am Exot Anim Pract.* 2017;20:87-110.

15. Raynaud A, Adrian M. Cutaneous lesions with papillomatous structure associated with viruses in the green lizard (*Lacerta viridis*). *Compt Rend Hebd Séanc Acad Sci.* 1976;283(7):845-847.

16. Zeigel RF, Clark HF. Histologic and electron microscopic observations on a tumor-bearing viper: Establishment of a C-type virus-producing cell line. *J Nat Cancer Inst.* 1971;46(2):309-321.

17. Cooper JE, Gschmeissner S, Holt PE. Viral particles in a papilloma from a green lizard (*Lacerta viridis*). *Lab Anim.* 1982;16:12-13.

18. Chandra AMMS, Jacobson ER, Munn RJ. Retroviral particles in neoplasms of Burmese pythons (*Python molurus bivittatus*). *Vet Pathol.* 2001;38:561-564.

19. Herbst LH, Lenz J, Van Doorslaer K, et al. Genomic characterization of two novel reptilian papillomaviruses, *Chelonia mydas* papillomavirus 1 and *Caretta caretta* papillomavirus 1. *J Virol.* 2009;338(1):131-135.

20. Herbst LH. Fibropapillomatosis of marine turtles. *Ann Rev Fish Dis.* 1994;4:389-425.

21. Jacobson ER, Buergelt C, Williams B, et al. Herpesvirus in cutaneous fibropapillomas of the green turtle *Chelonia mydas*. *Dis Aquat Org.* 1991;12:1-6.

22. Jacobson ER, Gaskin JM, Clubb SL, et al. Papilloma-like virus infection in Bolivian side-neck turtles. *J Am Vet Med Assoc.* 1982;181(11):1325-1328.

23. Marschang RE. Viruses infecting reptiles. *Viruses.* 2011;3(11):2087-2126.

24. Hellebuyck T, Pasmans F, Ducatelle R, et al. Detection of arenavirus in a peripheral odontogenic fibromyxoma in a red tail boa (*Boa constrictor constrictor*) with inclusion body disease. *J Vet Diagn Invest.* 2015;27(2):245-248.

25. Schillinger L., Selleri P, Frye FL. Lymphoblastic lymphoma and leukemic blood profile in a red-tail Boa (*Boa constrictor constrictor*) with concurrent inclusion body disease. *J Vet Diagn Invest.* 2011;23:159-162.

26. Summa NM, Guzman DSM, Hawkins MG, et al. Tracheal and colonic resection and anastomosis in a boa constrictor (*Boa constrictor*) with a T-cell lymphoma. *J Herpetol Med Surg.* 2015;25(3-4):87-99.

27. Stenglein MD, Sanders C, Kistler AL, et al. Identification, characterization and in vitro culture of highly divergent arenaviruses from boa constrictors and annulated tree boas: candidate etiological agents for snake inclusion body disease. *Mbio.* 2012;3(4):e00180-00212.

28. Rubin AI, Chen EH, Ratner D. Basal-cell carcinoma. *New Engl J Med.* 2005;335(21):2262-2269.

29. Wehner MR, Shive ML, Chren M-M, et al. Indoor tanning and non-melanoma skin cancer: systematic review and meta-analysis. *Brit Med J.* 2012;345:e5909.

30. Hauck ML. Tumors of the skin and subcutaneous tissues. In: Withrow SJ, Vail DM, Page RL, eds. *Small Animal Clinical Oncology.* 5th ed. St. Louis, MO: Elsevier;2013:305-320.

31. Lanckneus A, Martel A, Bosseler L, Hellebuyck T. Chromatophoromas in reptiles. *Vl Dierg Tijdschr.* 2016;86:5-15.

32. Hannon DE, Garner MM, Reavill DR. Squamous cell carcinomas in inland bearded dragons (*Pogona vitticeps*). *J Herpetol Med Surg.* 2011;21(4):101-106.

33. Heckers KO, Aupperle H. Pigment-forming tumors in reptiles: light regime and its dark sides. *Proc Annu Conf Assoc Reptilian Amphib Vet.* 2014;31-35.

34. Keck M, Zimmerman DM, Ramsay EC, et al. Renal adenocarcinomas in cape coral snakes (*Aspidelaps lubricus lubricus*). *J Herpetol Med Surg.* 2011;21(1):5-9.

35. Needle D, McKnight CA, Kiupel M. Chondroblastic osteosarcoma in two related spiny-tailed monitor lizards (*Varanus acanthurus*). *J Exot Pet Med.* 2013;22(3):265-269.

36. Cruz Cardona JA, Conley KJ, Wellehan Jr. JFX, et al. Incomplete ovariosalpingectomy and subsequent malignant granulosa cell tumor in a female green iguana (*Iguana iguana*). *J Am Vet Med Assoc.* 2013;239(2): 237-242.

37. Oros J, Tucker S, Fernandez L, et al. Metastatic squamous cell carcinoma in two loggerhead sea turtles *Caretta caretta. Dis Aquat Org.* 2004;58:245-250.

38. von Deetzen M-C, Müller K, Brunnberg L, Klopfleisch R. Non-metastatic squamous cell carcinoma in two Hermann's tortoises (*Testudo hermanni*). *Vet Sci Dev.* 2012;2(5):17-19.

39. Abou-Madi N, Kern TJ. Squamous cell carcinoma associated with a periorbital mass in a veiled chameleon (*Chamaeleo calyptratus*). *Vet Ophthalmol.* 2002;5(3):217-220.

40. Hellebuyck T, Ducatelle R, Van Caelenberg A, et al. Basal cell carcinoma in two Hermann's tortoises (*Testudo hermanni*). *J Vet Diagn Invest.* 2016;26:750-754.

41. Heckers KO, Aupperle H, Schmidt V, et al. Melanophoromas and iridophoromas in reptiles. *Dis Wildl Exot Species.* 2012;146:258-268.

42. Heckers KO, Schmidt V, Krastel D, et al. Malignes melanophoroma bei einer griechischen landschildkröte (*Testudo hermanni*). *Tiërartz Prax.* 2011;39(K):45-50.

43. Gregory CR, Harmon BG, Latimer KS, et al. Malignant chromatophoroma in a canebrake rattlesnake (*Crotalus horridus atricaudatus*). *J Zoo Wildl Med.* 1997;28:198-203.

44. Suedmeyer WK, Bryan JN, Johnson G, Freeman A. Diagnosis and clinical management of multiple chromatophoromas in an eastern yellowbelly racer (*Coluber constrictor flaviventris*). *J Zoo Wildl Med.* 2007;38:127-130.

45. Irizarry-Rovira AR, Wolf A, Ramos-Vara JA. Cutaneous melanophoroma in a green iguana (*Iguana iguana*). *Vet Clin Pathol.* 2006;35(1):101-105.

46. Simpson S. Pigmented cutaneous neoplasm in a central bearded dragon. *Proc Annu Conf Unusual Exot Pet Vet.* 2008;285-288.

47. Brot S, Sydler T, Nufer l, et al. Histologic, immunohistochemical and electron microscopic characterization of a malignant iridophoroma in a dwarf bearded dragon (*Pogona henrilawsoni*). *J Zoo Wildl Med.* 2015;46(3):583-587.

48. Bielli M, forlani A, Nardini G, et al. Mucinous melanophoroma in a northern red-bellied cooter (*Pseudemys rubriventris*). *J Exot Pet Med.* 2005;24:71-75.

49. Frye FL, Carney J, Harshbarger J, Zeigel R. Malignant chromatophoroma in a western terrestrial garter snake. *J Am Vet Med Assoc.* 1975;167:557-558.

50. Frye FL. Diagnosis and surgical treatment of reptilian neoplasma with a compilation of cases 1966-1993. *In Vivo.* 1994;8:885-892.

51. Liptak JM, Forrest LJ. Soft tissue sarcomas. In: Withrow SJ, Vail DM, Page RL, eds. *Small Animal Clinical Oncology.* 5th ed. St. Louis, MO: Elsevier;2013:356-380.

52. Lackovich JK, Brown DR, Homer BL, et al. Association of herpesvirus with fibropapillomatosis of the green turtle *Chelonia mydas* and the loggerhead turtle *Caretta caretta* in Florida. *Dis Aquat Org.* 1999;37:89-97.

53. Jones K, Ariel E, Burgess G, Read,M. A review of fibropapillomatosis in green turtles (*Chelonia mydas*). *Vet J.* 2015. In press.

54. Reavill DR, Schmidt RE. Lipomas in corn snakes (*Elaphe guttata guttata*): a series of four cases. *Proc Annu Conf Assoc Reptilian Amphib Vet.* 2003;60.

55. Goldschmidt MH, Hendrick MJ. Tumors of the skin and soft tissue. In: Meuten DJ, ed. *Tumors in Domestic Animals.* Ames, IA: Iowa State Press;2002:45-117.

56. Ritter JM, Garner MM, Chilton JA, et al. Gastric neuroendocrine carcinomas in bearded dragons (*Pogona vitticeps*). *Vet Pathol.* 2009;46:1109-1116.

57. Lyons JA, Newman SJ, Greenacre CB, Dunlap J. A gastric neuroendocrine carcinoma expressing somatostatin in a bearded dragon (*Pogona vitticeps*). *J Vet Diagn Invest.* 2010;22:316-320.

58. Mooij TS, Martel A, Bosseler L, et al. Atypische klinische presentatie van een metastatisch gastrisch neuroendocrien carcinoom bij een baardagame (*Pogona vitticeps*). *Vl Diergen Tijdschr.* 2014;83:293-298.

59. Well M, Rodiger KS. Cholangioma in a green iguana. *Kleintierprax.* 1992;93:415.

60. Liptak JM, Withrow SJ. Cancer of gastrointestinal tract. In: Withrow SJ, Vail DM, Page RL, eds. *Small Animal Clinical Oncology.* 5th ed. St. Louis, MO: Elsevier;2013:381-431.

61. Meyer DJ. The liver. In: Raskin RE, Meyer DJ, eds. *Canine and Feline Cytologie: A Color Atlas and Interpretation Guide.* 3rd ed. St. Louis, MO: Elsevier;2016:259-283.

62. Jacobson ER, Long PH, Miller RE, et al. Renal neoplasia of snakes. *J Am Med Vet Assoc.* 1986;189(9): 1134-1136.

63. Pirson S, Geerinckx L, Bosseler L, et al. Metastatic granulosa cell tumor in a mangrove snake (*Boiga dendrophila melanota*). *J Exot Pet Med.* 2017. In press.

64. Effron M, Griner L, Benirschke K. Nature and rate of neoplasia found in captive wild mammals birds and reptiles at necropsy. *J Nat Canc Inst.* 1977;59(1):185-198.

65. Onderka DK, Zwart P. Granulosa cell tumor in a garter snake (*Thamnophis sirtalis*). *J Wildl Dis.* 1978;14: 218-221.

66. Michaels SJ, Saneck R. Undifferentiated carcinoma in the ovary of a boa constrictor (*Boa constrictor ortoni*). *J Zoo Wildl Med.* 1988;19:237-240.

67. Solano-Gallego L, Masserdotti C. Reproductive system. In: Raskin RE, Meyer DJ, eds. *Canine and Feline Cytologie: A Color Atlas and Interpretation Guide.* 3th ed. St. Louis, MO: Elsevier;2016:313-352.

68. Anderson NL, Williams J, Sagartz JE, et al. Ovarian teratoma in a green iguana (*Iguana iguana*). *J Zoo Wildl Med.* 1996;27(1):90-95.

69. Gibbons P, Schiller C. What's your diagnosis: ovarian adenocarcinoma in a green iguana (*Iguana iguana*). *J Herpetol Med Surg.* 2000;10:34-38.

70. Levine B. Treatment of a malignant ovarian teratoma in a green iguana (*Iguana iguana*). *Exot DVM.* 2004;6(4):12-14.

71. Stacya BA, Vidalb JD, Osofskya A, et al. Ovarian papillary cystadenocarcinomas in a green iguana (*Iguana iguana*). *J Comp Pathol.* 2004;130:223-228.

72. Newman SJ, Brown CJ, Patnaik AK. Malignant ovarian teratoma in a red-eared slider (*Trachemys scripta elegans*). *J Vet Diagn Invest.* 2003;15:77-81.

73. Martorell J, Soto S, Barrera S, et al. Case report: Ovarian teratoma in a Mediterranean tortoise. *Compend Contin Educ Vet.* 2009;31(4):193-196.

74. Campbell TW. Clinical pathology. In: Mader DR, Divers SJ, eds. *Current Therapy in Reptile Medicine & Surgery.* St. Louis, MO: Elsevier;2014:70-92.

75. Barten SL, Fleming GJ. Current herpetologic husbandry and products. In: Mader DR, Divers SJ, eds. *Current Therapy in Reptile Medicine & Surgery.* St. Louis, MO: Elsevier;20142-12.

76. Helmick KE, Bennett RA, Ginn P, et al. Intestinal volvulus and stricture associated with a leiomyoma in a green turtle (*Chelonia mydas*). *J Zoo Wildl Med.* 2000;31(2):221-227.

77. Dumonceaux GA, Smith AJ, Garner MM. Anterior coelomic rhabdomyosarcoma in a central American banded gecko, *Coleonyx mitratus. Bull Assoc Reptilian Amphib Vet.* 1999;91:23-25.

78. Eyarefe OD, Antia RE, Oguntoye CO, et al. Rhabdomyosarcoma in a terrestrial tortoise (*Geochelone nigra*) in Nigeria: A case report. *J S Afr Vet Assoc.* 2012;83(1):300.

79. Cowan ML, Monks DJ, Raidal SR. Osteosarcoma in a woma python (*Aspidites ramsayi*). *Austr Vet J.* 2011;89(12):520-523.

80. Schmidt RE, Reavill DR. Metastatic chondrosarcoma in a corn snake (*Pantherophis guttatus*). *J Herpetol Med Surg.* 2012;22(3-4):67-69.

81. Honour SM, Ayroud M, Wheler C. Metastatic chondrosarcoma and subcutaneous granulomas in a grey rat snake (*Elaphe obsoleta obsoleta*). *Can Vet J.* 1993;34(4):238-240.

82. Schultze AE, Mason GL, Clyde VL. Lymphosarcoma with leukemic blood profile in a Savannah monitor lizard (*Varanus exanthematicus*). *J Zoo Wildl Med.* 1999;30:158-164.

83. Romagnano A, Jacobson ER, Boon GD, et al. Lymphosarcoma in a green iguana (*Iguana iguana*). *J Zoo Wildl Med.* 1996;27(1):83-89.

84. Folland DW, Johnston MS, Thamm DH. Diagnosis and management of lymphoma in a green iguana (*Iguana iguana*). *J Am Vet Med Assoc.* 2011;239(7):985-991.

85. Jankowski G, Sirninger J, Borne J, et al. Chemotherapeutic treatment for leukemia in a bearded dragon (*Pogona vitticeps*). *J Zoo Wildl Med.* 2011;42(2):322-325.

86. Bezjian M, Diep AN, Matos R, et al. Chinese box turtle (*Cuora flavomarginata*) with lymphoid leukemia characterized by immunohistochemical staining and cytochemical phenotyping. *Vet Clin Pathol.* 2013;42(3):368-373.

87. Ahmad M, Taqavi IH. Radiation induced leukemia and leukopenia in the lizard *Uromastyx hardwickii*. *Radiobiol Radiotherap.* 1978;19(3):353-360.

88. Banzato T, Hellebuyck T, Van Caelenberg A, et al. A review of diagnostic imaging of snakes and lizards. *Vet Rec.* 2013;173:43-49.

89. Hochleithner C, Holland M. Ultrasonography. In: Mader DR, Divers SJ, eds. *Current Therapy in Reptile Medicine & Surgery.* St. Louis, MO: Elsevier;2014:107-127.

90. Isaza R, Ackerman N, Schumacher J. Ultrasound-guided percutaneous liver biopsy in snakes. *Vet Radiol Ultrasound.* 2005;34(6):452-454.

91. Banzato T, Selleri P, Veladiano IA, et al. Comparative evaluation of the cadaveric and computed tomographic features of the coelomic cavity in the green iguana (*Iguana iguana*), black and white tegu (*Tupinambis merianae*) and bearded dragon (*Pogona vitticeps*). *Anat Histol Embryol.* 2013;42(6):453-460.

92. Banzato T, Russo E, Finotti I, et al. Ultrasonographic anatomy of the coelomic organs of boid snakes (*Boa constrictor imperator, Python regius, Python molurus molurus and Python curtus*). *Am J Vet Res.* 2012;73:634-645.

93. Meyer DJ. Chapter 1: The acquisition and management of cytology specimens. In: Raskin RE, Meyer DJ, eds. *Canine and Feline Cytologie: A Color Atlas and Interpretation Guide.* 3rd ed. St. Louis, MO: Elsevier;2016:1-15.

94. Kent MS. The use of chemotherapy in exotic animals. *Vet Clin North Am Exot Anim Pract.* 2004;7:807-820.

95. Spugnini EP, Baldi A. Electrochemotherapy in veterinary oncology: from rescue to first aid line therapy. *Methods Mol Biol.* 2014;23:21-38.

96. Brunner HM, Dutra G, Brunelli C, et al. Electrochemotherapy for the treatment of fibropapillomas in *Chelonia mydas*. *J Zoo Wildl Med.* 2014;45(2):213-218.

97. Siegel S, CroninKL. Palliative radiotherapy. *Vet Clin North Am Small Anim Pract.* 1997;27(1):149-155.

98. Mauldin GN, shiomitsu K. Principles and practice of radiation therapy in exotic and avian species. *Semin Avian Exot Pet Med.* 2005;14(3):168-174.

99. Leach MW, Nichols DK, Hartswell, et al. Radiation therapy of a malignant chromatophoroma in a yellow rat snake (*Elaphe obsoleta quadrivittata*). *J Zoo Wildl Med.* 1991;22:241-244.

100. Greenacre CB, Roberts R. Effect of strontium-90 on squamous cell carcinoma in an eastern box turtle (*Terrapene carolina*); discussion of alternative treatment modalities. *Proc 3rd Int Conf Vet Dis Reptile Amphib.* 2000.

Section 34

ARAV Medicine and Surgery Session

Heather Bjornebo, DVM, CertAqV, DABVP (Reptile and Amphibian Practice);
Kourtney Grimm, DVM
Session Chairs

Case Report of a Gastric Fistula and Repair in a Sonoran Desert Tortoise (*Gopherus morafkai*)

Ericka Meyers, DVM

Session #2073

Affiliation: From the VCA Valley Animal Hospital and Emergency Center, Tucson, AZ 85711, USA.

An approximately 10-year-old female Sonoran Desert tortoise presented for a pre-hibernation exam. On physical exam, the tortoise had a body condition score of 6/9 and had an unremarkable exam with the exception of some mud packed between the left pectoral and abdominal scute margins. The mud was cleaned to discover surrounding scutes that were devitalized. The scutes were trimmed away to reveal a copious amount of ingesta concealing the gastric cavity which had fistulated through the plastron. Chemistry and complete blood count were within normal limits, with the exception of a mild heterophila. The patient was sedated and the anatomy of the stomach was confirmed with an endoscope.The gastric tissue margins were debrided and sutured closed. Hibernation was prevented and the patient was started on enrofloxacin (5 mg/kg IM EOD; Bayer HealthCare LLC, Shawnee Mission, Kansas, USA) and meloxicam (0.2 mg/kg PO q24h; Boehringer Ingelheim Vetmedica, Inc, St, Joseph, MO, USA). Approximately 2 weeks postoperatively, the closure dehisced at which time a culture was obtained. Culture was negative for pathogens. A plastronotomy was recommended but declined by the owner. Closure was attempted once again with more aggressive debriding. After the second closure dehisced, tissue was submitted for histopathology which revealed chronic, active gastritis. Due to the owner's continued reluctance to perform a plastronotomy, three more attempts to debride and suture the fistula failed. A final surgical attempt to perform a partial gastric resection via plastronotomy proved successful. Postoperative medication included penicillin G (20,000 U/kg q24h; Butler Schein Animal Health, Dublin, OH, USA), enrofloxacin, and meloxicam. At 19 months postoperatively, the patient is thriving with no apparent residual complications. In conclusion, while the inciting cause of the fistula remains unknown, a plastronotomy was necessary to fully excise affected gastric tissue and permit healing.

Kitchen Creativity: Clinical Applications of an Airtight Food Storage Container as a Protective Wound Device

Lara Sypniewski, DVM, DABVP (Canine and Feline Practice),
Joao Brandao, LMV, MS,
Ian Kanda, RVT,
Mark Kramer, DVM,
Nicole Stec-Phillips, DVM

Session #2375

Affiliation: From the Center for Veterinary Health Sciences, Oklahoma State University, Stillwater, OK 74078, USA (Sypniewski, Brandao, Kanda), Avian and Exotic Animal Medical Center, Miami, FL 33143, USA (Kramer), South County Animal Hospital, Greenwood, AR 72938, USA (Stec-Phillips).

Using an airtight food storage container and reptile-safe, waterproof epoxy, shell defects may be adequately covered during healing, allowing chelonians to engage in natural behaviors. A 9-year-old intact breeding male African spurred tortoise (*Centrochelys sulcata*) underwent a plastrotomy to remove a urolith. Four months following the procedure, the plastrotomy site showed evidence of osteomyelitis. The site was debrided, the plastron resecured with an external fixation apparatus and covered with a modified plastic storage container. Wound management continued uninhibited by the apparatus. Three months after surgery, the external fixator was inadvertently removed and the protective device was utilized to maintain cleanliness as the plastron healed. Although showing signs of successful healing, the bandaging apparatus inhibited normal mating behavior. As such, the protective container was replaced with a contoured 3D printed shield to protect the plastrotomy site. A captive female river cooter (*Pseudemys concinna*) was presented for a bacterial shell infection leading to loss of one-quarter of the carapace with a resultant open wound. In order to protect the wound but allow access for future treatments, an airtight plastic food storage container was modified to cover the affected area. Wound management was performed using topical antibiotic ointments and antimicrobial baths. Housing in an aquatic environment led to a minor complication of water leakage due to an incomplete lid seal which resolved with lid replacement. These cases reflect the benefits and drawbacks of the clinical application a shell protective device used for management of damaged or surgically incised aquatic and terrestrial chelonian shells.

Risks and Complications Associated with Esophagostomy Tube Placement in Chelonian Patients

Joanna Hedley, BVM&S, DZooMed (Reptilian), DipECZM (Herpetology), MRCVS

Session #2160

Affiliation: From RVC Exotics Service, Beaumont Sainsbury Animal Hospital, Royal Veterinary College, London, UK.

Esophagostomy tube placement commonly is performed in chelonian patients, either to provide nutritional support during a period of anorexia or for administration of oral medications. Placement is usually a quick and simple procedure performed under a short general anesthetic or sedation. Studies in mammals indicate that tubes are associated with relatively few major complications and appear well tolerated by patients.[1-4] The aim of this study was to describe and quantify the risks and complications associated with esophagostomy tube placement in chelonians. A pilot survey was initially distributed among veterinarians in general practice to gain an understanding of techniques being used and complications encountered. During tube placement, reported complications included hemorrhage, perforation of the esophagus or stomach and extra-esophageal tube placement. Following tube placement, complications included tube obstruction, infection and tube displacement by the patient. Medical records from two exotics practices were then examined retrospectively. Reasons for esophagostomy tube placement, frequency and types of complications were recorded and analyzed in order to identify any potential risk factors. Complications were noted in < 62% of patients, with tube obstruction and tube displacement by the patient being the two most commonly seen problems. Both of these complications were easily managed and reflect a similar complication rate to that seen in feline studies.

References

1. Crowe DT, Devey JJ. Esophagostomy tubes for feeding and decompression: Clinical experience in 29 small animal patients. *J Am Anim Hosp Assoc.* 1997;33(5):393-403.

2. Ireland LM, Hohenhaus AE, Broussard JD, Weissman BL. A comparison of owner management and complication in 67 cats with esophagostomy and percutaneous gastrotomy feeding tubes. *J Am Anim Hosp Assoc.* 2003;39(3):241-246

3. Levine PB, Smallwood LJ, Buback JL. Esophagostomy tubes as a method of nutritional management in cats: A retrospective study. *J Am Anim Hosp Assoc.* 1997;33(5):405-410.

4. Whittington JK. Esophagostomy feeding tube use and placement in exotic pets. *J Exot Pet Med.* 2013;22(2):178-191.

Blood Transfusions in Reptiles: Why Don't We Do More?

Anke C Stöhr, DVM, and
Javier G Nevarez, DVM, Dipl ACZM, DipECZM (Herpetology)

Session #2106

Affiliation: From the Department of Veterinary Clinical Sciences, Louisiana State University, Skip Bertman Drive, Baton Rouge, LA 70803, USA.

Blood transfusion medicine is an important treatment option in mammals, but little information is available on its application in reptiles. The limited research performed to date has focused primarily on blood administration and storage. Few case reports have been published from crocodilians (*Osteolaemus tetraspis*), squamates (*Morelia spilota, Pogona barbata*), and chelonians (*Trachemys decorata, Terrapene carolina, Geochelone sulcata*) with varying outcomes. The authors recently performed successful blood transfusions in an alligator snapping turtle (*Macrochelys temminckii*) presented with anemia and hypoproteinemia (packed cell volume [PCV] 4%/total solids [TS] 0.4 g/dl) due to chronic shell infection. The animal responded to treatment with ceftazidime, iron dextran, thiamine, topical treatment with iodine, dry-docking, and two heterologous blood transfusions from a common snapping turtle (*Chelydra serpentina*). On a second case, a red-tailed boa (*Boa constrictor*) presented for recurrent dehiscence of a surgical wound. After several days of debridement, antibiotic and anti-inflammatory treatment, a homologous blood transfusion was performed perioperative due to hypoproteinemia and anemia (PCV 11%/TS 3.6 g/dl) to improve the outcome of the surgery. The wound healed without further dehiscence. Despite these promising case reports, the use of blood products in reptiles is currently not well understood and remains challenging. No information is available on donor selection, compatibility testing, proper blood handling and application, prevention/monitoring of transfusion reactions, safety of repeated blood transfusions, differences in the application of autologous, homologous and heterologous blood transfusions, survival time of transfused red blood cells, and blood component therapy. We currently are working on different research projects to increase our knowledge of reptilian blood transfusion medicine.

Biochemical Enzyme Tissue Specificity in Eastern Box Turtles (*Terrapene carolina carolina*)

Laura Adamovicz, DVM, and
Matthew C. Allender, DVM, MS, PhD, Dipl ACZM

Session #2217

Affiliation: From the Wildlife Epidemiology Laboratory, College of Veterinary Medicine, University of Illinois, Urbana, IL 61802, USA.

Plasma biochemical enzymes are commonly evaluated as part of a diagnostic work-up for exotic species, but their interpretation is complicated by a lack of knowledge about tissue of origin in many reptiles. This study evaluated tissue specificity of 6 biochemical enzymes (alanine aminotransferase, ALT; aspartate aminotransferase, AST; alkaline phosphatase, ALP; creatine kinase, CK; gamma-glutamyl transferase, GGT; and glutamate dehydrogenase, GLDH) in 10 tissues (skeletal muscle, cardiac muscle, lung, liver, gallbladder, pancreas, gastrointestinal tract, kidney, spleen, and reproductive tract) from 10 Eastern box turtles (*Terrapene carolina carolina*). Activity for CK was highest in skeletal muscle, cardiac muscle, and gastrointestinal tract. The GLDH and ALT activities were highest in liver, kidney, and gallbladder. The ALP and GGT activities were elevated in kidney and gastrointestinal tract, and AST was relatively non-specific, with significantly higher activity in the cardiac muscle, liver, kidney, skeletal muscle, and gallbladder compared to other tissues ($P < 0.05$). These results may help guide interpretation of plasma biochemistry panels in reptiles.

Pharmacokinetics of Terbinafine Following a Single Oral Administration in Red-Eared Sliders (*Trachemys scripta elegans*)

David Eshar, DVM, DABVP (Exotic Companion Mammal Practice), DipECZM (Small Mammal), DipECZM (Zoo Health Management),
Nili Avni-Magen, DVM, DipECZM (Zoo Health Management),
Avital Paz, DVM,
Butch KuKanich, DVM, PhD, DACVCP

Session #2153

Affiliation: From the Department of Clinical Sciences, College of Veterinary Medicine, Manhattan, KS 66502, USA (Eshar), the Tisch Family Zoological Gardens in Jerusalem, Jerusalem, Israel (Avni-Magen, Paz), Department of Anatomy and Physiology, College of Veterinary Medicine, Manhattan, KS 66502, USA (KuKanich).

Given the potential clinical use of terbinafine in turtles, the purpose of this study was to assess the pharmacokinetics of terbinafine in red-eared sliders (*Trachemys scripta elegans*, n = 6) after a single-dose (15 mg/kg) oral administration. Blood samples were collected from each animal at times 0, 1, 2, 4, 8, 24, and 48 hours post-treatment. Plasma terbinafine concentrations were quantified with ultra-performance liquid chromatography mass spectrometry. The time to maximum concentration was 1.26 hours. The half-life and mean plasma concentrations were 5.35 hours and 201.5 ng/ml, respectively. More pharmacokinetic studies are indicated to determine and effective oral dosing in red-eared sliders.

Pharmacokinetics and Oral Bioavailability of Trimethoprim Sulfamethoxazole (TMS) in the Green Iguana (*Iguana iguana*)

Spencer Kehoe, DVM,
Stephen Divers, Prof BVetMed, DZooMed, DipECZM (Herpetology),
DipECZM (Zoo Health Management), Dipl ACZM, FRCVS,
Joerg Mayer, DVM, MS, DABVP (Exotic Companion Mammal Practice),
DipECZM (Small Mammal), Dipl ACZM,
Jessica Comoli, DVM,
Robert Arnold, PhD

Session #2098

Affiliation: From the Department of Zoological Medicine and Surgery, College of Veterinary Medicine, University of Georgia, 2200 College Station Road, Athens, GA 30605, USA (Kehoe, Divers, Mayer, Comoli), Department of Drug Discovery & Development at the Specialized Pharmaceutical and Experimental Center for Translational Research and Analysis (SPECTRA), Harrison School of Pharmacy, Auburn University, 720 South Donahue Drive, Auburn, AL 36849, USA (Arnold).

The management of bacterial infections in reptiles is reliant on a small number of antibiotics based on limited pharmacokinetic and pharmacodynamic studies. At University of Georgia, of 141 bacterial isolates cultured from reptiles between 2005-2015, 91% of gram negatives and 62.5% of gram positives were sensitive to trimethoprim-sulfamethoxazole (TMS). This is a first-line broad spectrum antibiotic that has received very little attention and, to date, no peer-reviewed scientific studies evaluating the pharmacokinetics or oral bioavailability of this drug in reptiles. The objective of this study was to determine the pharmacokinetics of orally administered TMS using the green iguana (*Iguana iguana*) as a reptile model. We hypothesized that a 30 mg/kg oral dose of TMS would achieve adequate plasma concentrations in the green iguana. TMS was administered orally and intravenously at 30 mg/kg in a cross-over study design (with 14-day washout), to healthy adult iguanas. Plasma concentrations were then evaluated by high-performance liquid chromatography, and the oral bioavailability of TMS was then determined. The PK parameters were estimated using a non-compartmental analysis of naive averaged data. The IV trimethoprim results were: C_{max} of 12.8 μg/ml, $AUC_{(0\text{-infinity})}$ of 42.4 μg*h/ml, Vd of 0.36 ml/kg, and CL of 118 ml/h/kg. The PO trimethoprim results were; C_{max} of 5.89 μg/ml at T_{max} 8 hours with 100% bioavailability. The IV sulfamethoxazole results were; C_{max} of 41.8 μg/ml, $AUC_{(0\text{-infinity})}$ of 1120 μg*h/ml, Vd of 847 ml/kg, and CL 22.3 ml/h/kg. The PO sulfamethoxazole results were; C_{max} of 17.1 μg/ml at T_{max} 12 hours with a bioavailability of 76%.

Postprandial Changes in Biochemical Parameters in Bearded Dragons (*Pogona vitticeps*)

Christoph Mans, Dr med vet, Dipl ACZM, DipECZM (Zoo Health Management), and
Lily Parkinson, DVM

Session #2239

Affiliation: From the School of Veterinary Medicine, University of Wisconsin-Madison, 2015 Linden Drive, Madison, WI 53706, USA.

Previous studies investigating postprandial effects on blood biochemical analytes have been performed in aquatic (sea turtles) or semiaquatic turtles (red-eared sliders), but not in terrestrial reptiles; however, the formation and excretion of protein waste products differs greatly between aquatic and terrestrial reptile species, with terrestrial reptile producing significiantly more uric acid than aquatic reptiles. Therefore, the objective of this study was to determine if feeding has a significant effect on plasma biochemistry parameters, in particular in uric acid levels, in bearded dragons, a commonly kept terrestrial pet reptile. In a prospective, controlled, randomized non-crossover study 9 bearded dragons were used. Blood was collected in 6 animals immediately before and 4, 24 and 48 hours after feeding crickets at an amount equivalent to 1% of body weight. Three animals were used as the control group. Compared to baseline values, uric acid levels (median [25-75%]: 2.3 mg/dl [1.6-3.6 mg/dl]) were significantly higher by 24 hours after feeding (UA: 5.6 mg/dl [3.7-9 mg/dl] $P = 0.003$), but not by 4 hours. Uric acidc levels returned to pre-feeding values by 48 hours. In the control animals uric acid levels decreased over time. Bearded dragons show a pronounced post-prandial increase in plasma uric acid levels and therefore uric acid levels obtained in clinical cases or in research studies should be interpreted cautiously and should take a postprandial increase in consideration, if animals were fed within < 48 hours prior to blood collection.

Surgical Techniques for Facial Reconstruction in Common Snapping Turtles (*Chelydra serpentina*)

Leslie Reed, DVM

Session #2042

Affiliation: From The Wildlife Rehabilitation Center of Minnesota, 2530 Dale St N, Roseville, MN 55113, USA.

Common snapping turtles (*Chelydra serpentina*) in the wild frequently suffer from severe facial injuries as a consequence of being struck by a vehicle. The trauma oftentimes involves the mandible and maxilla of the beak, and may include the rostrum, surrounding facial bones and soft tissues of the oral cavity. Surgical intervention is usually necessary to repair these severe injuries, and proper timing and technique are crucial to a positive outcome. Common materials such as stainless steel wire and hardware screws, epoxy putty, medical grade glue, and suture may be used to successfully repair most facial injuries. The use of sterile intravenous catheters or feeding tubes can be utilized as temporary nares while injury to the rostrum undergoes healing. Given that proper postoperative care is practiced, most snapping turtles can be successfully released back to the wild 10-12 weeks post-trauma.

Section 35

ARAV Diagnostics Session

Pia Bartolini, CVT, VTS (CP-Exotics);
David Hannon, DVM, DABVP (Avian Practice)
Session Chairs

What Laboratories Do Veterinarians Use for Infectious Disease Diagnostics in Reptiles and Amphibians?

Rachel E. Marschang, PD, Dr med vet, DipECZM (Herpetology), FTÄ Mikrobiologie, ZB Reptilien,
Dana Varble, DVM,
Timothy H. Hyndman, BVMS, PhD,
Members of the ARAV Infectious Diseases Committee

Session #2116

Affiliation: From Laboklin GmbH & Co KG, Steubenstr 4, Bad Kissingen, Germany (Marschang), Association of Reptilian and Amphibian Veterinarians, 5003 SW 41st Blvd, Gainesville, FL 32608-4930, USA (Varble), College of Veterinary Medicine, School of Veterinary and Life Sciences, Murdoch University, Murdoch, Western Australia 6150, Australia (Hyndman).

Finding laboratories that offer diagnostic testing for the wide and increasing range of infectious agents found in reptiles and amphibians can be challenging. As an initial step in putting together a list to help clinicians tackle this difficult question, the Association of Reptilian and Amphibian Veterinarians (ARAV) surveyed its members from October 2017 to February 2018 to find out which laboratories are being used by clinicians. The short online survey received responses from 283 members. The majority (216, 76.3%) were clinicians. Other groups represented laboratory or industry employees, and people working with wildlife or with breeding facilities. Most responders listed multiple laboratories to which they submit samples, depending on their needs. A total of 90 laboratories were listed, 83 of which were clearly identifiable. Forty laboratories have offices in North America, 34 in Europe, 7 in Oceania, 3 in Asia, and 1 in Africa, a few have offices on multiple continents. Several of the laboratories listed could not be identified based on the short name provided. The types of laboratories used included large general veterinary diagnostic laboratories offering testing for a wide range of species, specialized laboratories offering testing for specific diseases, university and government laboratories, and laboratories specializing in pathology and histology. In addition to knowing where reptile and amphibian samples can theoretically be submitted for testing, it is important to have a framework for judging the potential quality of the testing offered. Little work has been done on standardizing testing for reptilian and amphibian infectious diseases. There are a number of criteria that can be used to evaluate a laboratory (Table 1).

Besides these criteria, another aspect to consider that does not directly reflect the quality of the laboratory, is whether it offers collaboration on interesting cases and also whether it publishes results of its work, making them available for peer review.

Not all laboratories that offer testing for reptiles and amphibians will fulfill all of the suggested criteria listed in Table 1 and different points will be more important in different cases. The laboratories on the list obtained during the survey are being contacted to provide more detail on reptile- and amphibian-specific testing. This list will be made available to the ARAV membership and will be updated regularly.

Disclosure statement: Two of the authors (REM and THH) are employed in laboratories that offer diagnostic testing for reptiles and amphibians.

Table 1. Potential criteria for evaluating a laboratory for infectious disease testing in reptiles and amphibians.

Criterion	Description	Comments
Controls		
Positive control	A sample known to be positive that is tested each time any sample is tested.	Provides a check that the test has worked when this sample is positive. Without a positive control, or any sample being positive, all negative results need to be interpreted with caution.
Negative control	A sample known to be negative that is tested each time any sample is tested.	Provides a check that there has not been contamination between samples and that the system is detecting the correct agent. Without a negative control, the interpretation of positive results is less certain because a positive sample may have contaminated negative samples or the test may not be detecting what it is supposed to detect.
Additional controls for PCR		
Internal control	Something that is added to the sample that can be detected.	Provides a check that the sample was processed correctly. Without this, the interpretation of negative results is less certain because the negative result may have been caused by an error in sample processing or the presence of inhibitory substances.
Detection of housekeeping gene/protein	Detection of material that should be present in all samples, irrespective of whether the sample contains the infectious agent.	Not only does this provide a check of sample processing (as per the internal control) but the sample quality also is assessed. Failure to detect a housekeeping gene/protein in a negative sample raises the possibility that the sample was too degraded for the infectious agent to be detectable.
Method for differentiating positive and negative results	All methods will need cut-offs and specific systems for differentiating between positive and negative results.	Depending on the test used, the methods used to determine whether a sample is positive or negative can vary. For serologic tests, this demarcation point theoretically should have been determined previously by controlled infection studies. For PCR, it can be based on the size or melting point of the product, or whether a signal was produced in more specific PCRs (for probe-based real-time PCRs). In some cases, additional testing will be necessary to confirm the result. Does the laboratory provide information on how a result was determined? How do they deal with equivocal results?
Sequencing (for PCR)	Determining the genetic identity of what was detected by a PCR.	Some PCRs are necessarily broadly reactive (eg, pan-adenovirus) so they can detect genetically diverse members of that group of organisms. Sequencing determines what was detected by the PCR. If a PCR is broadly reactive, then positive results should be verified by sequencing to make sure they are true positives (ie, a reptile adenovirus was detected and not a contaminating human adenovirus).
Transparency of testing methods	Was the test method used included on the results report? If so, is it a method that has been peer-reviewed?	Methods allow results to be better interpreted because it becomes clear what the test was designed to detect. Tests that use proprietary methods (ie, are unpublished and/or confidential) cannot be independently verified by other laboratories.

Criterion	Description	Comments
Additional controls for PCR *continued*		
Communication	How are the results communicated to the submitter?	Does the laboratory provide advice on interpreting results? Does the laboratory provide recommendations for additional testing based on the patient history and results obtained?
Accreditation (eg, ISO/IEC, NATA, AAVLD	A system for independently evaluating various aspects of a laboratory. Accreditation is dependent on these standards being met and quality control measures being in place.	Provides assurances that certain aspects of the testing are being done to pre-defined standards and that quality control measures are in place. It addresses technical competence and aspects such as scientific approach and documentation, objectivity and impartiality, and reproducibility. It generally only very marginally deals with reptile and amphibian specific testing. The absence of accreditation does not necessarily mean a laboratory is inadequate but it will usually mean that independent oversight of the laboratory does not happen.
Logistics	How the laboratory operates.	What are the costs of services? Does this represent value for money? How easy is submission and transport of samples to the laboratory? What is the typical turnaround time from sample receipt to results report? Does the laboratory store samples for future testing? If so, for how long? Are results kept confidential?
Comparability	Repeatability: The ability of the same laboratory to obtain the same result when testing the same sample several times. Reproducibility: Whether different laboratories testing the same sample arrive at the same result.	There are many reasons why different results can be obtained when samples are tested repeatedly or when a single animal is repeatedly sampled. Efforts to increase precision in reptile and amphibian veterinary diagnostic testing have been hampered by several factors, including lack of funding and inadequate volume of samples in some cases. Ring testing, which is used in many other areas of laboratory diagnostics, is not used currently in reptile and amphibian diagnostics. Laboratories should be aware of what others are doing and how results generally compare to others.

Determination of Leopard Gecko (*Eublepharis macularius*) Packed Cell Volume and Plasma Biochemistry Reference Intervals and Reference Values

Sarah Alberton, DMV,
Ophelie Cojean, DMV, IPSAV,
Rémi Froment, DMV, DACVP,
Édouard Maccolini, DMV, IPSAV,
Claire Vergneau-Grosset, DMV, IPSAV, CES, Dipl ACZM

Session #2337

Affiliation: From the Zoological Medicine Service and the Clinical Pathology Service, Veterinary Medicine Faculty, University of Montreal, 3200 rue Sicotte, Saint-Hyacinthe, Canada, J2S 2M2.

Despite their popularity in the pet trade, packed cell volume (PCV) and biochemistry reference intervals have yet to be determined in leopard geckos. Tail autotomy during venipuncture is a concern in this species.[1] Blood samples were obtained in 26 healthy subadults (14 females and 12 males) from the ventral coccygeal vein or cranial vena cava under general anesthesia with isoflurane. Reptile biochemistry panels were submitted to Idexx laboratories. Reference intervals and reference values (for parameters with less than 20 values available) were established following the American Society for Veterinary Clinical Pathology guidelines.[2] Data were computed using the Reference Value Advisor software.[3,4] Venipuncture was performed uneventfully in all individuals. No lymph contamination occurred macroscopically. Reference intervals were: 21-40% for PCV, 33-394 μmol/L for uric acid (0.5-6.6 mg/dl), 10-222 IU/L for aspartate transaminase, 0-3701 IU/L for creatine kinase, and 1-38 μmol/L for bile acids. Reference values were determined for the following parameters due to limited sample volumes: 2.4 to > 9.25 mmol/L for total calcium in females (9.6 to >37 mg/dl), 0.4-5.4 mmol/L for phosphorus (1.2-16.4 mg/dl), and 25-80 g/L for total proteins (2.5-8.0 g/dl). Biochemistry results in leopard geckos were comparable to reference values described in crested geckos (*Rhacodactylus ciliatus*).[5] Limitations of this study include small sample number, active folliculogenesis in females,[6] and variable venipuncture sites.[7]

Acknowledgments: Authors thank Montréal Gecko Léopard, Dr Hélène Rembeaux, Dr Maude Gauthier-Bouchard. This study was supported by an ARAV grant and Idexx laboratories.

References

1. Delorme SL, Lungu IM et Vickaryous MK. Scar-free wound healing and regeneration following tail loss in the leopard gecko, *Eublepharis macularius*. *Anat Rec.* 2012;295:1575-1595.

2. Friedrichs KR, Harr KE, Freeman KP, et al. ASVCP reference interval guidelines: Determination of de novo reference intervals in veterinary species and other related topics. *Vet Clin Pathol.* 2012;41:441-453.

3. École Nationale Vétérinaire de Toulouse: Reference Value Advisor. Available at: http://www.biostat.envt.fr/reference-value-advisor/. Accessed September 20, 2017.

4. Geffré A, Concordet D, Braun JP, Trumel C. Reference Value Advisor: A new freeware set of macroinstructions to calculate reference intervals with Microsoft Excel. *Vet Clin Pathol.* 2011;40:107-112.

5. Mayer J, Knoll J, Wrubel M, Mitchell A. Characterizing the hematologic and plasma chemistry profiles of captive crested geckos (*Rhacodactylus ciliatus*). *J Herpetol Med Surg.* 2011;21:68-76.

6. Campbell TW. Clinical pathology of reptiles. In: Mader DR, Divers SJ, eds. *Current Therapy in Reptile Medicine and Surgery.* St. Louis, MO: Elsevier Saunders;2014:70-92.

7. Ramon Lopez-Olvera J, Montane J, Marco I, et al. Effect of venipuncture site on hematologic and serum biochemical parameters in marginated tortoise (*Testudo marginata*). *J Wildl Dis.* 2003;39:830-836.

Relationship Between Ionized and Total Calcium Values in Diseased Client-Owned Reptiles and Their Diagnostic Sensitivity

Nicola Di Girolamo, DMV, MSc(EBHC), GPCert(ExAP), PhD, DipECZM (Herpetology),
Louise Lam,
Lenox Yim, BA,
Katriona Bradley, BVMS, MRCVS

Session #2158

Affiliation: From Tai Wai Small Animal and Exotic Hospital, 69-75 Chik Shun Street, Hong Kong.

Ionized calcium is considered the standard for assessment of calcium status in reptiles. We conducted a prospective diagnostic accuracy study. All reptiles that underwent blood work for clinical reasons from May 2017 to April 2018 had simultaneous measurement of various chemistry analytes including total calcium (VetScan VS2, Avian/reptile rotor) and ionized calcium (i-STAT, Cartridge CG8+) on the same sample. Diagnostic agreement was defined as both total and ionized calcium indicating a status of normo-, hypo- or hypercalcemia in the same patient based on pre-defined cutoffs. Of 129 reptiles, 56 (43.4%, 95%CI: 35.2 to 52.0%) had values of ionized and total calcium not in diagnostic agreement. Two of them had ionized hypocalcemia and total hypercalcemia, 6 of them had ionized hypocalcemia and total normocalcemia, 7 of them had ionized normocalcemia and total hypocalcemia, 29 of them had ionized normocalcemia and total hypercalcemia, 12 of them had ionized hypercalcemia and total normocalcemia. There was a weak significant correlation between total and ionized calcium ($r = 0.49$; $P < 0.001$), and a mean difference of 1.7 mmol/L (95%CI: 1.58 to 1.9; LoA: -0.01 to 3.49). Values of total and ionized calcium in reptiles have poor diagnostic agreement. The diagnostic disagreement affected almost half of the patients and was so important that in certain cases ionized calcium would have resulted in a diagnosis of hypocalcemia and total calcium in a diagnosis of hypercalcemia. Based on current available evidence, both measurements should be obtained for a complete understanding of calcium status in reptiles.

Section 36

ARAV Anesthesia Session

Pia Bartolini, CVT, VTS (CP-Exotics);
David Hannon, DVM, DABVP (Avian Practice)
Session Chairs

Evaluation of Four Different Sedation Protocols in Bearded Dragons (*Pogona vitticeps*)

Christoph Mans, Dr med vet, Dipl ACZM, DipECZM (Zoo Health Management), and
Anthony Barnstable, BS

Session #2243

Affiliation: From the School of Veterinary Medicine, University of Wisconsin-Madison, 2015 Linden Drive, Madison, WI 53706, USA.

Procedural sedation is increasingly used in reptiles in order to facilitate a variety of diagnostic and therapeutic procedures and to reduce the morbidity and mortality associated with general anesthesia; however, the studies evaluating sedation protocols in reptiles, and in particular in bearded dragons, one of the most common pet reptile species, are limited. In 2 separate prospective, randomized, controlled, blinded, complete crossover studies, 4 different injectable sedation protocols were evaluated in adult bearded dragons. In the first study, alfaxalone (20 mg/kg SC) was compared to dexmedetomidine-ketamine (0.1 and 20 mg/kg, SC). The alfaxalone protocol resulted in a more rapid onset of sedation, evident by loss of righting reflex within 5 minutes in most animals. Deep pain response was lost in most animals following sedation with dexmedetomdine-ketamine but in none of the animals following alfaxalone administration. In the second study, a completely reversible protocol consisting of dexmedetomidine-midazolam (DM, 0.1 and 1 mg/kg, SC) was compared to a partial reversible protocol consisting of dexmedetomidine-midazolam-ketamine (DMK, 0.1 and 1 and 10 mg/kg, SC). The addition of ketamine resulted in a significant increase in sedation depth, evidenced by loss of righting reflex in most animals in the DMK group compared to none of the animals in the DM group. Despite the inclusion of the non-reversible ketamine to the DMK protocol there was no significant difference in recovery times following administration of atipamezole and flumazenil in both groups 45 minutes after administration of the sedative drugs.

The Effect of Fraction of Inspired Oxygen in Alfaxalone-Sedated Bearded Dragon (*Pogona vitticeps*)

Christoph Mans, Dr med vet, Dipl ACZM, DipECZM (Zoo Health Management),
Cameron Ratliff, DVM, Lily Parkinson, DVM

Session #2149

Affiliation: From the School of Veterinary Medicine, University of Wisconsin-Madison, 2015 Linden Drive, Madison, WI 53706, USA.

In a prospective, randomized, controlled, blinded, complete crossover study the effects of supplementation with 100% O_2 compared to room air (21 % oxygen) in sedated spontaneously breathing bearded dragons was evaluated. Eight adult inland bearded dragons (*Pogona vitticeps*) were sedated with alfaxalone (20 mg/kg, SC) and received either 21% (room air equivalent) or 100% oxygen flow-by (1 liter per minute) via face mask with a 2-week washout period between treatments. Sedation parameters, cardiopulmonary parameters, venous blood gases, and post-sedation food intake were evaluated. Respiratory rate, heart rate, oxygen saturation and sedation quality were comparable between treatment groups. Venous blood gas analysis demonstrated higher total CO_2 and HCO_3 concentrations in the 21% oxygen group. Post-sedation food intake was not affected by inspired oxygen fraction provided during sedation. Based on the findings of this study, the fraction of inspired oxygen does not appear to have any clinically significant effects on physiologic parameters during and after sedation in bearded dragons. Therefore 100% oxygen supplementation can be considered in sedated bearded dragons without the risk of inducing hypoventilation. Likewise, the lack of 100% oxygen supplementation is unlikely to result in clinical significant consequences in healthy sedated bearded dragons.

Description of Reptile Anesthesia at a Veterinary Referral Hospital over a Period of Eighteen Years (1999-2017)

Kellyn M. Sweeley, DVM,
Jane E. Quandt, DVM, MS,
Stephen J. Divers, BVetMed, DZooMed, DipECZM (Herpetology),
DipECZM (Zoo Health Management), Dipl ACZM, FRCVS

Session #2125

Affiliation: From the Department of Small Animal Medicine and Surgery, College of Veterinary Medicine, University of Georgia, Athens, GA 30602, USA.

A retrospective study was undertaken to evaluate reptile anesthetic standards and protocols at a referral institution, and to evaluate anesthesia-related deaths for common patient variables that may have increased their risk. From May 1999 through December 2018, medical records of reptiles that underwent injectable and/or inhalant anesthesia at a veterinary referral hospital were reviewed. Only cases with complete medical and anesthesia records were included. Collected data included general patient details, diagnosis, procedure(s), and in-depth specifics regarding aspects of premedication, induction, maintenance, monitoring, and recovery including anesthesia-related deaths and necropsy results. Data were sorted into 3 groups (lizards, snakes, chelonians and 1 crocodilian) and described. Further medical record evaluation was given to reptiles that experienced anesthesia-related deaths. A total of 247 detailed anesthesia records were identified for 167 patients. Common findings among groups included opioid premedication, propofol induction, and isoflurane maintenance as the most commonly used agents, no major surgeries performed under injectable anesthesia alone, and mean recovery times around 42 minutes. Newly occurring trends were the use of alfaxalone for induction, norepinephrine constant rate infusion to maintain blood pressure, and epinephrine given on recovery. Those that experienced anesthesia-related death were commonly anemic with hepatic and/or renal pathology, a low body condition score, and already were considered high-risk for anesthesia. The practical application of this research is a better understanding of commonly used reptile anesthetic standards and protocols, the results of their use in practice, and an appreciation that reptile anesthesia be maintained at the same standards as small animal anesthesia.

A Systematic Study of Anesthetics and Analgesics in the Brown Anole Lizard (*Anolis sagrei*)

Ashley M. Rasys,
Stephen J. Divers, BVetMed, DZooMed, DipECZM (Herpetology),
DipECZM (Zoo Health Management), Dipl ACZM, FRCVS,
James D. Lauderdale, PhD,
Douglas B. Menke, PhD

Session #2143

Affiliation: From Department of Cellular Biology, University of Georgia, Athens, GA 30602, USA (Rasys, Lauderdale), Department of Genetics, University of Georgia, Athens, GA 30602, USA (Rasys, Menke), Department of Small Animal Medicine and Surgery, College of Veterinary Medicine, University of Georgia, Athens, GA 30602, USA (Divers).

Small reptiles are becoming increasingly common household pets, and the use of small reptiles in animal research is expanding. There is a pressing need to establish appropriate methods for sedation and pain management of small reptiles for the veterinary and research communities. The purpose of this study is to systematically test the efficacy of anesthetics and analgesics that are commonly used in mammals in the brown anole lizard, *Anolis sagrei*, which is emerging as a new animal model for studies of embryonic development. The drugs—alfaxalone, dexmedetomidine, hydromorphone, ketamine, and tribromoethanol—were administered intramuscularly or subcutaneously, alone or in combination at varying doses to 68 lizards, weighing 2.5-3.5 grams (8 anesthesia and 2 surgical trials; total of 14 groups; minimum of 4 lizards per group). Drug induction time, duration, anesthesia level, and adverse effects were all assessed. Differences of anesthesia level were observed between groups of lizards where injection site and method or drug combination were changed ranging from deep to light anesthesia. Alfaxalone and tribromoethanol given alone or when grouped with dexmedetomidine all induced a surgical plane of anesthesia while ketamine/hydromorphone/dexmedetomidine only produced moderate to light sedation. Hydromorphone when given with alfaxalone/dexmedetomidine resulted in deep anesthesia with respiratory depression. Following anesthesia trials, a surgical procedure (5-10 mm bilateral celiotomy) was successfully performed using the anesthetic combinations alfaxalone/dexmedetomidine or tribromoethanol/dexmedetomidine. This study represents a significant step towards understanding the effects of anesthetics/analgesics in lizards and the data generated here will benefit both veterinary medicine and research.

Acknowledgments: Funding sources include the National Science Foundation #1149453, National Institutes of Health T32GM007103, and ARCS Foundation 2017-18.

Anesthesia of Hatchling Sea Turtles Using Ketamine-Medetomidine-Tramadol

T. Franciscus Scheelings, BVSc, MVSc, MANZCVSc (Med Aust Wildl),
DipECZM (Herpetology),
Christopher Gatto, BSc, MSc,
Richard Reina, PhD

Session #2044

Affiliation: From the School of Biological Sciences, Monash University, Wellington Rd, Clayton, VIC, Australia, 3800.

In a prospective clinical trial, 10 hatchling green sea turtles (*Chelonia mydas*) were anesthetized using a combination of intramuscular ketamine-medetomidine-tramadol. Prior to anesthesia, resting cardiopulmonary parameters, cloacal temperature, and venous blood gas and biochemistry were obtained from hatchling green sea turtles. Animals were then anesthetised with ketamine (5 mg/kg), medetomidine (0.05 mg/kg), and tramadol (5 mg/kg) via intramuscular injection. Turtles were checked for depth of anesthesia at 5-minute intervals by recording reflexes (righting, palpebral, pinch, cloacal) and measuring heart rate, respiratory rate and cloacal temperature. After 20 minutes, a second venous blood sample was obtained for further blood gas and biochemical analysis, and the medetomidine was antagonized using atipamezole (5:1 medetomidine, 0.25 mg/kg). All turtles were successfully anesthetized with a mean time to induction of 3.4 minutes (± 1). In all animals, a loss of reflexes (except for palpebral reflex) and voluntary movement was observed for the entire 20 minutes. Anesthesia resulted in marked apnea for the duration of the procedure. Venous blood gas and biochemistry analysis indicated that a 20-minute period of apnea had no measureable effects on physiology. All turtles recovered uneventfully after atipamazole antagonization, with mean time to first breath 4.5 minutes (± 3.7), and mean recovery time 15.5 minutes (± 15.4). Intramuscular ketamine-medetomidine-tramadol, antagonized with atipamazole appears to be an effective anesthetic protocol in hatchling green sea turtles for short procedures; however, animals should be intubated and provided with supplemental oxygen

Armour-Plated Medicine: Tips, Tricks, and Treatments of Shingleback Lizards (*Tiliqua rugosa*)

Shane Simpson, BVSc(Hons), GCM(VP)

Session #2078

Affiliation: From Karingal Veterinary Hospital, Frankston, Victoria, 3199, Australia.

While commonly kept as pets in Australia, outside of their country of origin shingleback lizards (*Tiliqua rugosa*) are rare and often expensive acquisitions for prospective keepers. Veterinarians outside of Australia may see this species infrequently because of its rarity and cost. An understanding of shingleback ecology, reproduction, unusual anatomical features, and common conditions that afflict shingleback lizards is vital to allow appropriate husbandry advice to be given and medical treatments to be performed.

Section 37

ARAV Case Reports Session

James Bogan, DVM, DABVP (Canine and Feline Practice), CertAqV;
Erica Mede, CVT
Session Chairs

Clinical Presentation, Treatment and Outcome of a Case Series of Mediterranean Tortoises with Rhododendron (Grayanotoxin) Intoxication

Karina Mathes, DVM, Dr med vet, Cert Spec in Reptiles and Amphibians, DipECZM (Herpetology), and
Michael Fehr, DVM, Dr med vet, PhD, DipECZM (Small Mammals)

Session #2266

Affiliation: From the Clinic for Small Mammals, Reptiles and Birds, University of Veterinary Medicine Hannover, Buenteweg 9, D-30559 Hannover, Germany.

Three Mediterranean tortoises (1 spur-thighed [*Testudo graeca*] and 2 marginated tortoises [*T. marginata*]) were presented showing neurologic signs with a history of possible exposure to flowers and foliage of azaleas (*Rhododendron* spp.). In the case of the spur-thighed tortoise, the owner first reported that no toxicity should be possible because no poisonous plants grew in the outdoor enclosure; however, when taking a thorough medical history it was mentioned that some flowers from an azalea plant in the neighbor's garden might have blown into the tortoise pen. This information, combined with neurologic signs such as weakness and lethargy, extreme salivation and tonic extensor seizures of the hind limb, lead to the assumption of a possible grayanotoxin intoxication with parts of plants from the family Ericaceae. The animal was treated with subcutaneous prednisolone, furosemide, fluid therapy, and oral charcoal and water. Two days later, one marginated tortoise with similar symptoms and a history of possible ingestion of rhododendron flowers was presented and the same treatment started. This time multiple structures were radiographically visible in the stomach and an immediate gastroscopy for removal of flowers and leaves was carried out and was partially successful. A second marginated tortoise from the same owner presented with initial neurologic signs and multiple radiographically suspicious structures in the stomach. Gastroscopy was carried out immediately resulting in the successful removal of flowers and foliage of *Rhododendron* spp. followed by the same treatment as the other two animals received. All three tortoises survived.

Two Cases of *Encephalitozoon pogonae* Detected in *Pogona vitticeps* in Australia

Joshua Llinas BVSc, MVS,
John Mackie BVSc, PhD, DACVP,
Tim Hyndman BVMS, PhD

Session #2023

Affiliation: From Greencross Vets Jindalee, 1/111 Dandenong Road, Mount Ommaney, Queensland, Australia (Llinas), QML Vetnostics Pathology, 11 Riverview Place, Murarrie, Queensland, Australia (Mackie), and Murdoch University, Murdoch Drive, Murdoch, Western Australia, Australia (Hyndman).

Microsporidia are obligate, intracellular pathogenic fungi. In reptiles, they are most commonly reported in squamates. We report cases of systemic microsporidiosis from 2 collections of Australian inland bearded dragons (*Pogona vitticeps*). From the first collection, a dragon was presented with lethargy, inappetence and weight loss, while from the second collection a dragon presented for a routine wellness examination. Both dragons had palpable intracoelomic masses, which were assessed by ultrasound and removed surgically. Histologic evaluation of both masses revealed granulomatous inflammation within or adjacent to ovarian tissue, containing numerous aggregates of organisms consistent with microsporidia. The first dragon died shortly after surgery while the second dragon died 5 weeks after surgery, following 3 weeks of treatment with 25 mg/kg fenbendazole PO q48h. Necropsy samples from both dragons were collected and the microsporidian *Encephalitozoon pogonae* was detected in a variety of tissues and swabs (but not blood) by PCR and sequencing. Agamid adenovirus-1 (AgAdV-1) was detected in the dragon from the first collection but its significance was undefined. *E. pogonae* and AgAdV-1 were also detected in cloacal swabs from 3 asymptomatic dragons from the first collection that were in-contact with the affected dragon.

An Unusual Coelomic Mass in an Inland Bearded Dragon (*Pogona vitticeps*)

Mike Corcoran, DVM, DABVP (Reptile and Amphibian Practice), CertAqV

Session #2120

Affiliation: From the New England Aquarium, 1 Central Wharf, Boston, MA 02110, USA.

A 3.5-year-old female inland bearded dragon (*Pogona vitticeps*) presented for a second opinion on a coelomic mass. The mass was evaluated via ultrasound, cytology, histopathology and endoscopy. Eventually the mass was removed and submitted en mass for histopathology. Histopathology revealed that the mass was the fat pad with a granuloma from a bacterial infection. The animal recovered well and has not had recurrence of disease.

Use of Transesophageal Echocardiography to Observe Cardiovascular Flow Patterns in American Alligators During Isoflurane Anesthesia

Bonnie J. Gatson, DVM, DACVAA,
Nicole R. Furst, BS,
Simon Swift, MA, VetMB, CertSAC, Dipl ECVIM-CA (Cardiology), MRCVS,
Darryl J. Heard, BSc, BVMS, PhD, Dipl ACZM,
James FX Wellehan, DVM, MS, PhD, Dipl ACZM, Dipl ACVM (Virology),
Dipl ACVM (Bacteriology and Mycology)

Session #2074

2016 ARAV Conservation and Research Grant Winner

Affiliation: From the Department of Comparative, Diagnostic, and Population Medicine (Gatson, Heard, Wellehan) and the Department of Small Animal Clinical Sciences (Swift), University of Florida College of Veterinary Medicine, 2015 SW 16th Ave, Gainesville, FL 32608, USA.

The mechanism by which epinephrine hastens anesthetic recovery in alligators (*Alligator mississippiensis*) is unknown. Transesophageal echocardiography was used to observe alterations in cardiovascular flow patterns following epinephrine administration in isoflurane-anesthetized alligators. Using a crossover design, 10 alligators (5.7 ± 1.4 kg) underwent isoflurane anesthesia in oxygen for 90 minutes. Upon discontinuing isoflurane, alligators were randomly administered either intramuscular epinephrine (0.1 mg/kg) or equal volume saline. A blinded observer recorded the following times during recovery: return of spontaneous ventilation, return of palpebral reflex, return of withdrawal reflex, reaction to stimulation, spontaneous movement, and extubation that occurred following spontaneous ventilation and movement. Echocardiography was performed before anesthetic induction, immediately before discontinuing isoflurane, and following treatment administration. Cardiovascular flow patterns across the Foramen of Panizza, pulmonary outflow tract, and left aortic outflow tract were observed using color flow and spectral Doppler. After evaluating for normality, recovery parameters were analyzed using a paired two-tailed t-test; $P < 0.05$ was considered significant. Following epinephrine administration, biphasic flow across the pulmonary artery observed throughout systole prior to treatment was replaced by a single flow wave with a peak in mid-systole. Saline administration caused no change in flow profiles across the pulmonary artery. All recovery parameters occurred significantly faster following intramuscular epinephrine. Biphasic flow across the pulmonary valve during isoflurane anesthesia likely represents closure of the cog wheel valve mid-systole limiting pulmonary outflow. Continuous flow across the pulmonary valve following epinephrine administration suggests reduced pulmonary-to-systemic shunting, allowing faster recovery from isoflurane anesthesia in this species.

First-Degree Atrioventricular Block and Ataxia in a Central Bearded Dragon (*Pogona vitticeps*) with Atherosclerosis and Brain Spongiosis

Lionel Schilliger, Dr vet med, DipECZM (Herpetology),
DABVP (Reptile and Amphibian Practice),
Clément Paillusseau, Dr vet med,
Frédéric Gandar, Dr vet med,
Isabelle Desprez, Dr vet med,
Peggy Passavin, Dr vet med,
William Claude, Dr vet med,
Vittorio Saponaro, PhD, Dr vet med,
Marine Roche-Catholy, Dr vet med,
Valérie Chetboul, PhD, Pr vet med, Dipl ECVIM-CA

Session #2194

Affiliation: From the Veterinary Clinic of Auteuil Village, 35 rue Leconte de Lisle, 75016 Paris, France (Schilliger, Paillusseau, Gandar), Cardiology Unit of Alfort (UCA), Veterinary School of Alfort, 7 avenue du Géneral de Gaulle, 94700 Maisons-Alfort, France (Passavin, Claude, Saponaro, Roche-Catholy, Chetboul), Exotics Medicine Service, Veterinary School of Alfort, 7 avenue du Géneral de Gaulle, 94700 Maisons-Alfort, France (Desprez, Claude), INSERM U955, Equipe 03, 51 avenue du Maréchal de Lattre de Tassigny, 94010 Créteil cedex, France (Chetboul).

A 0.5-kg 5-year-old male bearded dragon (*Pogona vitticeps*) presented with a 2-month history of lethargy, anorexia and impaired locomotion. Upon physical examination, bradyarrhythmia (heart rate: 20 bpm; reference range 24-170) and ataxia were noted. Electrocardiography revealed a prolonged P-R interval (320 milliseconds; reference range 75-253), indicative of a first-degree atrioventricular (AV) block. On echocardiography, heart size was slightly above normal range without noticeable lesions of the myocardium and minor pericardial effusion was visible. Blood biochemistry (sodium, potassium, phosphate, calcium, uric acid, total cholesterol, fT4, T4, and liver enzymes) was unremarkable except for elevated aspartate aminotransferase (105 IU/L; reference range 1-53 IU/L). Adenovirus testing was negative by polymerase chain reaction. No abnormalities of the central nervous system (CNS) were detected by computed tomography. Following euthanasia, thickening of the arterial trunks and atrophy of the cerebellum were observed on necropsy. Histology showed multifocal atherosclerosis of efferent heart vessels, and multifocal spongiosis of brain tissue with cerebellar atrophy, mild hydrocephalus, degeneration of the optic chiasm, and multifocal sclerosis of arterioles. Although a common disease in captive psittacine birds, atherosclerosis is poorly documented in reptiles.[1-9] Risk factors in birds include stress, nonnative diet, lack of exercise, elevated plasma cholesterol, and hepatic lipidosis. Spongiosis of the CNS can be associated with inclusion body disease or sunshine virus infection in snakes, but has not been described in lizards. Neurologic signs observed in this patient were likely due to brain abnormalities, while heart enlargement and the AV block could be the consequences of atherosclerosis.

Acknowledgments: Authors thank Dr vet med Alexandra Nicolier, Dipl ECVP from Vet Diagnostics, 14 avenue Rockfeller, 69008, Lyon, France and Dr Jesse Bonwitt, BVSc, MSc, MRCVS.

References

1. Beaufrère H. Avian atherosclerosis: in parrots and beyond. *J Exot Pet Med.* 2013;22:336-347.

2. Finlayson R, Symons C, T-W-Fiennes R. Atherosclerosis: A comparative study. *Br Med J.* 1962;1(5277): 501-507.

3. Finlayson R, Woods SJ. Arterial disease of reptiles. *J Zool.* 2009;183(3):397-410.

4. Jensen B, Veterskov S, Boukens BJ, et al. Morpho-functional characterization of the systemic venous pole of the reptile heart. *Sci Rep.* 2017;7(1):6644.

5. Hyndman TH, Shilton CM, Doneley RJ, et al. Sunshine virus in Australian pythons. *Vet Microbiol.* 2012;161(1-2):77-87.

6. Ritchie B. Virology. In: Mader DR, ed. *Reptile Medicine and Surgery.* St Louis, MO: Elsevier Saunders;2006:391-417.

7. Schilliger L, Lemberger K, Chai N, et al. Atherosclerosis associated with pericardial effusion in a central bearded dragon (*Pogona vitticeps*). *J Vet Diag.* 2010;22:789-792.

8. Silverman S, Sanchez-Migallon Guzman D, Stern J, et al. Standardization of the two-dimensional transcoelomic echocardiographic examination in the central bearded dragon (*Pogona vitticeps*). *J Vet Cardiol.* 2016;18(2):168-178.

9. Tamukai K, Takami Y, Akabane Y, et al. Plasma biochemical reference values in clinically healthy captive bearded dragons (*Pogona vitticeps*) and the effects of sex and season. *Vet Clin Pathol.* 2011;40(3):368-373.

Clinical Bacterial Isolates and Antimicrobial Susceptibility Patterns from Reptiles (2005-2016)

Pak Kan Tang, BSc (Comp Path), BVetMed (Hons), MRCVS,
Susan Sanchez, BSc, MSc, PhD, CBiol, FSB,
Stephen J. Divers, BVetMed, DZooMed, DipECZM (Zoo Health Management), DipECZM (Herpetology), Dipl ACZM, FRCVS

Session #2083

Affiliation: From the School of Veterinary Medicine, University College Dublin, Belfield, Dublin 4, Ireland (Tang), Department of Infectious Disease (Sanchez) and Department of Small Animal Medicine and Surgery (Divers), College of Veterinary Medicine, University of Georgia, GA 30602, USA.

Antimicrobial agents remain the cornerstone of therapy for bacterial infection in both humans and animals; however, it is apparent that there is an increased prevalence of antimicrobial resistance, provoking a global threat to the public health. The aim of this retrospective study was to identify the nature of bacterial isolates and sensitivity patterns from clinical cases and determine an appropriate antimicrobial stewardship policy. This was achieved by analyzing the medical records at University of Georgia College of Veterinary Medicine to identify bacterial culture submissions and sensitivity results from reptiles presented between January 2005 and December 2016. Prevalence of bacterial genera and species were analysed. Results were subcategorised into gram-negative bacteria and gram-positive bacteria, and their susceptibility patterns against antimicrobial agents were reviewed. Samples (n = 131) were submitted, of which 49.6% (n = 65) were culture positive with a total of 138 identified organisms. *Pseudomonas* spp. (n = 20) and *Enterococcus* spp. (n = 18) were the most frequently identified gram-negative and gram-positive bacteria in reptiles, respectively. Our results demonstrated that aminoglycosides, particularly amikacin, are appropriate empirical antibiotic choices in the presence of gram-negative bacteria, while the presence of gram-positive bacteria may indicate that doxycycline is the most suitable first-line. This study emphasizes the importance of performing Gram smears in relation to empirical antimicrobial therapy. Our findings also demonstrate the complete ineffectiveness of ceftazidime against gram-positive bacteria and it has no superior efficacy against gram-negative bacteria compared with other antimicrobial agents (including trimethoprim-sulfa).

Diagnosis and Radiation Therapy of Myxosarcoma in a Russian Tortoise (*Agrionemys horsfieldii*)

Ber-In Lee, DVM, MS,
Chu-Lin Kang, DVM,
Chen-Hsuan Liu, DVM, PhD,
Chian-Ren Jeng, DVM, PhD,
Yuan-Peng Kuan, DVM, MS,
Pin-Huan Yu, DVM, PhD

Session #2112

Affiliation: From Institute of Veterinary Clinical Sciences (Lee, Kang, Kuan, Yu), and Institute of Molecular and Comparative Pathology (Liu, Jeng), School of Veterinary Medicine, National Taiwan University, No 1, Sec 4, Roosevelt Road, Taipei, 10617, Taiwan, ROC.

A male Russian tortoise (*Agrionemys horsfieldii*) of unknown age was presented for a solid mass between the right elbow and scapula. The patient showed no discomfort related to the mass. There was no remarkable hematologic and biochemical abnormality but fine needle aspiration suggested a spindle cell tumor. Surgical biopsy and CT scan revealed myxosarcoma dorsal to the plastron that extended to pectoral girdle. Surgical removal combined with stereotactic body radiation therapy was performed. Acute effect of radiation therapy was recorded after radiation therapy; however, the tortoise maintained good body condition in the following 3 months without delayed effect. Computed tomography performed 3 months after radiation therapy showed no sign of tumor recurrence. This is the first report of myxosarcoma managed by surgery and radiation therapy in tortoises with a successful outcome. Stereotactic body radiation therapy should be considered as one of the treatment options when managing myxosarcoma in tortoises.

Fatal Coagulopathy Secondary to Chronic Hepatopathy in Three Lizards

Christoph Mans, Dr med vet, Dipl ACZM, DipECZM (Zoo Health Management),
Katharine Kierski, BS,
Lily Parkinson, DVM

Session #2250

Affiliation: From the School of Veterinary Medicine, University of Wisconsin-Madison, 2015 Linden Drive, Madison, WI 53706, USA.

Coagulopathies associated with hepatopathies in reptiles have been poorly documented in the literature. Here we report 3 cases of pet lizards diagnosed with chronic hepatopathies which died or were euthanize due hemorrhage most likely associated with an underlying coagulopathy. Case 1 was a 12-year-old female green iguana (*Iguana iguana*), which was diagnosed with hepatopathy based on increased plasma bile acids and computed tomography. The animal died from severe hemorrhage secondary to a broken toe nail. Necropsy revealed severe hepatic fibrosis and hepatocellular carcinoma. Case 2 was a 9-year-old green iguana that presented for repeated episodes of hemoptysis and was diagnosed with hepatomegaly based on computed tomography (CT). One year after the initial diagnosis the animal acutely died and necropsy revealed acute pulmonary hemorrhage in addition to severe hepatic fibrosis and bile duct hyperplasia. Case 3 was a 4-year-old male Mali uromastyx, which was evaluated for anorexia and diagnosed with severe hepatic lipidosis on CT. Five weeks after diagnosis the patient presented for peracute onset of severe dyspnea and was euthanized. Necropsy revealed multiple blood clots obstructing trachea and primary bronchi, as well as severe hepatic lipidosis. In all 3 cases it is suspected that the underlying hepatopathies lead to secondary coagulopathies and hemorrhage. Therefore, coagulopathies should be considered as a complication of chronic hepatopathies in reptiles and appropriate treatment, such as vitamin K administration initiated. Computed tomography is a highly sensitive and non-invasive diagnostic modality to diagnose hepatopathies in reptiles.

Chronic Lymphocytic Leukemia in a Bearded Dragon (*Pogona vitticeps*)

Lacy Gilmer, DVM,
Drury Reavill, DVM, DABVP (Avian Practice), DABVP (Reptile and Amphibian Practice), DACVP,
William Vernau, BSc, BVMS, DVSc, PhD, DACVP,
Nicole I. Stacy, DVM, Dr Med Vet, DACVP (Clinical)

Session #2336

Affiliation: From Calvert Veterinary Center, 4100 Mountain Road, Ste C, Pasadena, MD 21122, USA (Gilmer), Zoo/Exotic Pathology Service, 6020 Rutland Drive #14, Carmichael, CA 95608-0515, USA (Reavill), Department of Pathology, Microbiology and Immunology, University of California Davis, School of Veterinary Medicine, Davis, CA 95616, USA (Vernau), the Department of Comparative, Diagnostic, and Population Medicine, University of Florida, College of Veterinary Medicine, 2015 SW 16th Ave, Gainesville, FL 32610, USA (Stacy).

A 3-year-old male bearded dragon (*Pogona vitticeps*) presented with weight loss, lethargy, and bilateral chemosis. Complete blood count (CBC) at presentation included marked lymphocytosis ($36.6 \times 10^3/\mu l$). Treatment was initiated with ceftazidime (20 mg/kg IM q72h), ciprofloxacin ophthalmic 0.3% solution (1 drop OU q12h), and nutritional supplementation. The patient improved in activity, appetite, and gained weight during the initial treatment period; however, a subsequent CBC at day 71 showed further increase of lymphocytosis ($209.4 \times 10^3/\mu l$) with a tentative diagnosis of lymphocytic leukemia by blood film evaluation. The lymphocyte population was predominantly composed of small well-differentiated lymphocytes with clumped chromatin and inapparent nucleoli and frequently exhibiting cytoplasmic hand mirror-like projections. Immunocytochemical immunophenotyping demonstrated CD3 expression with a small continuous rim of cytoplasmic positivity, consistent with a diagnosis of T-cell chronic lymphocytic leukemia (CLL). Prednisolone 3 mg/kg PO q24 was initiated on day 90 after presentation. On day 127 post-presentation, the lymphocytosis was markedly decreased ($11.9 \times 10^3/\mu l$), but severe periodontal disease had developed. Prednisone was reduced to 2 mg/kg PO q24h. Complete blood cell count on day 159 had a lymphocyte count within normal limits ($12.9 \times 10^3/\mu l$) and prednisone was further reduced to 1 mg/kg PO q24h. Ceftazadime was discontinued. Chemistry findings included hyperglycemia of 536 mg/dl and hyperuricemia of 7.2 mg/dl. The animal was clinically doing well until day 253 when he represented for acute lethargy and died on day 254. Initial necropsy results revealed bilateral pyelonephritis. Histopathology was pending at time of manuscript submission. This case describes the cytologic and immunocytochemical diagnosis of T-cell CLL in a bearded dragon, with initial successful management.

References

1. Tocidlowski ME, McNamara PL, Wojcieszyn JW. Myelogenous leukemia in a bearded dragon (*Acanthodraco vitticeps*). *J Zoo Wildl Med.* 2001;32(1):90-95.

2. Georoff TA, Stacy NI, Newton AN, et al. Diagnosis and treatment of chronic T-lymphocytic leukemia in a green tree monitor (*Varanus prasinus*). *J Herp Med Surg.* 2009;19:106-114.

3. Jankowski G, Sirninger J, Borne J, Nevarez JG. Chemotherapeutic treatment for leukemia in a bearded dragon (*Pogona vitticeps*). *J Zoo Wildl Med.* 2011;42(2):322-325.

Suspected Hyperviscosity Syndrome in Four Eastern Indigo Snakes (*Drymarchon corais couperi*)

Sarah Miller, DVM,
Elise E. B. LaDouceur, DVM, DACVP,
Michael M. Garner, DVM, DACVP,
Sam Rivera, DVM, MS, DABVP (Avian Practice), Dipl ACZM, DipECZM(Zoo Health Management)

Session #2162

Affiliation: From Zoo Atlanta, 800 Cherokee Avenue SE, Atlanta, GA 30315, USA (Miller, Rivera), Northwest ZooPath, 654 W Main St, Monroe, WA 98272, USA (LaDouceur, Garner), Charles River Laboratories, 15 Worman's Mill Rd, Frederick, MD 21701, USA (LaDouceur).

Four apparently healthy 3-year-old Eastern indigo snakes (*Drymarchon corais couperi*) died acutely during transport for release as part of a head start program. Gross necropsy revealed yellow coelomic fluid and multiple red, roughly spherical nodules or cysts on the mucosa of the lower esophagus in all animals. Total protein of the coelomic fluid was assessed in 3 animals (range 5.3-6.2 g/dl) and 1 sample was sent for further analysis. Coelomic fluid was a modified transudate with lymphocyte predominance and extracellular bacteria that was suspected to be contamination. Coelomic fluid culture was done in 3 animals. *Salmonella* sp. (n = 3), *Proteus mirablis* (n = 3), *Escherichia coli* (n = 2), and *Staphylococcus xylosus* were isolated, all with wide susceptibility patterns. Histologic exam revealed amphophilic to deeply eosinophilic fluid within vasculature, and extraluminal in the coelom, lungs, and eye. A hyperviscosity-like syndrome was suspected in all cases. A moderate adrenal stress response was also noted in three animals and mild erosive esophagitis was seen in one animal. Increased blood viscosity in hyperviscosity-like syndrome may be caused by pathologic changes in cellular or protein fractions, resulting in poor blood flow and perfusion, vascular leakage and organ congestion. In humans, typical clinical signs are mucosal bleeding, visual changes, and cardiorespiratory syndrome. This is uncommonly reported in veterinary medicine.

Section 38

ASSOCIATION OF REPTILIAN AND AMPHIBIAN VETERINARIANS

ARAV Nutrition and Gastroenterology Session

Anneliese Strunk, DVM, DABVP (Avian Practice);
Louisa Asseo, DVM, DABVP (Canine and Feline Practice)
Session Chairs

Feeding Trial and Health Assessment of Corn Snakes (*Pantherophis guttatus*) Fed an Insect-Based Sausage Diet

Kimberly Boykin, DVM,
Karina Butler-Perez, BS,
Mark A. Mitchell, DVM, MS, PhD, DipECZM (Herpetology)

Session #2067

Affiliation: From the Department of Veterinary Clinical Sciences, Louisiana State University, School of Veterinary Medicine, Baton Rouge, LA 70803, USA.

Processing whole vertebrate prey into sausage casings with various scents has been performed by herpetoculturists over the years in an attempt to reduce feeding costs and increase diet variety for carnivorous species. Diet options for captive snakes are limited due to the lack of prey available at the commercial level. Feeder mice are readily available, but these prey species are known to injure snakes when offered as live prey and their high nutritional content can lead to obesity in captive snakes. In addition, some reptile owners are disturbed by the idea of using whole vertebrate prey to feed their animals. Use of an alternative protein source, such as black soldier fly larvae, addresses many of these issues and has the potential to open the market for other sausage-based diets. Nine juvenile male corn snakes were housed individually under identical conditions for 4 months in this crossover feeding trial Each snake either received the experimental insect-based diet or a standard, frozen-thawed mouse diet for 2 months, before switching to the second diet for the remaining 2 months. All snakes received, on average, 2 meals per week with each meal equaling 15% of the animal's body weight. Physical examinations and growth parameters, including weight, snout-vent-length, tail length, and girth, were measured at the beginning and end of each 2-month period. Fecal contents were collected to assess apparent digestibility, and metabolism of the snakes was measured via respirometry. Diet acceptance also was compared by measuring the time from food offering to first strike and to swallowing.

Appetite Increase in an Aged Albino Burmese Python Using Photobiomodulation to Increase Parasympathetic Response

Janet Gordon Palm, DVM, CVCP

Session #2039

Affiliation: From Animobility Integrative Veterinary Services, PO Box 749, Fruita, CO 81521, USA.

Abstract: Low-level laser therapy (LLLT), also called photobiomodulation, has become an accepted healing modality in the areas of pain management, rehabilitation, wound healing, and fracture repair. The unique Class 2 low-power (5 mW; 7.5 mW) non-heat-producing, frequency-specific, visible-light wavelength (635 nm; 405 nm) laser is 1/100 of the power of a common Class 4 laser. The visible light does not scar retinal tissue, allows for a blink response, and thus there is no need for protective eyewear. The line-generated beams and the ability to vary the frequency of the pulsation of the light energy allows the targeting of various surface areas, affecting different tissues and systems. Individualized frequency settings are based on the vibrational frequency variations of different tissues and the studies of Nikola Tesla, Royal Raymond Rife, Paul Nogier, among others.[1] The acronym for frequency-specific low-level laser therapy (FSLLLT) has been used to describe this device. A unique ability of this combination of specifications of visible light and low power, with an energy density that affects the autonomic nervous system, allows for increasing appetite in anorexic animals.[2] Appetite stimulation was demonstrated with an albino Burmese python previously owned by Michael Jackson at Neverland Ranch, now residing at The Children's Nature Center in Grand Junction, CO. This appetite stimulation occurred on multiple occasions along with an increase in vitality throughout the 1.5 years he received maintenance LLLT. Numerous species (including but not limited to, horses, dogs, cats, birds, small mammals, large exotic cats, and wildlife), have been documented by the author to show similar increase in parasympathetic responses with concurrent decrease in sympathetic tone. Lowered heart rates and blood pressures have been recorded immediately post treatment on dogs, as well as decrease in digital pulse rates on laminitic horses. Calming of fractious animals, increasing appetite and gut motility, as well as allowing restorative sleep are the observed benefits. The ability to achieve these healing responses without sedation, from a short distance, and without need for eye protection, allows for a truly non invasive treatment protocol.[3]

Introduction

The purpose of this presentation is to show that there are non-invasive possibilities for increasing appetite and vitality through signaling the body's innate pharmacy through photobiomodulation. The safe use of a unique Class 2 laser system demonstrates an improved quality of life for an aged albino Burmese python, beyond what was expected.

The Class 2 non-heat-producing laser system has been used globally in the human field in the areas of rehabilitation medicine, sports performance, neurodegenerative disease, Alzheimers and traumatic brain injuries, as well as burn units, owing in part to its Food and Drug Administration clearance in acne and onychomycosis.[4] Veterinarians are able to utilize the benefits of the frequency-specific low-level laser therapy (FSLLLT) for cellular enhancement beyond pain and inflammation. Stem cell stimulation and nerve regeneration have been described in the literature, when low-power non-heat-generating lasers and LEDs are used.[5,6]

The ability to affect the autonomic nervous system is perhaps the most amazing response, of which few are aware. The author has experienced and documented lowered blood pressure within seconds of the completion of the therapy implemented using specific frequencies that can be programmed into the device. This author hopes to inspire more to investigate these responses further.

True photobiomodulation involves cellular enhancement using low-power lasers and LEDs without resultant thermal change within the tissues.[7] This term has erroneously been generalized to include high-power heat-producing therapy lasers, and thus confusion will occur until specific guidelines are enforced in future studies.

There is no question that higher-power Class 3b and 4 lasers demonstrate beneficial results in areas of joint and muscular rehabilitation as they inhibit pain with the additional heat increase; however, there exists the potential to cause thermal injury either due to improper device handling or to fragile endocrine (adrenals, pancreas) or gonadal tissues intolerant of this temperature variation. Proper training can minimize this risk; however, colleagues have reported smell of burning feathers, thermal burns to an equine spine, as well as animal discomfort, which are unacceptable risks when other non-invasive options are available. Suitable required eye protection can be cumbersome as well as impractical.

Materials and Methods

A Class 2 low-power (5 mW; 7.5 mW) non-heat-producing, frequency-specific visible-light wavelength (635 nm; 405 nm) laser system with line-generated beams, was used to perform recurring maintenance therapy on a geriatric albino Burmese python. This python along with another python were farmed out to a local rehabilitation facility as a result of the death of their famous owner, Michael Jackson. The pythons resided there until the facility fell upon hard times and the care of the inhabitants became questionable.

The pythons were transferred to the Children's Nature Center, a 501-C3 zoological and educational facility, in Fruita, CO in 2014. One did not survive. The other python had not eaten for 6 months, was encrusted in feces, and had a respiratory infection. With the help of a local herpetologist and veterinarian, the animal was able to adapt and eventually thrive in its new location. In 2016, the educational facility with its naturalistic habitats for its over 120 reptiles, arachnid, and aquatic inhabitants, was moved to a larger space in a mall setting in nearby Grand Junction. The python gradually became anorexic and was having prolonged sheds.

At the time the author became involved, an area veterinarian was preparing to pull blood samples to investigate possible metabolic causes. The dramatic response after the LLLT postponed this indefinitely as of this writing.

Low-level laser therapy (LLLT) treatments were performed using 3 180-second modules. A 5 mW laser unit with 2 diodes composed of 2 line-generated 635 nm wavelength beams was used to allow for simultaneous stimulation of nerve root as well as area of involvement. The frequencies selected were referenced from studies by Rife, using electric current, which is another form of electromagnetic energy.

1. Sympathetic overstimulation: Placing one diode that is preset with neurologic settings through the brain via the foremen magnum, and the other diode with settings for sympathetic nervous system, pain relief, over thoracolumbar spine where the sympathetic ganglia reside.

2. Parasympathetic stimulation: Placing the preset neurologic diode and the diode with the parasympathetic and colon settings both over the cranium and dorsal spine including sacral area.

3. Pain and inflammation: Includes settings for lymphatic drainage, as well as liver. Both diodes were placed over entire cranium, spine, and body.

These modules were repeated again 2 hours later.

Staff reported that the python struck at a rat that evening. Another treatment the following day resulted in the animal succeeding in ingestion of the freshly thawed rat that night.

Maintenance was discussed and the python was treated with only the pain and inflammation settings twice each month for 2 months, then once monthly. Four more anorexic episodes during the course of the year warranted similar treatment with the addition of sympathetic and parasympathetic modules. Appetite stimulation and increased vitality resulted after each dedicated LLLT application. The number of therapy applications varied from 1-4 visits, within a 2- to 4-day span. Staff members stated that the python's shed cycles appeared to be expedited as well and it seemed to be more active and interactive for days.

Results

The appetite and vitality of this geriatric python was observed to be increased on 4 separate occasions throughout the year within 24-72 hours of being treated with the Class 2 frequency-specific visible-light- wavelength, low-power laser. Also reported was normal shedding episodes had appeared to be more efficient in less time than what had been seen previously. In one episode, the python struck at and ate a thawed rat within minutes of completing the LLLT. The animal's responses to this non-invasive healing modality may pave the way for enhancing the quality of life for future animals, including other more fragile wildlife.

Discussion

Photobiomodulation is a promising versatile healing modality with an amazing ability to signal the body's own innate pharmacy to expedite healing in its many forms. The distinction in the terminology of true photobiomodulation is evolving to include only those light therapies that do not produce a thermal response. Clearly there are many unanswered questions about how quantum medicine truly works. A quote from astrophysicist Dr. Neil de Grasse Tyson reminds us in that "The universe is under no obligation to make sense to us." Consequently, the ability of photobiomodulation in the form of FSLLLT to affect the autonomic nervous system is perhaps the most amazing response of which few are aware. With documented variations in vital statistics in multiple species, including lowered blood pressure within seconds of the completion of the therapy implemented using specific frequencies programmed into the device, this author hopes to inspire more to investigate these responses further. The lack of need for eye protection, ability to enhance tissue regeneration systemically, and the ability to treat broad areas from a short distance without sedation, while in motion, offers exciting possibilities for sensitive wildlife.[8-16]

References

1. Oschman J. *Energy Medicine: The Scientific Basis.* 2nd ed. New York, NY: Elsevier Churchill Livingstone;2015.

2. Inman W. Advanced laser therapy. Available at: https://vomtech.com/. Accessed July 1, 2018.

3. Gordon Palm J. Evidence supporting low power laser therapy: Increasing applications increases success. *Proc Erchonia Seminars.* 2016.

4. Food and Drug Administration. Laser surgical instrument for use in general and plastic surgery and in dermatology. Available at: https://www.erchonia.com/wp-content/uploads/2017/07/K153164_FFS_Onychomycosis_6_3_2016.pdf. Accessed July 1, 2018.

5. Neira R, Jackson R, Dedo D, Ortiz CL, et al. Low-level laser-assisted lipoplasty appearance of fat demonstrated by MRI on abdominal tissue. *Am J Cosm Surg.* 2001;18:3.

6. Gianni F, Soares DM, Barreto MP, Barboza CA. Effect of LLLT on mesenchymal stem cell proliferation: A systemic review. *Lasers Med Sci.* 2015;30(8):2189-2194.

7. Anders J, Lanzafame RJ, Arany P. Low-level light/laser therapy versus photobiomodulation therapy. *Photomed Laser Surg.* 2015;33(4):183-184.

8. Mester E, Jaszagi-Nagi E. The effects of laser radiation on wound healing and collagen synthesis. *Studia Biophys*. 1973;35:227-230.

9. Hamblin MR, Ferraresi Cleber, Huang Y-Y, et al. Low-level light therapy: Photobiomodulation. Bellingham, WA; Spie Press;2018.

10. Hode L, Tuner J. *Laser Phototherapy: Clinical Practice and Scientific Background.* Coeymans Hollow, NY: Prima Books;2014.

11. Yu W, et al. Improvement of host response to sepsis by photobiomodulation. *Lasers Surg Med.* 1997;21(3); 262-268.

12. McDonald RS, Gupta S. 405nm light exposure of osteoblasts and inactivation of bacterial isolates from arthroplasty patients: potential for new disinfection applications? *Eur Cell Mater.* 2013;25:204-214.

13. Jann H. Equine wound healing: Influence of low level laser therapy on an equine metacarpal wound healing model. *Photon Lasers Med.* 2012;117-122.

14. Joensen J, Demmink JH, The thermal effects of therapeutic lasers with 810 and 904nm wavelengths on human skin. *Photomed Laser Surg.* 2011;29(3):145-153.

15. Denegar C, Saliba E, Saliba S. *Biologic Effects of Laser Therapy; Therapeutic Modalities for Musculoskeletal Injuries.* 3rd ed. Champaign, IL: Human Kinetics;2016.

16. Karu T. *Ten Lectures on Basic Science of Laser Phototherapy.* Coeymans Hollow, NY:Prima Books;2007.

Evaluation of Vitamin A Gut Loading in Black Soldier Fly Larvae

Kimberly Boykin, DVM,
Karina Butler-Perez, BS,
Mark A. Mitchell, DVM, MS, PhD, DipECZM (Herpetology)

Session #2069

Affiliation: From the Department of Veterinary Clinical Sciences, Louisiana State University, School of Veterinary Medicine, Baton Rouge, LA 70803, USA.

Black soldier fly (BSF) larvae are an excellent source of calcium for insectivores; however, previous studies have identified that they lack appreciable amounts of fat-soluble vitamins (A, D_3, and E). In order to make BSF larvae a more complete food item, fat soluble vitamins should either be provided via gut loading or with a multi-vitamin dusting supplement. The purpose of this study was to identify factors associated with gut loading vitamin A into BSF larvae and to develop feeding recommendations for a more consistent gut loading process. Factors that were addressed include the vitamin A concentration added to the diet, length of time given to gut load, moisture content of the diet, and density of larvae during feeding. Diets and larvae were analyzed for dry matter, crude protein, fat, fiber, trace minerals, and vitamin A concentration. Larval vitamin A concentrations increased in a linear fashion with increasing dietary vitamin A. Moisture content of the diet ($P = 0.003$, $F = 12.624$) and larval density ($P = 0.005$, $F = 11.054$) were both found to be significant factors contributing to the larvae's gut loading capacity. Based on initial results and vitamin A recommendations from the National Research Council for rats and poultry, gut loading recommendations for BSF larvae when fed to insectivorous reptiles and amphibians are as follows: vitamin A concentration of diet between 15,000-20,000 μg/kg, a gut loading time period of 24 hours, moisture content between 55-65% moisture, and a larval density between 0.1-1 larvae per each gram of moist substrate.

Digestibility of Black Soldier Fly Larvae when Fed to Leopard Geckos (*Eublepharis macularius*)

Kimberly Boykin, DVM,
Karina Butler-Perez, BS,
Mark A. Mitchell, DVM, MS, PhD, DipECZM (Herpetology)

Session #2068

Affiliation: From the Department of Veterinary Clinical Sciences, Louisiana State University, School of Veterinary Medicine, Baton Rouge, LA 70803, USA.

Black soldier fly (BSF) larvae have been marketed as an excellent way to provide calcium to reptiles and amphibians without the hassle and variability of dusting or gut loading; however, previous studies have indicated that digestibility and nutrient availability is poor due to the larvae's tough exoskeleton. Breaking the exoskeleton prior to feeding has been suggested to increase digestibility and absorption of key nutrients. In addition, fasted larvae are not considered to be nutritionally complete meals as they are deficient in fat soluble vitamins (A, D_3, and E). The need for vitamin supplementation in either the form of gut loading or multi-vitamin dusting still exists. In this feeding and digestibility trial, 24 adult male leopard geckos were housed individually under identical conditions and fed 1 of 3 diets for 60 days: diet 1 consisted of fasted and intact BSF larvae; diet 2 consisted of intact vitamin A gut-loaded larvae; and diet 3 consisted of vitamin A gut-loaded larvae that were cut in half prior to feeding. Samples of the larvae and fecal output from the animals were collected daily and apparent digestibility was calculated for dry matter, protein, fat, fiber fractions, and minerals (Ca, P, Mg, Na, K, Cu, Fe, Mn, Mo, Zn). Growth measurements, biochemistry profiles, and serum and liver vitamin A concentrations were compared among each of the 3 diet groups.

Nutrition and Husbandry Conditions of European Tortoises (*Testudo* spp.) in Captivity

Petra Koelle, DVM, Dr med vet, PrivDoz, Cert Spec for Fish including Ornamental Fish, Cert Spec for Reptiles, Cert Spec for Pet Nutrition consultation,
Sven Reese, DVM, Dr med vet, PrivDoz, Cert Spec for Anatomy, Cert spec for Information Technology,
Thomas Bauer, DVM

Session #2114

Affiliation: From the Clinic of Small Animal Internal Medicine, University of Munich (LMU), Veterinaerstrasse 13, D-80539 Muenchen, Germany (Koelle, Bauer), University of Munich (LMU), Veterinaerstrasse 13, D-80539 Muenchen, Germany (Reese).

A survey was developed by online or print questionnaire (German language) to evaluate the current population structure, husbandry conditions, diet regime, and health status of captive *Testudo* species in Germany. Respondents (n =1,075) kept 8,486 Palaearctic tortoises. The Hermann's tortoise (*T. hermanni)* was the most kept species (76%). Nearly 62% of the respondents acquired their tortoise from breeders, while reptile fairs (5.0%) did not play an important role. More than two thirds of the respondents evaluated the quality of the information they obtained from breeders and from the internet as good, while only 51% evaluated the information from pet shops as good. More than 75% of the respondents housed their tortoises in an outdoor enclosure containing a greenhouse or cold frame, which is considered as the most species-appropriate way of husbandry. A diet consisting of more than 80% grasses and weeds were fed by 67.7% of the respondents during the summer vegetation period. A diet containing > 60% annual and perennial grasses and weeds and hay or hay products was fed by 92.4%; however, 4.2% fed a diet of > 80% salad, 0.2% a diet of > 80.0% fruits, and 0.3% a diet of > 80% processed or non-vegetarian food. A tortoise with a diagnosed disease was owned by 8.2% of the respondents. Pyramidal growth syndrome (PGS)[1-3] is considered an indicator for the quality of tortoise husbandry. Only 18.5% of the respondents kept tortoises with distinct PGS of the levels 2, 3 and 4 as identified by means of photos in the questionnaire. According to this study, the likelihood of tortoises developing PGS is high in animals kept in a terrarium and/or fed a diet of < 80% grasses and weeds in summer. The highest predisposition to develop PGS was found in Hermann's tortoises.

References

1. Koelle P. *Die Schildkröte: Heimtier und Patient.* Stuttgart, Germany: Enke Verlag;2009.

2. Stancel CF, Dierenfeld ES, Schoknecht PA. Calcium and phosphorus supplementation decreases growth, but does not induce pyramiding, in young red-eared sliders, *Trachemys scripta elegans*. *Zoo Biol.* 1998;17:17-24.

3. Wiesner C, Iben C. Influence of environmental humidity and dietary protein on pyramidal growth of carapaces in African spurred tortoises (*Geochelone sulcata*). *J Anim Physiol Anim Nutr.* 2003;87:66-74.

How Dimension in *Testudo* spp. Influences Weight Loss During Hibernation

Emanuele Lubian, DVM, GPCert(ExAP),
Edoardo Bardi, DVM, GPCert(ExAP),
Alessandro Vetere, DVM, GPCert(ExAP),
Gianluca Deli, DVM, GPCert(ExAP)

Session #2198

Affiliation: From MyPetClinic, Milano, Italy (Lubian, Bardi), Ambulatorio veterinario Gaggiano, Milano, Italy (Lubian, Vetere), Ospedale Veterinario Universitario Piccoli Animali, Università degli Studi di Milano, Italy (Lubian, Bardi), Ospedale Veterinario Gregorio VII, Roma, Italy (Deli).

Hibernating animals lose some of their weight during hibernation. In Mediterranean tortoises, people often prefer to avoid hibernation during first years of their life to prevent death. This study monitored 120 chelonians 1-45 years of age, genus *Testudo* (in particular *T. graeca*, *T. marginata*, *T. Hermanni*), which were maintained in the same conditions during the summer season and at 6 °C during the entire winter months (totaling 5 months). The health status of each animal was confirmed through clinical examination before the winter. Weights from each animal were taken every month. If weight loss exceeded 15% of prehibernation body weight the animal was awakened. Results of this study demonstrate that the decrease in weight is inversely proportional to the dimensions of the animal. Authors agree that the percentage of weight decrease decreases with the increasing body mass/body surface ratio. This study supports the idea that young animals are more vulnerable during hibernation than adults; however, the authors don't suggest avoiding hibernation, but monitoring it.

References

1. Gilles-Baillien M. Seasonal variations in weight and hematological data of *Testudo hermanni hermanni*. *Arch Int Physiol Biochim.* 1973;81(4):723-32.

2. Lubian E, Vetere A, Floridi C, Deli G, Bardi E, Millefanti M. Hematological and serological values changes in different seasons in Testudo hermanni boettgeri. *Proc ICARE.* 2017;574.

Acquisition of the Microbiota in Sea Turtles

T. Franciscus Scheelings, BVSc, MVSc, MANZCVSc (Med Aust Wildl),
DipECZM (Herpetology),
Rob Moore, PhD,
Marcel Klaassen, PhD,
Richard Reina, PhD

Session #2043

Affiliation: From the School of Biological Sciences, Monash University, Wellington Rd, Clayton, VIC, Australia, 3800 (Scheelings and Reina), RMIT University, Plenty Rd, Bundoora, VIC, Australia, 3083 (Moore), and Centre for Integrative Ecology, Deakin University, Waurn Ponds, VIC, Australia, 3216 (Klaassen).

A diverse array of microorganisms, known as the microbiota, reside on and within all metazoans. Acquisition of the microbiome during birth is critical for priming the gastrointestinal tract and immune system of mammals. Given the reliance of animals on their symbiotic micro-organisms, it is highly likely that initial microbial colonisation of reptile gastrointestinal tracts also plays a significant role in development of their immune systems; however, it is not known how microbes are acquired by hatchlings in order to prime them for healthy gut function and immunity. Using an experimental protocol, combining laboratory and field data, we examined how environment and maternal microbes influence the microbiota of hatchling green sea turtles (*Chelonia mydas*). Colonic swabs were collected from nesting female turtles from Heron Island, Australia and their entire clutch of eggs was collected from the turtle into a sterile plastic bag. A small proportion of these eggs were opened using sterile technique and the contents of the egg were sampled. We then separated the remainder of the clutch into two treatment groups. Some eggs were buried in artificial nests on the island while others were taken back to the laboratory to incubate in commercially available sand. Swabs were collected from hatchling turtles as they emerged from the eggs from both treatment groups. Analysis of microbial composition was achieved using 16s rRNA gene sequencing. We then compared the microbiota of mother and hatchlings to determine how the hatchling microbes had been acquired. We will discuss these results in the context of how they improve our knowledge of sea turtle biology and their implications for artificial incubation of eggs.

Diet Influences the Blood Nutritional Parameters and Microbiome of Rehabilitating Green Sea Turtles

Jennifer C.G. Bloodgood, MS, PhD,
Terry M. Norton, DVM, Dipl ACZM,
Lisa A. Hoopes, PhD,
Sonia M. Hernandez, DVM, PhD, Dipl ACZM

Session #2229

Affiliation: From the Daniel B. Warnell School of Forestry and Natural Resources, University of Georgia, 180 E Green St, Athens, GA 30602, USA and Southeastern Cooperative Wildlife Disease Study, College of Veterinary Medicine, University of Georgia, 589 DW Brooks Dr, Athens, GA 30602, USA (Bloodgood, Hernandez), Georgia Sea Turtle Center, Jekyll Island Authority, 214 Stable Rd, Jekyll Island, GA 31527, USA (Norton), Georgia Aquarium, 225 Baker St NW, Atlanta, GA 30313, USA (Hoopes).

Threatened and endangered green turtles (*Chelonia mydas*) admitted for rehabilitation are challenging to feed, as this species exhibits a unique shift with age from an omnivorous to primarily herbivorous foraging strategy. They often are fed non-herbivorous, high-protein, low-fiber diets to counteract poor appetite and emaciation, even when presumably in the herbivorous life-stage. This can result in gastrointestinal pathology and obesity, delaying recovery and release. To understand the impact of diet on health and recovery, blood nutritional parameters in green turtles undergoing rehabilitation at the Georgia Sea Turtle Center ($n = 34$) were compared to free-ranging turtles captured in Florida ($n = 34$). Blood samples were collected from free-ranging turtles at a single time point, and from rehabilitation turtles at 3 timepoints: admission (primarily seafood diet), mid-rehabilitation (mixed seafood and herbivorous diet), and recovery (primarily herbivorous diet). Fecal samples for microbiome analysis were also collected from rehabilitation animals ($n = 19$) over the 3 time points. Rehabilitation turtles at recovery had significantly higher plasma cholesterol, triglycerides, and phosphorous, and lower vitamin D, calcium, and magnesium than free-ranging turtles (all $P < 0.01$). They also exhibited significantly altered microbiome composition, including fewer Firmicutes that aid in digestion of their natural diet of seagrass and algae. A diet that more closely mimics the wild diet is clearly important for green turtles during rehabilitation. Formulation of a novel, nutritionally complete gel diet based on our findings should enhance recovery of turtles in rehabilitation and help maintain the nutritional health of animals in long-term captivity.

Reference Values for Complete Blood Count, Phosphorus, and Total and Ionized Magnesium and Calcium in Red-Footed Tortoises (*Chelonoidis carbonaria*)

Kyra J. Berg, DVM,
Orlando Diaz-Figueroa, DVM, MS, DABVP (Avian Practice),
Matthew Schexnayder, DVM,
Britton J. Grasperge, DVM PhD DACVP,
Javier G. Nevarez, DVM, PhD, Dipl ACZM, DipECZM (Herpetology)

Session #2100

Affiliation: From the Department of Veterinary Clinical Sciences, Louisiana State University, School of Veterinary Medicine, Skip Bertman Drive, Baton Rouge, LA 70803, USA (Berg, Nevarez), Department of Pathobiological Sciences, Louisiana State University, School of Veterinary Medicine, Skip Bertman Drive, Baton Rouge, LA 70803, USA (Schexnayder, Grasperge), and Lake Howell Animal Clinic, 856 Lake Howell Rd, Maitland, FL 32751, USA (Diaz).

Abstract: Calcium and phosphorus imbalance results in nutritional secondary hyperparathyroidism (NSHP) in reptiles. Magnesium is also known to regulate calcium but its role in NSHP is unknown. Hematologic analysis from apparently healthy red-footed tortoises (*Chelonoidis carbonaria*) was performed to establish values for complete blood count, total and ionized magnesium, total and ionized calcium, and phosphorus. Objectives included determination of associations between total and ionized magnesium, and total calcium and total magnesium.

Introduction

Calcium and phosphorus play a role in nutritional secondary hyperparathyroidism (NSHP) in reptiles, with chelonians commonly affected with this condition. Magnesium, the second-most abundant intracellular cation in the body, is a co-factor and activator for the renal enzyme that converts Vitamin D to its final active form.[1] Hypomagnesemia may result in increased intracellular calcium ion concentration, decreasing the circulating ionized calcium in peripheral blood.[2] Magnesium needs and deficiencies have been described in humans and cattle but little information is available regarding magnesium physiology in reptiles.

Methods

Venipuncture was performed in 40 apparently healthy juvenile red-footed tortoises (*Chelonoidis carbonaria*) kept in uniform husbandry conditions. Weight and external carapacial temperature were recorded. The first objective was to establish reference values for packed cell volume, slide estimate white blood cell count, and white blood cell differential; and plasma phosphorus, total and ionized calcium, and total and ionized magnesium. The second objective was to determine if there were any associations between the analytes. Statistics were performed according to standardized guidelines for establishing reference ranges.[3] Descriptive statistics revealed normal distribution of the complete blood counts, chemistry values, and morphometrics for the population as a whole. Data was not normally distributed when split by gender.

Results

The results of the study may improve diagnostic and medical management of ill tortoises, and provide new insight into relationships and possible correlations between calcium and magnesium in reptiles. This information will ultimately stimulate development of future research studies on calcium and magnesium, and efficacy of magnesium supplementation in reptiles.

References

1. Kirchgessner MS, Tully TN, Nevarez J, et al. Magnesium therapy in a hypocalcemic African grey parrot. *J Avian Med Surg.* 2012;26(1):17-21.

2. Alam MK, Begum N, Begum S. Relationship of serum ionized calcium and magnesium concentration with parasympathetic nerve function in type 2 diabetes mellitus. *J Bangladesh Soc Physiol.* 2016;11(2):70-73.

3. Friedrichs KR, Harr KE, Freeman KP, Szladovitz B, Walton RM, Barnhart KF, Blanco-Chavez J. ASVCP reference interval guidelines: determination of de novo reference intervals in veterinary species and other related topics. *Vet Clin Pathol.* 2012;41(4):441-453.

Nutritional Assessment of an Experimental Cricket Diet and Crickets on the Health of Leopard Geckos

Laure Molitor, MS, DVM, and
Mark Mitchell, DVM, MS, PhD, DipECZM (Herpetology)

Session #2295

Affiliation: From the Department of Veterinary Clinical Sciences, School of Veterinary Medicine, Louisiana State University, Skip Bertman Drive, Baton Rouge, LA 70803, USA.

Leopard geckos (*Eublepharis macularis*) are one of the most popular pet reptiles kept in captivity. The purpose of this study was to generate evidence-based nutritional data using an experimental cricket gutloading diet with the goal decreasing the prevalence of secondary nutritional hyperparathyroidism by creating a nutritionally balanced diet for these animals. A randomized paired design was utilized. Juvenile leopard geckos were matched by weight and sex, then randomly divided into 2 groups: treatment (12 geckos) and control (12 geckos). Geckos assigned to the treatment group were provided crickets supplemented with the experimental diet. Groups of 12 crickets were placed in small transport cages and were gutloaded for 4-6 hours before being offered to each subject in the treatment group. Geckos in the control group were provided crickets fasted for at least 24 hours. The total weight of crickets offered every other day was 5% of the leopard gecko's body weight. Over the course of the study, body weights were measured weekly. Computed tomography scans were collected at baseline (day 0), 45 days, and 90 days after initiating the study. Houndsfeld unit density collected from the mandibular cortical bone was used to compare the 2 groups. The experimental cricket diet had 5.9% calcium dry matter, while the diet of the non-supplemented crickets had 0.11% calcium. The calcium to phosphorus ratios for the diets were 10:1 and 1:10, respectively. Vitamin A concentration on an as is basis in the experimental cricket diet and non-supplemented crickets was 5000 IU/kg and < 1000 IU/kg.[1] Vitamin D concentrations on an as is basis in the experimental diet and fasted crickets were 1000 IU/kg and < 250 IU/kg, respectively.[1] While the geckos consumed both diets, the experimental cricket diet had calcium, vitamin A, and vitamin D concentrations that were considered more physiologically appropriate for geckos.

References

1. Finke MD. Gut loading to enhance the nutrient content of insects as food for reptiles: A mathematical approach. *Zoo Biol.* 2003;22:147-162.

Section 39

ASSOCIATION OF REPTILIAN AND AMPHIBIAN VETERINARIANS

ARAV Poster

Femoral Head and Neck Excision Arthroplasty for Management of Coxofemoral Degenerative Joint Disease with Pseudogout in a Red-Eared Slider (*Trachemys scripta elegans*)

Yen Tao Lee, BVSc, MSc,
Yun Shan Chiu, BVSc, MSc,
Li Hsung Chiang, BVSc, MSc

Session #2139

Affiliation: From the Brave Vet Exotic Animal Veterinary Hospital, 227, Minquan W Rd, Datong Dist, Taipei 10357, Taiwan.

A 10-year-old red-eared slider (*Trachemys scripta elegans*) was presented with a 1-year history of multiple joint swelling that including bilateral stifles and elbows. The patient's right hind limb was suddenly lame 3 days prior to admission. Right coxofemoral luxation was observed in dorsoventral radiographs. Femoral head and neck excision arthroplasty (FHNEA) via a caudal approach and surgical debridement in other swelling joints were performed. Caseous materials were found at the right coxofemoral joint and other joints during the surgery. There was no growth of microorganisms by bacterial culture of the surgical site. The caseous materials sent for pathologic examination were confirmed as calcium deposits. Metastatic calcinosis (pseudogout) was diagnosed. Function of the right hind limb was improved 1.5 months after surgery. In this case, the cause of the coxofemoral luxation remains unclear, but metastatic calcinosis and degenerative joint disease may be considered as one of the etiologies. Although FHNEA has been performed in a wide range of mammalian and avian species, there are limited reports in the literature involving reptiles.[1-4] This is a successful case of the use of a caudal approach to perform FHNEA and recover the limb function in a red-eared slider. Future studies are required to survey the etiology of coxofemoral luxation and the postoperative outcome of FHNEA in different surgical approaches in reptiles.

References

1. Chitty J, Raftery A. *Essentials of Tortoise Medicine and Surgery.* Oxford, UK: Wiley-Blackwell; 2013.

2. Naylor AD. Femoral head and neck excision arthroplasty in a leopard tortoise (*Stgmochelys pardalis*). *J Wildl Dis.* 2013;44(4):982-989.

3. Barten SL, Bennett RA. Treatment of chronic coxofemoral luxation by femoral head and neck excision arthroplasty in a white-throated monitor (*Varanus albigularis*). *Bull Assoc Reptilian Amphib Vet.* 1996;6:10-13.

4. McArthur M, Hernandez-Divers S. Surgery. In: McArthur SR, Wilkinson R, Meyer J, eds. *Medicine and Surgery of Tortoises and Turtles.* Oxford, UK: Blackwell Publishing Ltd;2004:403-464.

Index

Symbols

A

B

C

D

E

F

G

H

I

J

K

L

M

N

O

P

Q

R

S

T

U

V

W

X

Z

Section 42

ARAV Poster Addendum

Gap in pagination due to formatting issues.

Pages 1017-1030

Cryptosporidium varanii: Only for Lizards?

Anke C Stöhr, DVM,
Sean M Perry, DVM,
Mark Mitchell, DVM, MS, PhD, DipECZM (Herpetology)

Session #2107

Affiliation: From the Department of Veterinary Clinical Sciences, Louisiana State University, Skip Bertman Drive, Baton Rouge, LA 70803, USA.

Cryptosporidium varanii and *Cryptosporidium serpentis* are the most common causes of intestinal and gastric cryptosporidiosis reported in lizards and snakes, respectively. The fatal course of intestinal disease caused by *C. varanii* has been repeatedly reported in lizards, but it has been hypothesized previously that the pathogenicity is low in snakes. A 3-year-old, female Western hognose (*Heterodon nasicus*) in good body condition was presented due to severe, acute, hemorrhagic diarrhea shortly after shipment to their new owners. A large amount of flagellates was found in a direct fecal smear. Despite treatment, the animal died within 24 hours. Pathohistology identified multisystemic soft tissue mineralization, acute myocardial necrosis, ulcerative hemorrhagic colitis with intralesional gram-negative and gram-positive bacilli, and apically attached protozoa. *Salmonella* sp. was cultured from the intestine and *C. varanii* was identified via PCR. Vitamin D toxicity had been suggested, but no dietary source could be identified. *Cryptosporidium* has been previously found to be associated with enteritis in different snake species, but no molecular identification of the detected *Cryptosporidium* was done in these cases; however, a genetically distinct *Cryptosporidium* genotype has been found in Japanese grass snakes (*Rhabdophis tigrinus*) associated with enteritis. Immunosuppression due to another underlying disease and the recent transport of the Western hognose may have favored the fatal outcome in this animal; however, this case report indicates the pathogenicity of *C. varanii* in snakes and stresses the importance of PCR and sequencing to understand disease etiology, epidemiology and outcome.

Reptarenavirus Detection in a Group of Wagler's Pit Vipers (*Tropidolaemus wagleri*)

Janosch Dietz, Dr med vet,
Ekaterina Kolesnik, med vet,
Kim O. Heckers, Dr med vet, FTA Pathologie,
Rachel E. Marschang, PD Dr med vet, DipECZM (Herpetology), FTÄ Mikrobiologie

Session #2119

Affiliation: From Laboklin GmbH & Co. KG, Steubenstr 4, D-97688 Bad Kissingen, Germany.

Reptarenaviruses have been determined to cause inclusion body disease (IBD) in boas and pythons. IBD-like disease has also been described in colubrids in two reports and in palm vipers (*Bothriechis marchi*) in one report,[1] but no virus was detected in those reports, which were published before the discovery of reptarenaviruses. Eight Waglers`s pit vipers (*Tropidolaemus wagleri*) from a mixed collection of snakes died with signs of respiratory disease within a course of six months, four were submitted for necropsy. A mild to moderate infiltration with heterophils was found in the tracheal mucosa and pulmonary interstitium of all four. The small and large intestines of two snakes were moderately to severely inflamed and a high concentration of flagellates was found in the colons of both. Multiple nematodes with associated severe chronic granulomatous hepatitis and perihepatitis were found in the liver parenchyma and capsule of one snake. No intracellular inclusions were observed in any of the tissues examined. Tissues from all four snakes were tested for the presence of ferla, adeno, reo, nido, and reptarenaviruses by PCR. Reptarenaviruses were detected in mixed tissues from two of the snakes, all other PCRs were negative. Sequences of the PCR products from both animals were nearly identical to one another and to the corresponding sequence of *Giessen reptarenavirus* from a boa constrictor (*Boa constrictor*). This is the first description of reptarenavirus detection in a viperid species. The importance of the virus infection for disease in these animals requires further study.

References

1. Raymond JT, Garner MM, Nordhausen RW, Jacobson EJ. A disease resembling inclusion body disease of boid snakes in captive palm vipers (*Bothriechis marchi*). *J Vet Diagn Invest.* 2001;13:82-86.

Congenital Pulmonary Unilateral Aplasia in a Common Snapping Turtle Admitted to a Wildlife Rehabilitation Center

Renee Schott, DVM

Session #2122

Affiliation: From the Wildlife Rehabilitation Center of Minnesota, 2530 Dale St N Roseville, MN 55113, USA.

A common snapping turtle (C*helydra sepentina*) was found on the ice of a frozen lake in Minnesota and brought to a wildlife rehabilitation center. The admission physical exam was within normal limits except for a large carapacial deformity. A computed tomography scan was performed and a 3-D model was reconstructed; it was determined the carapacial deformity was due to congenital unilateral pulmonary aplasia.

References

1. Schachner ER, Sedlmayr JC, Schott R, et al. Pulmonary anatomy and a case of unilateral aplasia in a common snapping turtle (*Chelydra serpentina*): Developmental perspectives on cryptodiran lungs. *J Anat.* 2017;231:835-848.

Establishing Amphibian and Reptile Tissue Culture Cell Lines to Improve Disease Diagnostics

Robert J. Ossiboff, DVM, PhD, DACVP

Session #2142

Affiliation: From Aquatic, Amphibian, and Reptile Pathology Program, Department of Comparative, Diagnostic, and Population Medicine, College of Veterinary Medicine, University of Florida, 2015 SW 16th Ave, Gainesville, FL 32610-0123, USA.

Tissue culture cell lines are an extremely important in vitro reagent. They are an integral component in the discovery, characterization, and diagnosis of obligate intracellular pathogens. Though cell lines from domesticated and laboratory animals are abundant and readily available, there is a paucity of cell lines from other species. This is particularly true for the non-mammalian taxa, including amphibians and reptiles (herpetofauna). While there are approximately 17,000 species of amphibians and reptiles, only three cell lines, two reptile and one amphibian, are available from the American Type Culture Collection (ATCC), the largest collection of human and animal cell lines in the world. Such a lack of appropriate reagents is a major limiting factor in the diagnosis and characterization of disease in reptiles and amphibians. Establishment of continuously dividing tissue culture cell lines was attempted from primary and embryonic reptile and amphibian tissues. Optimal conditions and techniques for cell line establishment were determined, cell lines were generated from turtle, snake, and lizard hosts, and were used to amplify known and novel reptilian viral pathogens, including ophidian nidoviruses. Increased availability to cell lines will improve herpetofaunal disease diagnostics, particularly in regard to pathogen isolation. The ultimate goal of this project is to establish a wide-ranging panel of in vitro cell lines at critical points along the evolutionary tree of reptiles and amphibians, and to make these reagents available to researchers and diagnosticians globally.

Cytology Scrapings of Oral Cavities of 31 Specimens of *Testudo horsfieldii* with Dyspnea and Lack of Appetite: Eosinophilic Intranuclear Inclusions Found in Keratinocytes of 14 Specimens

Alessandro Vetere, DVM, GPCert(ExAP),
Emanuele Lubian, DVM, GPCert(ExAP),
Erika Treves, DVM

Session #2169

Affiliation: From Ambulatorio Veterinario Gaggiano, Via Galvani 42, Gaggiano, 20083, Italy.

Scrapings of the oral cavity were performed in 31 specimens of *Agrionemys horsfieldii* between the months of April 2015 to October 2017. The specimens were presented for lack of appetite, lethargy and/or nasal and ocular discharge. Upon examination, 14 of the 31 subjects presented with dyspnea, ocular discharge and white oral necrotic/diphtheritic plaques. Ulceration of the mucosa occured during the removal of the plaques (group 1). At the initial presentation, 8 of the subjects showed only severe dehydration and depression (group 2). The last 9 specimens were alert and reactive to stimuli (group 3) but presented loss of appetite and ocular and conjunctival discharge.

During every clinical exam 2 samples also were taken: one for the cytology and another was an oral swab without media transport. Both were sent to a laboratory to test for herpesvirus and *Mycoplasma agassizi* through PCR. Cytology showed eosinophilic intranuclear *inclusions,* multinuclear syncytium, acantholysis, and a number of other alterations that can be attributed to a herpes infection in 6 specimens of group 1 and 8 specimens of group 3. Moreover the subjects that were tested through PCR turned out to be positive for Testudinis herpesvirus 1 (TeHV1) and 1 subject was positive for *Mycoplasma agassizii.* In conclusion, an early diagnosis became possible by means of a rapid cytological exam (cytobrush) during the clinical exam in the subjects that showed symptoms compatible with herpesvirus while the PCR exam was still pending.

References

1. Cooper JE, Gheschmeissner S, Bone RD. Herpes-like virus particles in necrotic stomatitis of tortoises. *Vet Rec.* 1988;123:554.

2. Cowan ML, Raidal SR, Peters A. Herpesvirus in a captive Australian Krefft-s river turtle (*Emydura macquarii kreftii*). *Aust Vet J.* 2015;93:46-49.

3. Heldstab A, Bestetti G. Herpesviridae causing glossitis and meningoencefalitis in land tortoises (*Testudo hermanni*). *Herpetopatologia.* 1976;1:5.

4. Hauser B, Mettler F, Rübel A. Herpesvirus-like infection in two young boas. *J Comp Pathol.* 1983;93: 515-519.

5. Jacobson ER, Mansell JL, Sundberg JP, et al. Cutaneous fibropapillomas of green turtles (*Chelonia mydas*). *J Comp Pathol.* 1989;101:39-52.

6. Jacobson ER, Gaskin JM, Roelke M, et al. Conjunctivitis, tracheitis, and pneumonia associated with herpesvirus infection in green sea turtles. *J Am Vet Med Assoc.* 1986;189:1020-1023.

7. Lucke B. Studies on tumors in cold blooded vertebrates. *Annu Rep Tortugas Laboratory Carnegie Institute*; 1938.

8. Marschang RE. Evidence for a new herpesvirus serotype associated with stomatitis in Afghan tortoises (*Testudo horsfieldii*). *Proc Annu Conf Assoc Reptilian Amphib Vet.* 1999;77.

9. Origgi F, Jacobson ER, Romero CH, et al. Diagnostic tools for herpesvirus detection in chelonians. *Proc Annu Conf Assoc Reptilian Amphib Vet.* 2000;127.

10. Origgi FC. *Testudinid herpesviruses:* A review. *J Herpetol Med Surg.* 2012;22:42-54.

Prevalence and Genetic Characterization of *Cryptosporidium* sp. from Captive Reptiles in France: Risks for Manipulators?

William Claude, Dr vet med,
Marie-Thérèse Poirel,
Slimania Benabed,
Lionel Zenner, Dr vet med,
Magalie Rene-Martellet, Dr vet med

Session #2173

Affiliation: From the Veterinary School of Alfort, 7 avenue du Géneral de Gaulle, 94700 Maisons-Alfort, France.

Cryptosporidiosis is well known in reptiles but there is a lack of data regarding the prevalence in some countries in Europe. This study took place in France between 2016 and 2017 and included the analysis of 153 samples from 98 snakes, 35 lizards, and 20 tortoises belonging to 41 different species. Most of them came from pet stores, zoo, and private owners. Only 131 samples were valid and examined for the presence of *Cryptosporidium* sp. Oocysts were detected by direct smear with saccharose assay, then a two-step nested PCR at the small subunit (SSU) rRNA locus after purification with a commercially available kit as described previously.[1] The nested PCR revealed that 9 samples (6.9%) were positive for *Cryptosporidium* sp. and among them 2 samples (1.5%) were also positive with saccharose assay. The sequencing of amplified fragments, which was possible for 8 positive samples, was done in the laboratory of Vaulx-en-Velin (Rhône, France). The sequencing resulted in the identification of reptilian specific *Cryptosporidium* sp. such as *C. saurophilum*, *C. serpentis* and *C. ducismarci*. Reptile nonspecific *Cryptosporidium* oocyst were also detected in 1 tortoise (*Testudo graeca*), identified as a strain of *Cryptosporidium hedgehog*. This reptile non-specific *Cryptosporidium* showed a 5/785 nucleotide mismatch to *C. parvum* and *C. hominis*. Its zoonotic potential is unknown for now and it was never reported in a human case of cryptosporidiosis but its similarity with zoonotic strains of *Cryptosporidium* should be taken into consideration to avoid a possible public health implication.

References

1. Xiao L, Singh A, Limor J, et al. Molecular characterization of *Cryptosporidium* oocysts in samples of raw surface water and wastewater. *Appl Environ Microbiol.* 2001;67(3):1097-1101.

Novel Atadenovirus in Veiled Chameleons

Cathy Johnson-Delaney, DVM,
Drury R. Reavill, DVM, DABVP (Avian Practice), DABVP (Reptile and Amphibian Practice), DACVP,
April J. Childress,
James F. X. Wellehan, DVM, MS, PhD, Dipl ACZM, Dipl ACVM (Virology, Bacteriology/Mycology)

Session #2189

Affiliation: From NW Zoological Supply, 3614 Smith Ave, Everett, WA 98201, USA (Johnson-Delaney), Zoo/Exotic Pathology Service, 6020 Rutland Drive #14, Carmichael, CA 95608-0515, USA (Reavill), University of Florida, College of Veterinary Medicine, 2015 SW 16th Ave, Gainesville, FL, 32608, USA (Childress, Wellehan).

Multiple juvenile veiled chameleons (*Chamaleo calyptratus*) belonging to a commercial breeding colony clinically presented with blepharospasm, dehydration, anorexia and wasting. Death would occur within 3 days of clinical signs despite symptomatic treatment. Bacterial cultures did not reveal a pathogen. One juvenile was submitted for a complete post mortem. On histologic examination the primary lesion was a multifocal moderate to severe subacute hepatitis with intranuclear inclusions. Eosinophilic to amphophilic intranuclear inclusions also were identified in the epithelium of the oral mucosa and esophagus. The conjunctivitis identified was not associated with a specific disease etiology. Due to the pneumonia, hepatitis, and conjunctivitis identified, chlamydial immunohistochemistry was performed. Replicate sections were negative with stains for *Chlamydia* genus. Paraffin-embedded formalin-fixed tissues were submitted for pan-adenoviral PCR and sequencing. Bayesian and maximum likelihood hylogenetic analysis confirmed the presence of a novel atadenovirus, here termed veiled chameleon adenovirus 1.

References

1. Corsaro D, Venditti D. Emerging chlamydial infections. *Crit Rev Microbiol.* 2004;30:75-106.

2. Menendez-Conejero R, Nguyn TH, Singh AD, et al. Structure of a reptilian adenovirus reveals a phage tailspike fold stabilizing a vertebrate virus capsid. *Structure.* 2017;25:1562-1573.

3. Wellehan JFX, Johnson AF, Harrach B, et al. Detection and analysis of six lizard adenoviruses by consensus primer PCR provides further evidence of a reptilian origin for the atadenoviruses. *J Virol.* 2004;78:13366-13369.

High-Grade Tick Infestation in European Tortoises Re-Imported From Turkey and Resulting Health Issues

Sabine Oefner, and
Markus Baur

Session #2199

Affiliation: From the Reptile Rescue Center Munich, Kaulbachstraße 37, 80539 Munich, Germany.

The Reptile Rescue Center Munich accompanied a re-import of 2500 European tortoises of the species *Testudo hermanni*, *Testudo graeca*, and *Testudo marginata* from a private owner, who had to leave Turkey due to political reasons. The animals were moved back to Germany and were housed in 2 large greenhouses (1500 square meters in total) as an interim solution for 6 months. On arrival most of the *Testudo graeca* and some of the *Testudo marginata* showed a high-grade infestation with ticks (*Hyalomma aegyptium*). The *Testudo hermanni* were unaffected and no ticks were found on these animals. All affected *Testudo graeca* showed mild to severe skin problems like abscesses, old scar tissue, fresh wounds and signs of dermatitis as well as signs of icterus (yellow skin, yellowish discolored mucous membrane). Blood samples were drawn from some of the animals for virus testing and blood values. The values for all liver enzymes were in the reference range, further investigation of possible liver disease was declined by the owner. Blood smears of the affected animals showed infection with different blood parasites (mostly hemogregarine and plasmodia). As ticks are known vectors for several diseases like Crimean-Congo hemorrhagic fever, a thorough investigation was engineered to identify the blood parasites and possible transmissible diseases to evaluate health risks for the tortoises and for humans. Results for this investigation are still ongoing and will be published later.

Hydropericardium in a *Boa constrictor imperator* that Died After Severe Enteritis

Alessandro Vetere, DVM, GPCert(ExAP),
Emanuele Lubian, DVM, GPCert(ExAP),
Diana Binanti, DVM, PhD, Dipl ECVP

Session#2213

Affiliation: From Ambulatorio Veterinario Gaggiano, Via Galvani 42, Gaggiano, 20083, Italy (Vetere,Lubian), AbLab S.r.l.s, Via Privata Massa Neri 13, Sarzana,19038 (SP), Italy (Binanti).

A 6-year-old, 4-kg albino male *Boa constrictor imperator* was presented for a single episode of diarrhea of 1-day duration. During the clinical examination, the management history was determined to be correct for this species (preferred optimal temperature zone, nutrition, cage size, and humidity). At the examination there were no abnormalities except for moderate dehydration. A cloacal swab was performed for bacterial culture (aerobic and anaerobic plus antibiogram) and a coprologic examination of fecal material presented by the owner was also carried out (direct smear and flotation), which was negative for parasites. A blood smear was subsequently performed in order to carry out a differential leukocyte count that showed a moderate heterophilia. Pending the outcome of the antibiogram, subcutaneous rehydrating therapy and antibiotic therapy was initiated every 72 hours for 3 treatments. After 5 days from the first visit the animal died 1 hour after a regular meal. The post-mortem exam was performed 2 hours later and did not show any evident changes in the ceolomic organs, with the exception of a moderate pericardial effusion. Meanwhile, the bacterial culture grew an *Escherichia coli*, mucoid strain. The organs of the ceolomic cavity were fixed in formalin and then sent to the laboratory for histopathology. The only organs showing pathologic changes were the heart with marked and widespread peripheral edema and a large portion of the colon, which was affected by subacute heterophilic enteritis and diffuse lymphocytic enteritis.

References

1. Barten SL. Cardiomyopathy in a king snake (*Lampropeltis calligaster rhombomaculata*). *Vet Med Small Anim Clin.* 1980;75:125-129.

2. Jacobson ER, Homer B, Adams W. Endocarditis and congestive heart failure in a burmese python (*Python molurus bivittatus*). *J Zoo Wildl Med.* 1991;22:245-248.

3. Mader DR. Fluid therapy. In: Mader DR, ed. *Reptile Medicine and Surgery.* 2nd ed. Philadelphia, PA: Elsevier Saunders;2006:539-547.

4. Schilliger L, Vanderstylen D, Pietrain J, et al. Granulomatous myocarditis and coelomic effusion due to *Salmonella enterica arizonae* in a Madagascarian Dumeril's boa (*Acrantophis dumerili*, Jan. 1860). *J Vet Cardiol.* 2003;5:43-45.

5. Selleri P, Di Girolamo N. Cardiac tamponade following cardiocentesis in a cardiopathic boa constrictor imperator (*Boa constrictor imperator*). *J Small Anim Pract.* 2012;53:487.

Acute Wild Garlic Intoxication in an Aldabra Giant Tortoise and African Spurred Tortoises

Tom Hellebuyck, Dr med vet, DipECZM (Herpetology),
Jules Simard, med vet,
Norbert van de Velde, med vet,
Lise Geerinckx, med vet

Session #2259

Affiliation: From the Clinic for Poultry, Exotic Companion Animals, Wildlife and Experimental Animals, Department of Pathology, Bacteriology and Avian Diseases, Faculty of Veterinary Medicine, Ghent University, Salisburylaan 133, B-9820, Merelbeke, Belgium (Hellebuyck, Simard, van de Velde), Dierenartsencentrum Trigenio, Dorsel 38, B-2560, Nijlen, Belgium (Geerinckx).

This report documents acute wild garlic (*Allium ursinum*) intoxication in tortoises. A group of 4 captive tortoises, consisting of 1 Aldabra giant tortoise (*Aldabrachelys gigantea*) and 3 African spurred tortoises (*Centrochelys sulcata*), was fed a large quantity of wild garlic foliage. Approximately 12 hours later, 3 tortoises developed acute and pronounced lethargy, anorexia and signs of hemolytic anemia. The severity of the clinical signs in the individual tortoises seemed to be correlated with the volume of wild garlic ingested. At initial presentation, approximately 24 hours after being fed wild garlic, 1 African spurred tortoise died. Supportive care as well as the administration of activated charcoal was initiated in the remaining tortoises. The Aldabra giant tortoise died 5 days following presentation, but the other 2 tortoises recovered. Gross pathologic, histopathologic, blood biochemistry and hematologic findings largely compared to what has been documented in other species with *Allium* spp. intoxication.[1,2] To the authors' knowledge this is the first report of *Allium ursinum* intoxication in reptiles. Although the development of intoxication due to the ingestion of wild garlic in tortoises seems to be dose dependent, veterinarians and reptile pet owners should be aware of the potential toxic effects of feeding this herb to tortoises. Further toxicologic studies would be required to reveal possible species-dependent effects in tortoises.

References

1. Cope RB. Allium species poisoning in dogs and cats. *J Vet Med.* 2005;100:562-566.

2. Salgado BS, Monteiro LN, Rocha NS. Allium species poisoning in dogs and cats. *J Venom Anim Toxins incl Trop Dis.* 2001;17(1):4-11.

A Preliminary Study of Hematologic Values in Two Endangered Turtle Species: *Indotestudo travancorica* and *Batagur kachuga*

E. Bardi, DVM, GPCert(ExAP),
E. Lubian, DVM, GPCert(ExAP),
V. Aquaro, DVM,
M. Masi, DVM,
N. Whitaker, Biologist,
S. Romussi, DVM, PhD

Session #2260

Affiliation: From Ospedale Veterinario Universitario Piccoli Animali, Università degli Studi di Milano, Italy (Bardi, Lubian, Romussi), MyPetClinic, Milano, Italy (Bardi, Lubian), SIC Bosco Difesa Grande, Gravina di Pugli Bari, Italy (Aquaro), Centro Veterinario Specialistico, Roma, Italy (Masi), Madras Crocodile Bank Trust, Centre for Herpetology, Tamil Nadu, India (Whitaker).

The Indian subcontinent is incredibly rich in biodiversity and is home to many species of herpetologic and conservation interest. The aim of this study was to provide the first assessment of hematologic values in travancore tortoise (*Indotestudo travancorica*) and red-crowned roofed turtle (*Batagur kachuga*), two endangered species on which medical literature is still lacking. Between May and June, 19 specimens of *I. travancorica* and 17 of *B. kachuga* were sampled. Both populations were housed by the Madras Crocodile Bank Trust, Centre for Herpetology (Tamil Nadu, India). From each animal, a blood sample was obtained from the dorsal coccygeal vein and stored in a lithium-heparin tube. In both species a complete red and white blood cell count (RBC and WBC) was performed with the Natt & Herrik method using a Neubauer chamber. For *I. travancorica,* hematocrit values were assessed by centrifugation of microcapillaries. Due to the small sample size, only descriptive statistics were applied. In our results we could not detect a difference between two sexes for hematologic values in the observed species. The only difference noted was in total WBC count between young and adult animals in *B. kachuga*, with the higher count in the adult population; however, without further data, this difference cannot be accurately attributed to age.

References

1. Bielli M, Nardini G, Di Girolamo N, Savarino P. Hematological values for adult eastern Hermann's tortoise (*Testudo hermanni boettgeri*) in semi-natural conditions. J Vet Diagn Invest 2015;27(1):68-73.

2. Lubian E, Vetere A, Floridi C, et al. Hematological and serological values changes in different seasons in *Testudo hermanni boettgeri. Proc ICARE.* 2017.

3. Campbell TW. Hematologic techniques in lower vertebrates. In: Campbell TW, ed. *Exotic Animal Hematology and Cytology* 4th ed. Hoboken, NJ: John Wiley & Sons;2015:199-205.

4. Teare JA. Physiological data reference values project. Apple Valley, MN: International Species Information System;2010.

Surgical Removal of a Tweezer Foreign Body in a Bearded Dragon

Erica Giles, DVM

Session #2317

Affiliation: From Hermitage Veterinary Hospital, 1500 North Hermitage Rd, Hermitage, PA 16148, USA.

Bearded dragons (*Pogona vitticeps*) are known for the inquisitive nature and voracious appetite, making them a prime candidate for foreign body ingestion. This poster will present a case from general practice of a bearded dragon with a metal foreign body which created two body wall fistulas. The patient for this case was an adult male bearded dragon, specific age unknown, that was surrendered to the clinic for treatment. The previous owner reported a pair of "feeding tweezers" had gone missing a month prior to them noticing an object protruding from the dragon's side. Up to surgery the patient was bright and alert, still eating, still defecating, active and in good body condition. Blood work prior to surgery and anesthesia showed no significant findings. A fecal sample came back positive for *Isospora* and the patient was treated with a course of ponazuril (30 mg/kg PO q48h for 3 doses; Diamondback Drugs, Scottsdale, AZ, USA). Surgery involved a standard paramedian approach. The tweezers were found to be within the stomach. Removal of the tweezers was performed by creating a gastrotomy incision and removal via gentle traction. The tweezers had perforated the stomach, gone through the ceolomic wall and skin, creating two fistula at both of the sharp end points of the tweezers. The stomach was gently separated from the body wall with a scalpel and cotton tip applicators. The gastric mucosal edges were freshened then closed in a simple continuous pattern using 5-0 PDO suture. The coelom was throughly flushed with warm saline. The body wall was closed in two layers; an internal, coelomic membrane/muscle belly layer and an external, skin layer. The coelomic membrane was closed using in a simple interrupted pattern using 5-0 PDO. The skin was closed using 3-0 Nylon in an interrupted, everting pattern. Both fistula were closed in a similar manner to the body wall. Closure was routine. Recovery was uneventful. Culture was collected at time of surgery and *Klebsiella pneumoniae*, *Proteus mirabilis*, and *Escherichia coli* were grown and all were sensitive to ceftazidime. Post-operatively, the patient received Ceftazidime (20 mg/kg IM q72h, Sargent, Schaumburg, IL, USA), meloxicam (0.3 mg/kg SC q48h; Eloxiject, Henry Schein, Dublin, OH, USA), tramadol (5 mg/kg PO q12h; Hometown Apothecary, New Brighton, PA, USA) and Emeraid IC omnivore (10-20 ml/kg PO q12-24h; Emeraid LLC, Cornell, IL, USA). At the time of this presentation, the patient had gained weight, was eating, defecating, and doing very well overall.

References and Resources

1. Carpenter J, Marion C. *Exotic Animal Formulary.* St. Louis, MO: Elsevier Saunders;2017.

2. Bennet RA, Lock BA. Nonreproductive surgery in reptiles. *Vet Clin North Am Exot Anim Pract.* 2000;3(3):715-732.

Transesophageal Echocardiographic Imaging in Healthy American Alligators

Nicole Furst,
Bonnie Gatson, DVM, Dipl ACVAA,
Simon Swift, MA, VetMB, CertSAC, Dipl ECVIM-CA (Cardiology), MRCVS

Session #2319

Affiliation: From the Department of Comparative, Diagnostic, and Population Medicine (Gatson) and the Department of Small Animal Clinical Sciences (Swift), College of Veterinary Medicine, University of Florida, 2015 SW 16th Ave, Gainesville, FL 32610-0123, USA.

Abstract: Various techniques have been described to characterize the unique cardiovascular system of crocodilians. In an effort to develop non-invasive imaging techniques to further study the crocodilian heart, transthoracic and transesophageal echocardiography was compared in 10 healthy American alligators (*Alligator mississippiensis*). Transthoracic echocardiography was obtained prior to induction of general anesthesia with mild restraint. Alligators were then induced with isoflurane in 100% oxygen and transesophageal images were obtained following 90 minutes of anesthesia. The overall image quality and the ability to observe the right outflow tract of the heart was significantly better with transesophageal imaging. Biphasic flow across the pulmonary artery was observed throughout systole. Flow across the Foramen of Panizza occurred in a bidirectional manner only during diastole, as the foramen was closed by the pulmonic and aortic valve leaflets during systole. The major disadvantage of transesophageal echocardiography is that general anesthesia is required to obtain images. In conclusion, transesophageal echocardiography is easy to perform and provides excellent image quality in American alligators.

Introduction

Crocodilians have an extraordinarily complex cardiovascular system that is unique among reptilian species. Unlike other reptiles, they have a 4-chambered heart with 2 aortas, 1 emerging from each ventricle, that are connected just outside the heart base via the Foramen of Panizza.[1,2] Pulmonary outflow is controlled by a cog-wheel valve that exists immediately proximal to the pulmonary artery.[3] This valve and several connections between the right and left aorta, including the Foramen of Panizza, permit various extracardiac shunting patterns to occur.[1,4] Various techniques have been used to characterize these shunting patterns including in situ heart preparations,[3] gas probes,[5] flow probes, and transthoracic echocardiography.[4] Using non-invasive imaging modalities such as echocardiography is valuable for evaluating the cardiovascular system of these animals while optimizing their welfare in a research setting. Having a better understanding of normal echocardiographic parameters will also improve the clinical management of crocodilians. The use of transesophageal echocardiography to visualize the heart may be useful in reptilian species where access for transthoracic imaging may be limited or challenging. The purpose of this study was to evaluate these two echocardiographic techniques in American alligators (*Alligator mississippiensis*) and describe normal echocardiographic parameters in this species.

Materials and Methods

Animals

This study was approved by the Institutional Animal Care and Use Committee at the University of Florida (Protocol# 201609263). Ten 5-year-old American alligators weighing between 3.9 and 8.7 kg (mean 5.7 ± 1.4 kg) were obtained for this study from a private institution. Three to four animals were housed together at a time for 5 days in a climate controlled indoor enclosure containing a small pool of water with a temperature of 78 °F overnight and 80 °F during the day. The alligators were fasted for at least 2 days before transport from the private institution and they were not fed throughout the duration of this study. During recovery from general anesthesia, alligators were placed in an individual enclosure without water until deemed fully recovered by a veterinarian when they were returned to the original enclosure.

Restraint and preparation

For all procedures, animals were fastened to a customized wooden board with multiple fabric restraining straps in sternal recumbency and remained restrained in this manner until extubation. A hollow plastic gag was placed in the mouth of each animal between the incisors and secured with electrical tape around the jaws. Transthoracic echocardiography was performed without sedation prior to induction of anesthesia while the animals were restrained. For transesophageal echocardiography, alligators were intubated using either a 4.0 or 4.5 mm uncuffed endotracheal tube and anesthetized using isoflurane at 5% in 100% oxygen at 2 L/min for induction and 3% in 100% oxygen at 1 L/min for maintenance. Padding was placed over the eyes to decrease visual stimulation. The transesophageal probe was advanced through the mouth gag into the esophagus to obtain images (Fig 1).

Figure 1. Preparation and restraint of alligators for transesophageal echocardiography included a customizable wooden board with restraint straps, mouth gag placed between the incisor teeth, electrical tape around the snout to secure the mouth gag, and padding over the eyes to lessen visual stimulation. The transesophageal probe is inserted through the mouth gag into the esophagus while the alligators are endotracheally intubated.

Imaging modalities and technique

Transthoracic echocardiography was performed using a 6 MHz probe and a GE Vivid E9 echocardiography machine at 125 fps, and at a depth of 3-4cm. Transesophageal imaging was obtained using a 6 MHz transesophageal probe that was advanced caudally along the esophagus until the heart was visualized. All echocardiography was performed by a boarded veterinary cardiologist. Color flow was used to identify flow across the mitral and aortic valves and spectral Doppler was used to show flow patterns.

Results

During transthoracic echocardiography, short and long axis views of the heart were obtained. The four chambers and the right aorta could be identified but the pulmonary artery and left aorta could not be reliably distinguished. Significant craniocaudal movement of the heart was apparent in the thoracic cavity and made image acquisition difficult during the cardiac cycle especially with the normal resting crocodilian heart rate ranging from 17-44 bpm (mean 31 ± 7.4 bpm). The blood was at such an oblique angle to flow that flow velocities could not be meaningfully obtained. The flow across the Foramen of Panizza could be visualized and, as the alligator red cells are echolucent, flow could be seen. Although the flow velocities across the valves could not be estimated, the profiles were evident on spectral Doppler. Biphasic flow across the pulmonary artery could be seen with flow at the start and end of systole (Fig 2A). There was no flow in mid systole presumably due to closure of the cog-tooth like valve at the base of the pulmonary artery.

The image quality of the transesophageal technique was improved compared to the thoracic imaging. Similar chambers were appreciated, but the left aorta and pulmonary artery could be identified much more reliably (Fig 3). Color flow showed blood flow through the Foramen of Panizza, mainly from the right aorta to the left aorta but on occasions did occur in a bidirectional manner confirmed with spectral Doppler. Flow across the foramen occurred during diastole as the pulmonic and aortic valve leaflets closed the foramen when they were open during systole. Flow from the pulmonary artery could also be seen and both color flow and spectral Doppler confirmed the biphasic flow with an early and late systolic peak (Fig 2B).

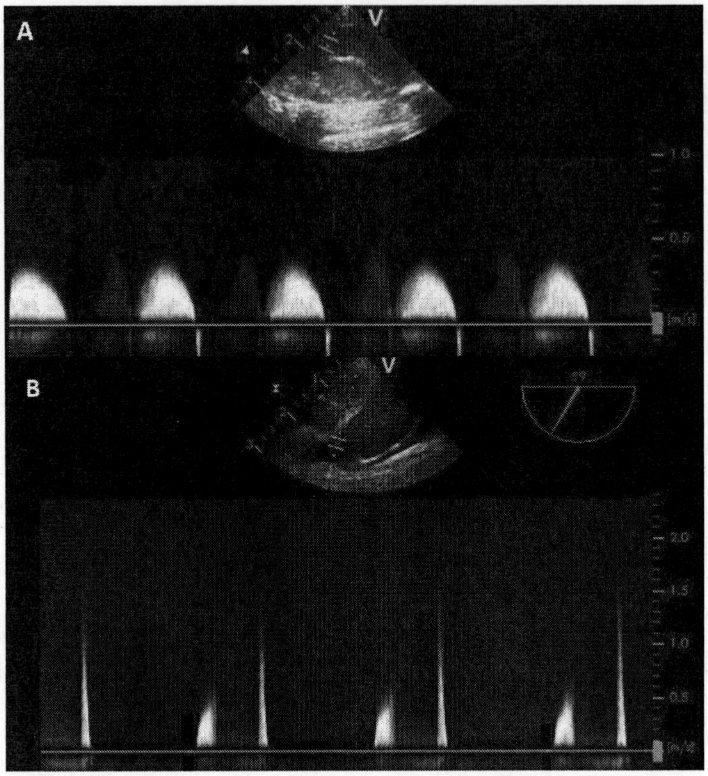

Figure 2. Biphasic flow through the pulmonary artery using A) transthoracic and B) transesophageal echocardiography.

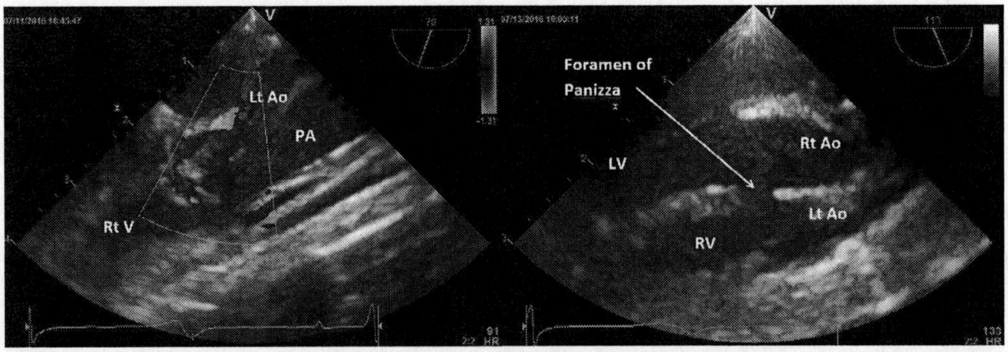

Figure 3. Transesophageal echocardiographic images in long-axis view of alligators during isoflurane anesthesia. (Lt Ao, left aorta; LV, left ventricle; PA, pulmonary artery; Rt Ao, right aorta; RV, right ventricle.)

Discussion

Compared to traditional transthoracic echocardiography, transesophageal image quality was superior. The entire right outflow tract can be seen with transesophageal echocardiography whereas the pulmonary artery and left aorta could not be reliably distinguished using the transthoracic method. These structures are important to identify when evaluating shunting patterns in crocodilians.

Though the transesophageal images gave a much clearer picture of the heart and associated structures, logistically it is more complicated. Alligators must be sedated at minimum or undergo general anesthesia to perform transesophageal echocardiography without fear of damage to the probe or harm to the animal (ie, esophageal tear). Unfortunately, flow measurements and pressure gradients across valves were difficult to obtain by either method. Further studies are needed to validate non-invasive flow measurements across valves, great vessels, and the Foramen of Panizza as well as chamber sizes, contractility, and measurements of cardiac output in crocodilian species.

Acknowledgments: Funding for this study was provided by grants from the University of Florida Faculty Development Fund and the Association of Reptilian and Amphibian Veterinarians. The authors would like to thank the St. Augustine Alligator Farm for graciously providing research animals.

References

1. Grigg G. Central cardiovascular anatomy and function in crocodilia. In: *Physiological Adaptations in Vertebrates: Respiration, Circulation, and Metabolism.* New York, NY: Marcel Dekker;1991:339-354.

2. Panizza B. Sulla structura del cuore e sulla circolazione del sangue del Crocodilus lucius. *Biblioth Ital.* 1883;87:87-91.

3. Franklin CE, Axelsson M. Physiology: An actively controlled heart valve. *Nature.* 2000;406:847-848.

4. Malvin GM, Hicks JW, Greene ER. Central vascular flow patterns in the alligator *Alligator mississippiensis. Am J Physiol.* 1995;269:1133-1139.

5. Findsen A, Crossley DA II, Wang T. Feeding alters blood flow patterns in the American alligator (*Alligator mississippiensis*). *Comp Biochem Physiol A Mol Integr Physiol.* 2018;215:1-5.

Standardization of Giant Amazon River Turtle (*Podocnemis expansa*) Glycemic Parameters

Kauê C. Ribeiro,
Lohany P. Franco,
Lucas C. M. Mendonça,
Patrícia P. Perin,
Eugenio G. Araujo, DVM, MSc, PhD,
Luciana B. M. Araujo, DVM, MSc, PhD

Session #2333

Affiliation: From Preventive Veterinary Medicine (Ribeiro, Franco, Mendonça, Perin, Araujo) and Veterinary Pathology (Eugenio Araujo), School of Veterinary Medicine and Animal Science of Federal University of Goiás, Av. Esperança, S/N, Campus Universitário, Goiania, Goiás, Brazil, 74690-900.

Acquiring physiologic data on wild animals in situ broadens knowledge and allows a more precise evaluation of well-being and sanitary conditions of captive specimens. The use of portable medical devices that enable the verification of these parameters immediately after the collection of the biological specimen increases the accuracy of the obtained data. We harvested and fractionated blood from 21 captive-bred specimens of *Podocnemis expansa* to validate the use of a portable glucometer (Accu-Check®). One drop was immediately applied on a dosing strip, a fraction subjected to centrifugation to obtain serum, and three other fractions added to sodium citrate, sodium heparin or potassium fluoride to observe the action of each anticoagulant on the reading. In addition, we obtained further serum and plasma samples and maintained them under refrigeration (4 °C) or frozen (-20 °C) for 48 hours until spectrophotometry was performed to verify if storage temperature interfered with sample analysis. The data obtained using the portable glucometer differed statistically from those obtained in spectrophotometry regardless of the anticoagulant used or the sample storage temperature ($P > 0.05$). There was no difference in the glucose concentration of samples kept refrigerated or frozen, as well as in plasma samples obtained with sodium heparin and potassium fluoride. The results obtained by spectrophotometry ranged within the reference values determined for the species.

Cardiac Tamponade Secondary to Hemopericardium following Cardiocentesis in a Ball Python (*Python regius*)

Jessica R. Comolli, DVM,
Brittany McHale, DVM, MPH,
Stephen Divers, BVetMed, DZooMed, DipECZM (Herpetology, Zoo Health Management), Dipl ACZM, FRCVS,
Spencer Kehoe, DVM,
Jörg Mayer, DVM, MS, DABVP (Exotic Companion Mammal Practice), DipECZM (Small Mammal), Dipl ACZM

Session #2403

Affiliation: From the Department of Small Animal Medicine and Surgery (Zoological Medicine) (Comolli, Divers, Kehoe, Mayer) and the department of Pathology (McHale), College of Veterinary Medicine, University of Georgia, 2200 College Station Rd, Athens, GA 30602, USA.

A 2-year-old female ball python was presented to the University of Georgia Zoological Medicine Service for evaluation of an oral discharge. The snake was anesthetized for diagnostic imaging, a tracheal wash, and blood collection. Blood was obtained via cardiocentesis with a 25-gauge needle using standard protocol. Immediately following blood collection a swelling was noted in the region of the heart. The snake was unable to be recovered from anesthesia and its heart subsequently stopped based on Doppler assessment. The cause of death at necropsy was determined to be acute cardiac tamponade secondary to hemopericardium, with no additional histologic lesions noted within the heart. This is the first report of cardiac tamponade following cardiocentesis in a snake without a diagnosed cardiopathy.

Association of Reptilian & Amphibian Veterinarians
5003 SW 41st Blvd
Gainesville FL 32608-4930
USA

ISBN 978-1-5108-7528-9

9 781510 875289

CURRAN ASSOCIATES INC.
proceedings.com